Springer Proceedings in Mathematics & Statistics

Volume 119

More information about this series at http://www.springer.com/series/10533

Springer Proceedings in Mathematics & Statistics

This book series features volumes composed of selected contributions from workshops and conferences in all areas of current research in mathematics and statistics, including OR and optimization. In addition to an overall evaluation of the interest, scientific quality, and timeliness of each proposal at the hands of the publisher, individual contributions are all refereed to the high quality standards of leading journals in the field. Thus, this series provides the research community with well-edited, authoritative reports on developments in the most exciting areas of mathematical and statistical research today.

Joachim Escher • Elmar Schrohe • Jörg Seiler •
Christoph Walker
Editors

Elliptic and Parabolic Equations

Hannover, September 2013

Springer

Editors
Joachim Escher
Leibniz University of Hannover
Institute for Applied Mathematics
Hannover
Germany

Elmar Schrohe
Leibniz University of Hannover
Institute for Analysis
Hannover
Germany

Jörg Seiler
Department of Mathematics
University of Torino
Torino
Italy

Christoph Walker
Leibniz University of Hannover
Institute for Applied Mathematics
Hannover
Germany

ISSN 2194-1009 ISSN 2194-1017 (electronic)
ISBN 978-3-319-12546-6 ISBN 978-3-319-12547-3 (eBook)
DOI 10.1007/978-3-319-12547-3

Library of Congress Control Number: 2015938443

Springer Cham Heidelberg New York Dordrecht London

Printed on acid-free paper

Springer is part of Springer Science+Business Media (www.springer.com)

Preface

The present volume originates from the international *Workshop on Elliptic and Parabolic Equations*, held on 10–12 September, 2013 at the *Leibniz Universität Hannover*. It represents a collection of refereed research papers and survey articles written by eminent scientists. Recent advances on various topics from different fields in elliptic and parabolic partial differential equations are included such as singular Riemannian manifolds, spectral analysis on manifolds, nonlinear dispersive equations, Brownian motion and kernel estimates, Euler equations, porous medium type equations, pseudodifferential calculus, free boundary problems, and bifurcation analysis.

Our thanks go to all participants of the workshop and the authors who contributed to this volume for making both successfully happen.

We gratefully acknowledge the financial support provided by the *Leibniz Universität Hannover* and the *Deutsche Forschungsgemeinschaft* and we thank the team of *Springer Verlag* for their help in realizing this volume.

Hannover and Torino
July 2014

Joachim Escher
Elmar Schrohe
Jörg Seiler
Christoph Walker

Contents

Uniformly Regular and Singular Riemannian Manifolds 1
Herbert Amann
1 Introduction 1
2 Notations and Definitions 14
3 Preliminaries 17
4 Uniformly Regular Riemannian Manifolds 22
5 Characteristics 24
6 Model Cusps and Funnels 28
7 Submanifolds of Euclidean Spaces 31
8 Singular Ends 37
References 42

Eigenvalue Estimates on Bakry–Émery Manifolds 45
Nelia Charalambous, Zhiqin Lu and Julie Rowlett
1 Introduction 45
2 Eigenvalue Estimates 46
2.1 One-dimensional Collapse 46
2.2 Maximum Principle and Gradient Estimates 47
2.3 Lower Bounds for the First Eigenvalue 48
2.4 Lower Bounds of Higher Eigenvalues 50
2.5 Upper Bounds 50
3 Sobolev Inequalities 52
4 Heat Kernel Estimates and the Noncompact Case 55
4.1 The Essential Spectrum 56
4.2 Proof of Theorem 6 57
References 59

A Note on the Local Well-Posedness for the Whitham Equation 63
Mats Ehrnström, Joachim Escher and Long Pei
1 Introduction 63
2 General Preliminaries and the Case of Initial Data on $\mathbb{R}$ 64
3 The Periodic Case 73
References 74

On the Lifetime of a Conditioned Brownian Motion in Domains Connected Through Small Gaps 77
Matthias Erven and Guido Sweers
1 Introduction 77
2 The Link Between Brownian Motion and Kernel Functions 80
3 Main Results 83
4 Estimates on Green Functions and Poisson Kernels 84
4.1 Recalling and Adjusting Known Estimates 85
4.2 Estimates for Higher-Order Green Function Derivatives 87
4.3 Asymptotic Estimates for G_{Ω_ℓ} 88
5 The Case that Both x and y Lie in A 91
6 The Main Proof: x Lies in A and y in B 92
6.1 The Halfway Statement 92
6.2 Comparing G_{Ω_ℓ} by Poisson Kernels 95
6.3 3G Asymptotics 101
7 Appendix: Sub- and Supersolution for a Special Dirichlet Problem 106
References 108

Analyticity of Rotational Water Waves 111
Joachim Escher and Bogdan-Vasile Matioc
1 Introduction 111
2 Two-Dimensional Periodic Flows on Water of Finite Depth 112
2.1 The Governing Equations and Equivalent Formulations 112
2.2 Weak Interpretations of the Height Function Formulation 118
2.3 Regularity Results for Periodic Flows on Water of Finite Depth 119
2.4 Regularity of Periodic Waves with Stagnation Points 122
3 Solitary Waves on Water of Finite Depth 123
3.1 Regularity of Solitary Waves 123
3.2 An Intrinsic Characterization of the Symmetric Solitary Waves 125
4 Deep-water Waves 133
5 Periodic Stratified Flows in Water of Finite Depth 134
References 135

Degenerate and Singular Porous Medium Type Equations with Measure Data 139
Ugo Gianazza
1 Introduction 139
2 A Priori Estimates for Weak Solutions: The Degenerate Case 142
3 A Priori Estimates for Weak Solutions: The Singular Case 145
4 Existence of Very Weak Solutions when $m > 1$ 148
5 Existence of Very Weak Solutions when $\frac{(N-2)_+}{N} < m < 1$ 152
References 156

Aspects of the Mathematical Analysis of Nonlinear Stratified Water Waves 159
David Henry and Bogdan-Vasile Matioc
1 Introduction 159
1.1 The Mathematical Model 160
2 Existence Results for Flows Admitting Stagnation Points 163
2.1 Stratified Flows with Linear Density Distribution and Constant Bernoulli Function 164
2.2 Stratified Flows with More General Density Distribution and Bernoulli Function 167
3 Existence Results for Flows Without Critical Points 169
3.1 Gravity Stratified Waves 169
3.2 Capillary-Gravity Stratified Waves 171
4 Stratification for Gerstner-Type Flows 172
5 Qualitative Properties for General Stratified Flows 173
References 175

A Calculus of Abstract Edge Pseudodifferential Operators of Type ρ, δ 179
Thomas Krainer
1 Introduction 179
2 Abstract Wedge Sobolev Spaces 180
3 Abstract Edge Pseudodifferential Operators 186
4 Boundedness and Compactness 197
5 Trace Class Operators 201
6 Appendix: Banach Operator Ideals in Hilbert Spaces 205
References 207

Boundary Value Problems for Elliptic Wedge Operators: The First-Order Case 209
Thomas Krainer and Gerardo A. Mendoza
1 Introduction 209
2 Setup 210
3 Some Considerations 211
4 Ellipticity, the Wedge Cotangent Bundle and the Structure Ring 212
5 Indicial and Normal Families 213
6 The Kernel Bundle of a Holomorphic Fredholm Family 215
7 Elliptic Systems of Variable Order 222
8 Boundary Value Problems for First-Order Elliptic Wedge Operators 228
References 231

The Time Singular Limit for a Fourth-Order Damped Wave Equation for MEMS 233
Philippe Laurençot and Christoph Walker
1 Introduction 233
1.1 State of the Art 235
1.2 The Time Singular Limit 239

2 A Lower Bound on the Maximal Existence Time ... 240
3 The Time Singular Limit $\gamma^2 \longrightarrow 0$ ... 244
References ... 245

Compositions in the Edge Calculus ... 247
B.-Wolfgang Schulze
1 Introduction ... 247
2 Edge Quantisation ... 248
2.1 Mellin Operators and the Cone Algebra ... 248
2.2 Smoothing Mellin Plus Green Edge Symbols ... 252
2.3 Edge-Degenerate Operators and Mellin-Edge Quantisation 255
3 Composition Results ... 256
3.1 Properties of Edge Symbols ... 256
3.2 Compositions ... 260
3.3 Conclusions and Remarks ... 263
4 Auxiliary Material ... 264
4.1 Kegel Spaces and Other Aspects of the Cone Calculus ... 264
4.2 Edge Spaces and Symbols with Twisted Symbolic Estimates 267
4.3 Asymptotics ... 269
References ... 270

On Bifurcation for Semilinear Elliptic Dirichlet Problems on Shrinking Domains ... 273
Nils Waterstraat
1 Introduction ... 273
2 Bifurcation of Critical Points for Essentially Positive Functionals . 276
2.1 Self-Adjoint Fredholm Operators and Their Spectra ... 277
2.2 The Bifurcation Theorem ... 279
3 Elliptic Dirichlet Problems on Shrinking Domains ... 283
4 Corollaries and Examples ... 288
References ... 290

Contributors

Herbert Amann Mathematical Institute, University of Zurich, Zurich, Switzerland

Nelia Charalambous Department of Mathematics and Statistics, University of Cyprus, Nicosia, Cyprus

Mats Ehrnström Department of Mathematical Sciences, Norwegian University of Science and Technology, Trondheim, Norway

Matthias Erven Mathematical Institute, University of Cologne, Cologne, Germany

Joachim Escher Institute for Applied Mathematics, Leibniz University of Hannover, Hannover, Germany

Ugo Gianazza "F. Casorati" Department of Mathematics, University of Pavia, Pavia, Italy

David Henry School of Mathematical Sciences, University College Cork, Cork, Ireland

Thomas Krainer Department of Mathematics and Statistics, Pennsylvania State University, Altoona Campus, Altoona, PA, USA

Philippe Laurençot Toulouse Institute of Mathematics, University of Toulouse, France

Zhiqin Lu Department of Mathematics, University of California, Irvine, CA, USA

Bogdan-Vasile Matioc Institute for Applied Mathematics, Leibniz University of Hannover, Hannover, Germany

Gerardo A. Mendoza Department of Mathematics, Temple University, Philadelphia, PA, USA

Long Pei Department of Mathematical Sciences, Norwegian University of Science and Technology, Trondheim, Norway

Julie Rowlett Mathematical Sciences, Chalmers University of Technology and the University of Gothenburg, Göteborg, Sweden

B.-Wolfgang Schulze Institute for Mathematics, University of Potsdam, Potsdam, Germany

Guido Sweers Mathematical Institute, University of Cologne, Cologne, Germany

Christoph Walker Institute for Applied Mathematics, Leibniz University of Hannover, Hannover, Germany

Nils Waterstraat Institute for Mathematics, Humboldt University of Berlin, Berlin, Germany

Uniformly Regular and Singular Riemannian Manifolds

Herbert Amann

Abstract A detailed study of uniformly regular Riemannian manifolds and manifolds with singular ends is carried out in this paper. Such classes of manifolds are of fundamental importance for a Sobolev space solution theory for parabolic evolution equations on noncompact Riemannian manifolds with and without boundary. Besides pointing out this connection in some detail, we present large families of uniformly regular and singular manifolds which are admissible for this analysis.

1 Introduction

The principal object of our concern is an in-depth study of evolution equations on noncompact Riemannian manifolds. We are particularly interested in establishing an optimal local existence theory for quasilinear parabolic initial boundary value problems in a Sobolev space framework. For this we need, in the first instance, a good understanding of fractional order L_p-Sobolev spaces, including sharp embedding and trace theorems, etc. Although fractional order Sobolev spaces can be invariantly defined on any noncompact Riemannian manifold, it is not possible to establish embedding and trace theorems in this generality. For these to hold one has to impose restrictions on the underlying manifold near infinity.

In our paper [2] we have introduced the class of uniformly regular Riemannian manifolds and shown, in particular, that fractional order Sobolev spaces on such manifolds possess all the properties alluded to above. (Also see [1] for complements and extensions to anisotropic settings.) This class encompasses the well-studied case of complete Riemannian manifolds without boundary and bounded geometry. Of course, in the study of boundary value problems manifolds with boundary are indispensable. In our previous papers [1], [2], and [3] we have presented examples

H. Amann (✉)
Mathematical Institute, University of Zurich, Winterthurerstr. 190, 8057 Zürich, Switzerland,
e-mail: herbert.amann@math.uzh.ch

J. Escher et al., (eds.), *Elliptic and Parabolic Equations,* Springer Proceedings in Mathematics & Statistics 119, DOI 10.1007/978-3-319-12547-3_1

of manifolds with boundary which are uniformly regular. Yet proofs have not been included. The reason being that it needs quite a bit of argumentation to establish these claims. It is the purpose of this paper to close this gap and carry out a detailed study of uniformly regular Riemannian manifolds. Some of the main results and their ramifications are explained in the following.

Let (M,g) be a smooth m-dimensional Riemannian manifold with boundary (which may be empty). Unless explicitly stated otherwise, $m \in \mathbb{N}^\times := \mathbb{N}\backslash\{0\}$. An atlas $\mathfrak{K}$ for M is said to be uniformly regular if it consists of normalized charts[1], has finite multiplicity, all coordinate changes have uniformly bounded derivatives of all orders, and if it is shrinkable. By the latter we mean that there is a uniform shrinking of all chart domains such that the result is an atlas as well. The normalization of the local charts also means that they are well adapted to the boundary in a natural precise sense. The shrinkability assumption is the most restrictive one. For example, the open unit ball in $\mathbb{R}^m$, endowed with the Euclidean metric $|dx|^2$, does not possess a uniformly regular atlas.

Let M be equipped with a uniformly regular atlas $\mathfrak{K}$. The metric g is called uniformly regular if its local representation $\kappa_* g$ is equivalent to the Euclidean metric of $\mathbb{R}^m$ and has bounded derivatives of all orders, uniformly with respect to $\kappa \in \mathfrak{K}$. Then (M,g) is said to be *uniformly regular* if it possesses a uniformly regular atlas $\mathfrak{K}$ and g is uniformly regular. Vaguely speaking, this means that M has an atlas whose coordinate patches are all "of approximately the same size." The concept of uniform regularity is independent of the particular choice of the atlas $\mathfrak{K}$ in a natural sense (made precise in (2.5)). Fortunately, in practice a specific atlas is rarely needed. It suffices to know that there exists one.

We denote by c constants ≥ 1 whose actual value may vary from occurrence to occurrence; but c is always independent of the free variables in a given formula, unless a dependence is explicitly indicated.

On the set of all nonnegative functions, defined on some nonempty set S, whose specific realization will be clear in any given situation, we introduce an equivalence relation $\sim$ by writing $f \sim g$ iff there exists c such that $f/c \leq g \leq cf$. Here inequalities between symmetric bilinear forms are understood as inequalities between the corresponding polar forms. By $\mathbf{1}$, more precisely $\mathbf{1}_S$, we denote the constant function $S \to \mathbb{R},\ s \mapsto 1$.

Now we present some examples to illustrate the extent of the concept of uniform regularity.

Examples 1.1. *(a)* $\mathbb{R}^m = (\mathbb{R}^m, |dx|^2)$ *and closed half-spaces thereof are uniformly regular.*

(b) Every compact Riemannian manifold is uniformly regular.

(c) If (M,g) is a Riemannian submanifold with compact boundary of a uniformly regular Riemannian manifold, then it is uniformly regular also.

(d) Complete Riemannian manifolds without boundary and bounded geometry are uniformly regular.

[1] Precise definitions of all concepts used in this introduction without further explanation are found in the main body of this paper—in Sect. 2, in particular.

(e) Products of uniformly regular manifolds are uniformly regular.

(f) If (M_1,g_1) and (M_2,g_2) are isometric, then (M_1,g_1) is uniformly regular iff (M_2,g_2) is so.

Proof. For (a) see (3.3). Statements (b)–(d) are proved in Sect. 4. Assertion (e) is a particular instance of Theorem 3.1. Claim (f) follows from Lemma 3.4. □

There are also Riemannian manifolds with singular ends which are uniformly regular. To explain this in more detail we need some preparation. We fix $d \geq m-1$ and suppose that (B, g_B) is an $(m-1)$-dimensional Riemannian submanifold of $\mathbb{R}^d$. Then, given $\alpha \geq 0$,

$$\left\{ (t, t^{\alpha} y) \ ; \ t > 1, \ y \in B \right\} \subset \mathbb{R} \times \mathbb{R}^d = \mathbb{R}^{1+d} \tag{1.1}$$

is the *infinite model α-funnel over B*. We denote it by $F_\alpha(B)$. It is an infinite cylinder if $\alpha = 0$, and an infinite (blunt) cone if $\alpha = 1$. Note that $F_\alpha(B)$ is an m-dimensional submanifold of $\mathbb{R}^{1+d}$. If $\partial B \neq \emptyset$, then $\partial F_\alpha(B) = F_\alpha(\partial B)$.

In Fig. 1 there is depicted part of a (rotated) three-dimensional model funnel $F_{1/2}(B)$ with a compact base B having two connected components and three boundary components.

For $0 \leq \alpha \leq 1$ we endow $F_\alpha(B)$ with the metric $g_{F_\alpha(B)}$ induced by the embedding $F_\alpha(B) \hookrightarrow \mathbb{R}^{1+d}$ so that $\bigl(F_\alpha(B), g_{F_\alpha(B)}\bigr)$ is a Riemannian submanifold of $(\mathbb{R}^{1+d}, |dx|^2)$. An open subset V of M is an *infinite α-funnel over B* if (V, g) is isometric to $\bigl(F_\alpha(B), g_{F_\alpha(B)}\bigr)$. It is a *tame end of* (M,g) if it is an infinite α-funnel with α belonging to $[0,1]$ and B being compact.

Suppose $\{V_0, V_1, \ldots, V_k\}$ is a finite open covering of M with $V_i \cap V_j = \emptyset$ for $1 \leq i < j \leq k$ such that V_i is a tame end for $1 \leq i \leq k$ and V_0, $V_0 \cap V_1, \ldots,$ $V_0 \cap V_k$ are relatively compact in M. Then (M,g) is said to have (finitely many) *tame ends.*

Theorem 1.2. *If (M,g) has tame ends, then it is uniformly regular.*

Proof. Section 8. □

As aforementioned, our motivation for the study of uniformly regular Riemannian manifolds stems from the theory of parabolic equations. To explain their role in the present environment we consider a simple model problem. We set

$$\mathscr{A} u := -\operatorname{div}(a \cdot \operatorname{grad} u) \ , \tag{1.2}$$

with a being a symmetric positive definite $(1,1)$-tensor field on (M,g) which is bounded and has bounded and continuous first order (covariant) derivatives. This is expressed by saying that $\mathscr{A}$ is a regular uniformly strongly elliptic differential operator. This low regularity assumption for a is of basic importance for treating quasilinear problems in which a depends on u.

We assume that $\partial_0 M$ is open and closed in ∂M and $\partial_1 M := \partial M \backslash \partial_0 M$. Then we put

$$\mathscr{B}_0 u := u \text{ on } \partial_0 M\ , \quad \mathscr{B}_1 u := (\nu \,|\, a \bullet \operatorname{grad} u) \text{ on } \partial_1 M,$$

where these operators are understood in the sense of traces and ν is the inward pointing unit normal vector field on $\partial_1 M$. Thus $\mathscr{B} := (\mathscr{B}_0, \mathscr{B}_1)$ is the Dirichlet boundary operator on $\partial_0 M$ and the Neumann operator on $\partial_1 M$.

Suppose $0 < T < \infty$. We write $M_T := M \times [0,T]$ for the space time cylinder. Moreover, $\partial = \partial_t$ is the "time derivative," $\Sigma_T := \partial M \times [0,T]$ the lateral boundary, and $M_0 = M \times \{0\}$ the "initial surface" of M_T. Then we consider the problem

$$\partial u + \mathscr{A} u = f \text{ on } M_T\ , \quad \mathscr{B} u = 0 \text{ on } \Sigma_T\ , \quad u = u_0 \text{ on } M_0. \tag{1.3}$$

The last equation is to be understood as $\gamma_0 u = u_0$ with the "initial trace" operator γ_0 .

Of course, $\partial_0 M$ or $\partial_1 M$ or both may be empty. In such a situation obvious interpretations and modifications are to be applied.

We are interested in an optimal L_p-theory for (1.3). To describe it we have to introduce Sobolev–Slobodeckii spaces. We always assume that $1 < p < \infty$. The Sobolev space $W_p^k(M)$ is then defined for $k \in \mathbb{N}$ to be the completion of $\mathscr{D}(M)$, the space of smooth functions with compact support, in $L_{1,\mathrm{loc}}(M)$ with respect to the norm

$$u \mapsto \Big(\sum_{j=0}^{k} \| \, |\nabla^j u|_{g_0^j} \|_{L_p(M)}^p \Big)^{1/p} . \tag{1.4}$$

Here $\nabla = \nabla_g$ is the Levi-Civita covariant derivative and $|\cdot|_{g_0^j}$ the $(0,j)$-tensor norm naturally induced by g. Thus $W_p^0(M) = L_p(M)$. If $s \in \mathbb{R}^+ \backslash \mathbb{N}$, then the Slobodeckii space $W_p^s(M)$ is defined by real interpolation:

$$W_p^s(M) := \big(W_p^k(M), W_p^{k+1}(M) \big)_{s-k,p}\ , \qquad k < s < k+1\ , \quad k \in \mathbb{N}\ .$$

Although these definitions are meaningful on any Riemannian manifold, they are not too useful in such a general setting since they may lack basic Sobolev type embedding properties, for example. The situation is different if we restrict ourselves to uniformly regular Riemannian manifolds. The following theorem is a consequence of the results of our paper [2] to which we direct the reader for details, proofs, and many more facts.

Theorem 1.3. *Let (M,g) be uniformly regular. Then the Sobolev–Slobodeckii spaces $W_p^s(M)$, $s \geq 0$, possess the same embedding, interpolation, and trace properties as in the classical Euclidean case. They can be characterized by means of local coordinates.*

We denote by $W_{p,\mathscr{B}}^s(M)$ the closed linear subspace of all $u \in W_p^s(M)$ satisfying $\mathscr{B} u = 0$ whenever s is such that this condition is well-defined (cf. [3, (1.4)]).

We set

$$A := \mathscr{A} \,|\, W_{p,\mathscr{B}}^2(M)\ ,$$

considered as an unbounded linear operator in $L_p(M)$ with domain $W^2_{p,\mathscr{B}}(M)$. Then (1.3) can be expressed as an initial value problem for the evolution equation

$$\dot{u} + Au = f \text{ on } [0,T]\,, \quad u(0) = u_0$$

in $L_p(M)$.

Now we are ready to formulate the basic well-posedness result in the present model setting.

Theorem 1.4. *Let (M,g) be uniformly regular and $p \notin \{3/2,\ 3\}$. Suppose that $\mathscr{A}$ is regularly uniformly strongly elliptic. Then (1.3) has for each*

$$(f,u_0) \in L_p\big([0,T],L_p(M)\big) \times W^{2-2/p}_{p,\mathscr{B}}(M)$$

a unique solution

$$u \in L_p\big([0,T],W^2_{p,\mathscr{B}}(M)\big) \cap W^1_p\big([0,T],L_p(M)\big).$$

The map $(f,u_0) \mapsto u$ is linear and continuous.
Equivalently: $-A$ generates an analytic semigroup on $L_p(M)$ and has the property of maximal regularity.

For this theorem we refer to [3] where nonhomogeneous boundary conditions and lower order terms are treated as well and further references are given. Analogous theorems apply to higher order problems and parabolic equations operating on sections of uniformly regular vector bundles.

On the surface, Theorem 1.4 looks exactly the same as the very classical existence and uniqueness theorem for second order parabolic equations on open subsets of $\mathbb{R}^m$ with smooth compact boundary (e.g., O.A. Ladyzhenskaya, V.A. Solonnikov, and N.N. Ural'ceva [23, Chapter IV] and R. Denk, M. Hieber, and J. Prüss [9]). However, it is, in fact, a rather deep-rooted vast generalization thereof, since it applies to any uniformly regular Riemannian manifold.

In this connection we have to mention the work of G. Grubb [15] who established a general L_p theory for parabolic pseudo-differential boundary value problems (also see Sect. IV.4.1 in [16]). It applies to a class of noncompact manifolds, called "admissible," introduced in G. Grubb and N.J. Kokholm [17]. It is a subclass of the above manifolds with tame ends, namely a family of manifolds with conical ends. Earlier investigations of pseudo-differential operators on manifolds with conical ends are due to E. Schrohe [27] who employs weighted Sobolev spaces.

Recently, a maximal regularity theory for parabolic differential equations on Riemannian manifolds without boundary and cylindrical ends has been presented by Th. Krainer [22]. This author uses a compactification technique to "reduce" the problem to a compact Riemannian manifold $(\tilde{M},\tilde{g})$, where $\tilde{g}$ is the cusp metric $dt^2/t^4 + g_Y$ in a collar neighborhood of the boundary Y of $\tilde{M}$. Then the theory of cusp pseudo-differential operators is applied in conjunction with the general $\mathscr{R}$-boundedness theory of maximal regularity for parabolic evolution equations. The final result is then formulated in the Sobolev space setting for $(\tilde{M},\tilde{g})$ which involves rather complicated weighted norms. In contrast, our result Theorem 1.4 uses the Sobolev space setting of (M,g) only. Due to Theorem 1.2, it applies to manifolds with cylindrical ends, in particular.

There is a tremendous amount of literature on heat equations on complete Riemannian manifolds without boundary and bounded geometry. Most of it concerns heat kernel estimates and spectral theory (see, for example, E.B. Davies [8] or A. Grigor'yan [13] and the references therein). By imposing further structural conditions, as the assumption of non-negative Ricci curvature, for instance, heat kernel estimates lead to maximal regularity results for the Laplace–Beltrami operator (e.g. M. Hieber and J. Prüss [18], A.L. Mazzucato and V. Nistor [24]. Also see A. Grigor'yan and L. Saloff-Coste [14] and L. Saloff-Coste [26]). Due to Example 1.1(d) our Theorem 1.4 applies in this setting without any additional restriction on the geometry of (M,g).

Let now (M,g) be a Riemannian manifold, which is not uniformly regular. It is said to be *singular of type* ρ if $\rho \in C^\infty\big(M,(0,\infty)\big)$ and $(M,g/\rho^2)$ is uniformly regular. Any such ρ is a *singularity function* for (M,g). We assume that ρ is bounded from above. Then $\inf\rho = 0$ and (M,g) is said to be *singular near* $\rho = 0$. In order for $\rho \in C^\infty\big(M,(0,\infty)\big)$ to qualify as a singularity function it has to satisfy structural conditions naturally associated with (M,g) (see (2.7)). Below we describe a large class of singularity functions which are closely related to the geometric structure near the "singular ends" of (M,g), that is, the behavior of (M,g) "near infinity."

Suppose (M,g) is singular of type ρ. We set $\hat{g} := g/\rho^2$ and $(\hat{M},\hat{g}) := (M,g/\rho^2)$. Then we can apply the preceding results to the uniformly regular Riemannian manifold $(\hat{M},\hat{g})$. Since $\hat{g}$ is conformally equivalent to g (and ρ satisfies appropriate structural conditions), we can express the Sobolev–Slobodeckii spaces $W_p^s(\hat{M})$, which are constructed by means of $\nabla_{\hat{g}}$, in terms of weighted Sobolev–Slobodeckii spaces on M. More precisely, we define $W_p^{k,\lambda}(M;\rho)$ for $k \in \mathbb{N}$ and $\lambda \in \mathbb{R}$ by replacing (1.4) in the definition of $W_p^k(M)$ by

$$u \mapsto \Big(\sum_{j=0}^{k} \big\|\rho^{\lambda+j}\,|\nabla^j u|_{g_0^j}\big\|_{L_p(M)}^p\Big)^{1/p}.$$

Furthermore,

$$W_p^{s,\lambda}(M;\rho) := \big(W_p^{k,\lambda}(M;\rho), W_p^{k+1,\lambda}(M;\rho)\big)_{s-k,p}, \qquad k<s<k+1, \quad k\in\mathbb{N},$$

and $L_p^\lambda(M;\rho) := W_p^{0,\lambda}(M;\rho)$. Then, see [4],

$$W_p^s(\hat{M}) \doteq W_p^{s,-m/p}(M;\rho), \qquad s \ge 0,$$

where $\doteq$ means: equal except for equivalent norms. In [4] it is also shown that

$$W_p^s(\hat{M}) \to W_p^{s,\lambda}(M;\rho), \quad u \mapsto \rho^{-\lambda+m/p}u$$

is an isomorphism. With its help we can transfer all properties enjoyed by the Sobolev–Slobodeckii spaces $W_p^s(\hat{M})$ to the weighted spaces $W_p^{s,\lambda}(M;\rho)$ (direct proofs, not using this isomorphism, are given in [2]).

There are also simple relations between the differential operators div and grad on (M,g) and $\operatorname{div}_{\hat{g}}$ and $\operatorname{grad}_{\hat{g}}$ on $(\hat{M},\hat{g})$, respectively. In fact, setting $\hat{a} := \rho^{-2}a$ we find (cf. [3, (5.19)])

$$\operatorname{div}(a \operatorname{grad} u) = \operatorname{div}_{\hat{g}}(\hat{a}\cdot \operatorname{grad}_{\hat{g}} u) + (u\hat{a}\cdot\rho^{-1}\operatorname{grad}_{\hat{g}}\rho \mid \operatorname{grad}_{\hat{g}} u)_{\hat{g}} .$$

Note that Theorem 1.4 applies to the operator

$$\hat{\mathscr{A}} u := -\operatorname{div}_{\hat{g}}(\hat{a}\cdot\operatorname{grad}_{\hat{g}} u)$$

provided it is regularly uniformly strongly elliptic on $(\hat{M},\hat{g})$. This is equivalent to the assumption that (1.2) be *regularly uniformly strongly ρ-elliptic*. By this we mean that the following conditions are satisfied:

$$\begin{aligned} &\text{(i)} \qquad \big(a(q)\cdot X \,|\, X\big)_{g(q)} \sim \rho^2(q)\,|X|^2_{g(q)}, \ \ X\in T_qM, \ \ q\in M .\\ &\text{(ii)} \qquad |\nabla a|_{g^2_1} \le c\rho . \end{aligned}$$

An elaboration of these facts leads to the following optimal well-posedness result for degenerate parabolic equations on singular manifolds. It is a special case of Theorem 5.2 of [3].

Theorem 1.5. *Let (M,g) be singular of type ρ and $p \notin \{3/2,\ 3\}$. Suppose $\mathscr{A}$ is regularly uniformly strongly ρ-elliptic and $\lambda\in\mathbb{R}$. Then problem (1.3) has for each*

$$(f,u_0)\in L_p\big([0,T],L_p^\lambda(M;\rho)\big)\times W_{p,\mathscr{B}}^{2-2/p,\lambda}(M;\rho)$$

a unique solution

$$u\in L_p\big([0,T],W_{p,\mathscr{B}}^{2,\lambda}(M;\rho)\big)\cap W_p^1\big([0,T],L_p^\lambda(M;\rho)\big) .$$

The map $(f,u_0)\mapsto u$ is linear and continuous.

Equivalently: let

$$A^\lambda := \mathscr{A} \,|\, W_{p,\mathscr{B}}^{2,\lambda}(M;\rho) .$$

Then $-A^\lambda$ generates a strongly continuous analytic semigroup on $L_p^\lambda(M;\rho)$ and has the property of maximal regularity.

In order to render this theorem useful we have to provide sufficiently large and interesting classes of singular manifolds. This is the aim of the following considerations.

Let (B,g_B) be as in definition (1.1). If we choose there $\alpha<0$, then we call the resulting Riemannian submanifold of $\mathbb{R}^{1+d}$ *infinite model α-cusp over B* and denote it by $C_{\infty,\alpha}(B)$ and its metric by $g_{C_{\infty,\alpha}(B)}$. Similarly as for funnels, an open subset V

of M is an *infinite α-cusp over B* of (M,g) if (V,g) is isometric to an infinite model α-cusp $\big(C_{\infty,\alpha}(B), g_{C_{\infty,\alpha}(B)}\big)$. It is *smooth* if B is compact.

We consider the following conditions:

$$
\left.\begin{array}{rl}
\text{(i)} & (\mathscr{M},g) \text{ is an } m\text{-dimensional Riemannian manifold}\,. \\
\text{(ii)} & \mathbf{V} \text{ is a finite set of pairwise disjoint infinite smooth } \alpha\text{-cusps } V \\
& \text{of } (\mathscr{M},g)\,. \\
\text{(iii)} & V_0 \text{ is an open subset of } \mathscr{M} \text{ such that } \{V_0\}\cup\mathbf{V} \text{ is a covering of } \mathscr{M} \\
& \text{and } (\mathscr{M},g) \text{ is uniformly regular on}^2\ V_0\,. \\
\text{(iv)} & \boldsymbol{\Gamma} \text{ is a finite set of pairwise nonintersecting compact connected} \\
& \text{Riemannian submanifolds } \Gamma \text{ of } \mathscr{M} \text{ without boundary and} \\
& \text{codimension at least 1 such that } \Gamma\subset\partial\mathscr{M} \text{ if } \Gamma\cap\partial\mathscr{M}\neq\emptyset \\
& \text{and } \Gamma\cap V=\emptyset \text{ for } \Gamma\in\boldsymbol{\Gamma} \text{ and } V\in\mathbf{V}\,. \\
\text{(v)} & \beta_\Gamma\ge 1 \text{ for } \Gamma\in\boldsymbol{\Gamma}.
\end{array}\right\} \qquad (1.5)
$$

We set

$$\boldsymbol{\alpha} := \{\alpha_V\ ;\ V\in\mathbf{V}\}\,,\quad \boldsymbol{\beta} := \{\beta_\Gamma\ ;\ \Gamma\in\boldsymbol{\Gamma}\}\,,\quad \mathscr{S} := \textstyle\bigcup_{\Gamma\in\boldsymbol{\Gamma}}\Gamma\,,$$

and

$$(M,g) := \big(\mathscr{M}\backslash\mathscr{S},\ g\,|\,(\mathscr{M}\backslash\mathscr{S})\big)\,.$$

Then $\boldsymbol{\alpha}$, respectively $\boldsymbol{\beta}$, is the *cuspidal weight* (vector) for $\mathbf{V}$, respectively $\boldsymbol{\Gamma}$, and $\mathscr{S}$ the *(compact) singularity set* of M. Furthermore, (M,g) is said to be a *Riemannian manifold with smooth cuspidal singularities of type* $[\mathbf{V},\boldsymbol{\alpha},\boldsymbol{\Gamma},\boldsymbol{\beta}]$.

If $\mathbf{V}=\emptyset$ and $\boldsymbol{\Gamma}=\emptyset$, then $\mathscr{M}=V_0$ and $(\mathscr{M},g)=(M,g)$ is uniformly regular. Thus we assume henceforth that $\mathbf{V}\cup\boldsymbol{\Gamma}\neq\emptyset$. If $\mathbf{V}=\emptyset$, then $V_0=\mathscr{M}$ and $(\mathscr{M},g)$ is uniformly regular. By the preceding results this is the case, in particular, if $\mathscr{M}$ is compact or $(\mathscr{M},g)$ has tame ends. However, (M,g) is not uniformly regular.

By its definition, (M,g) is a Riemannian submanifold of the ambient manifold $(\mathscr{M},g)$. In turn, the latter is obtained from (M,g) by setting $\mathscr{M} := M\cup\mathscr{S}$ and defining $g_{\mathscr{M}}$ by smooth extension of g. The crucial point of this procedure is that $(\mathscr{M},g_{\mathscr{M}})$ is a Riemannian manifold as well. To avoid technical subtleties we prefer to take $(\mathscr{M},g)$ as initial object. Due to the intimate connection between $(\mathscr{M},g)$ and (M,g) there is often no need to mention $(\mathscr{M},g)$ explicitly.

For $V\in\mathbf{V}$ we fix $q=q_V\in\bar V\backslash V$ and set

$$\delta_V=\delta_{V,q} := 1+\mathrm{dist}_{\mathscr{M}}(\cdot,q)\colon V\to[1,\infty) \qquad (1.6)$$

where $\mathrm{dist}_{\mathscr{M}}$ is the Riemannian distance in $(\mathscr{M},g)$. Note that $\sup\delta_V=\infty$ and $\mathrm{dist}_{\mathscr{M}}(\cdot,q)=\mathrm{dist}_M(\cdot,q)$ on V.

For $\Gamma\in\boldsymbol{\Gamma}$ there exists an open neighborhood $\mathscr{U}_\Gamma$ of Γ in $\mathscr{M}$ with $\mathscr{U}_\Gamma\cap\tilde\Gamma=\emptyset$ for $\tilde\Gamma\in\boldsymbol{\Gamma}$ satisfying $\tilde\Gamma\neq\Gamma$, and such that $\mathrm{dist}_{\mathscr{M}}(\cdot,\Gamma)$ is a well-defined smooth function. Then $U_\Gamma := \mathscr{U}_\Gamma\backslash\Gamma$ is open in M and the restriction δ_Γ of $\mathrm{dist}_{\mathscr{M}}(\cdot,\Gamma)$ to U_Γ is smooth and everywhere positive.

---

[2] Cf. the localized definitions in Sect. 2.

The following theorem is the main result of this paper as far as singular manifolds are concerned. Its proof is given in Sect. 8. Here and in similar situations obvious interpretations have to be used if either $\mathbf{V}$ or $\boldsymbol{\Gamma}$ is empty.

Theorem 1.6. *Let* (M,g) *be a Riemannian manifold with smooth cuspidal singularities of type* $[\mathbf{V},\boldsymbol{\alpha},\boldsymbol{\Gamma},\boldsymbol{\beta}]$. *Fix* $\rho \in C^\infty\big(M,(0,1]\big)$ *such that* $\rho \sim \mathbf{1}$ *on* V_0, $\rho \sim \delta_V^{\alpha_V}$ *on* $V \in \mathbf{V}$, *and* $\rho \sim \delta_\Gamma^{\beta_\Gamma}$ *near* $\Gamma \in \boldsymbol{\Gamma}$. *Then* (M,g) *is singular of type* ρ.

Corollary 1.7. *Theorem 1.5 applies whenever* (M,g) *is a Riemannian manifold with cuspidal singularities.*

It follows from the considerations in the main body of this paper that the special choice of ρ is of no importance. In fact: if ρ is replaced by $\tilde\rho$ with $\tilde\rho \sim \rho$, then $(M,\ g/\rho^2)$ and $(M,\ g/\tilde\rho^2)$ are equivalent in the sense defined in Sect. 2. In particular, the Sobolev–Slobodeckii spaces $W_p^s(M;\rho)$ and $W_p^s(M;\tilde\rho)$ differ only by equivalent norms. Thus merely the behavior of ρ "near infinity along V," for $V \in \mathbf{V}$, and near $\Gamma \in \boldsymbol{\Gamma}$, where ρ approaches zero, does matter. Notably, this shows that the choice of $q_V \in \bar V \backslash V$, as well as the special form of ρ on compact subsets of M, is irrelevant.

It remains to explain why the naming "manifold with smooth cuspidal singularities" has been chosen. This is clear if $\boldsymbol{\Gamma} = \emptyset$, but needs elucidation otherwise. The following considerations contribute to it. But first we introduce some notation.

For $d \in \mathbb{N}^\times$ we denote by $\mathbb{B}^d$ the open unit ball in $\mathbb{R}^d$, by $\mathbb{S}^{d-1}$ its boundary, the unit sphere, and by $\mathbb{H}^d := \mathbb{R}^+ \times \mathbb{R}^{d-1}$ the closed right half-space, where $\mathbb{R}^0 := \{0\}$. Then $\mathbb{B}_+^d := \mathbb{B}^d \cap \mathbb{H}^d$ and $\mathbb{S}_+^{d-1} := \mathbb{S}^{d-1} \cap \mathbb{H}^d$ are the right half-ball and half-sphere, respectively. Note that $\partial\mathbb{B}_+^d = \{0\} \times \mathbb{B}^{d-1} \cong \mathbb{B}^{d-1}$ and $\partial\mathbb{S}_+^{d-1} = \{0\} \times \mathbb{S}^{d-2}$ if $d \geq 2$ and $\partial\mathbb{S}^0 = \emptyset$. Lastly, $\dot{\mathbb{B}} := \mathbb{B}\backslash\{0\}$ for $\mathbb{B} \in \{\mathbb{B}^d, \mathbb{B}_+^d\}$.

Suppose $1 \leq \ell \leq m$ and $\mathbb{S} \in \{\mathbb{S}^{\ell-1}, \mathbb{S}_+^{\ell-1}\}$. Given $\alpha \geq 1$,

$$C_\alpha(\mathbb{S}) = C_{\alpha,\ell}(\mathbb{S}) := \big\{\, (t, t^\alpha y)\ ;\ 0 < t < 1,\ y \in \mathbb{S} \,\big\} \subset \mathbb{R}^{1+\ell} \tag{1.7}$$

is an ℓ-dimensional submanifold of $\mathbb{R}^{1+\ell}$ and

$$\varphi_\alpha \colon C_\alpha(\mathbb{S}) \to (0,1) \times \mathbb{S}\,, \quad (t, t^\alpha y) \mapsto (t,y)$$

is the "canonical stretching diffeomorphism." Observe that $\partial C_\alpha(\mathbb{S}) = \emptyset$ if $\mathbb{S} = \mathbb{S}^{\ell-1}$ or $\ell = 1$, and $\partial C_\alpha(\mathbb{S}_+^{\ell-1}) = C_{\alpha,\ell-1}(\mathbb{S}^{\ell-2})$ otherwise.

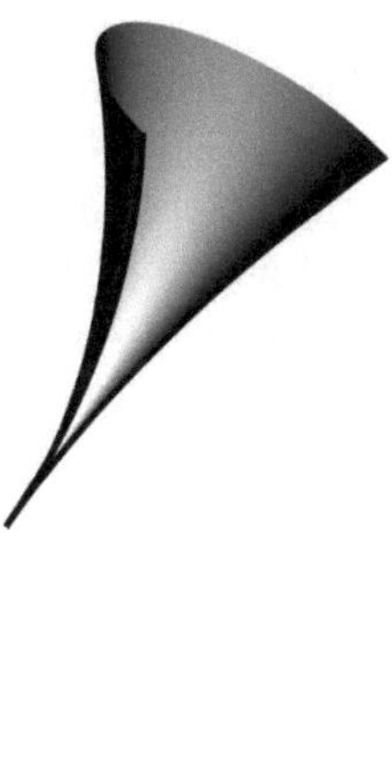

$C_\alpha(\mathbb{S})$ is a (blunt) *model α-cusp*, respectively *cone* if $\alpha = 1$, which is *spherical* if $\mathbb{S} = \mathbb{S}^{\ell-1}$ and *semi-spherical* otherwise. In Fig. 2 there is depicted a (rotated) semi-circular model 2-cusp in $\mathbb{R}^3$. Its boundary consists of two disjoint one-dimensional generators.

We endow $C_\alpha = C_\alpha(\mathbb{S})$ with the Riemannian metric g_{C_α} induced by the natural embedding $C_\alpha \hookrightarrow \mathbb{R}^{1+\ell}$. Then g_{C_α} is equivalent to the pull-back by φ_α of the metric $dt^2 + t^{2\alpha} g_{\mathbb{S}}$ of $(0,1) \times \mathbb{S}$, where $g_{\mathbb{S}}$ is the standard metric induced by $\mathbb{S} \hookrightarrow \mathbb{R}^\ell$.

Assume (Γ, g_Γ) is an $(m-\ell)$-dimensional compact connected Riemannian manifold without boundary. Then $W_\alpha := C_\alpha \times \Gamma$, whose metric is $g_{W_\alpha} := g_{C_\alpha} + g_\Gamma$, is a *model* (α,Γ)-*wedge* which is also called *spherical* if C_α is so, and *semi-spherical* otherwise. If $m = \ell$, then Γ is a one-point space, W_α is naturally identified with C_α, and all references to and occurrences of Γ are to be disregarded. Thus every cusp is a wedge also.

Let U be open in M. Then (U,g), more loosely: U, is a *spherical*, respectively *semi-spherical, cuspidal end of type* (α,Γ) *of* (M,g) if there exists an isometry Φ_α from (U,g) onto a spherical, respectively semi-spherical, model (α,Γ)-wedge (W_α, g_{W_α}). In this case U is *represented by* $[\Phi_\alpha, W_\alpha, g_{W_\alpha}]$ or, simply, by Φ_α.

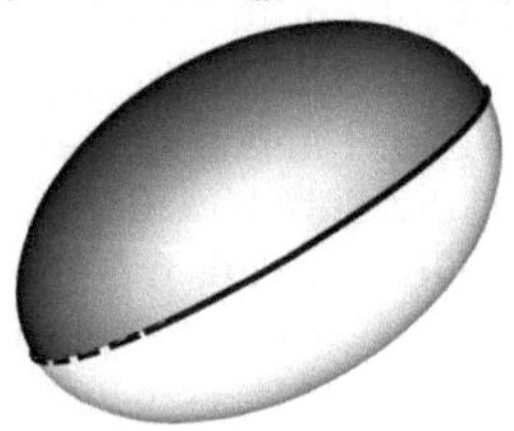

Now we return to the setting of Theorem 1.6 and consider a particular simple constellation. Namely, we assume that M is obtained from a three-dimensional ellipsoid $\mathcal{M}$ in $\mathbb{R}^3$ by removing an equator Γ. Its metric g is induced by the natural embedding $\mathcal{M} \hookrightarrow \mathbb{R}^3$. In this case $\mathbf{V} = \emptyset$ and $\boldsymbol{\Gamma} = \{\Gamma\}$.

On one component, $\partial_0 M$, of the boundary of M we put Dirichlet conditions (e.g. on the dark side of Fig. 3) and Neumann conditions on the other one, $\partial_1 M$. Note that $\partial_0 M$ and $\partial_1 M$ meet in $\mathcal{M}$ along Γ, but "do not see each other" in M. In other words, $\partial_0 M$ and $\partial_1 M$ are both open and closed in ∂M.

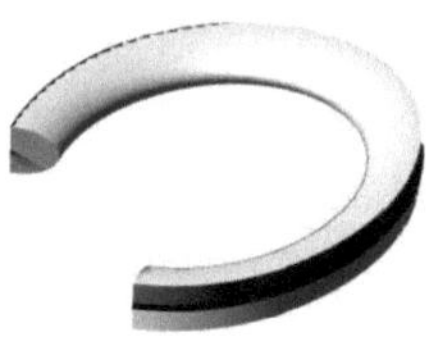

We consider a tubular neighborhood U of Γ in M and represent it as $\dot{\mathbb{B}}^2_+ \times \Gamma$ by means of the tubular diffeomorphism $\tau\colon U \to \dot{\mathbb{B}}^2_+ \times \Gamma$ (see Sect. 8 for details). A part of it is depicted in Fig. 4 in which the curve along the flat side represents Γ $(= \{0\} \times \Gamma)$, which does not belong to $\tau(U)$, however.

Let

$$\pi\colon \dot{\mathbb{B}}^2_+ \to (0,1) \times \mathbb{S}^1_+, \quad x \mapsto (|x|,\ x/|x|)$$

be the polar coordinate diffeomorphism. Then, given $\alpha \geq 1$, the composition

$$U \xrightarrow{\ \tau\ } \dot{\mathbb{B}}^2_+ \times \Gamma \xrightarrow{\ \pi \times \mathrm{id}_\Gamma\ } (0,1) \times \mathbb{S}^1_+ \times \Gamma \xrightarrow{\ \varphi_\alpha^{-1} \times \mathrm{id}_\Gamma\ } C_\alpha(\mathbb{S}^1_+) \times \Gamma \tag{1.8}$$

defines a diffeomorphism Φ_α from U onto the semi-circular model (α,Γ)-wedge $W_\alpha = C_\alpha(\mathbb{S}^1_+) \times \Gamma$. We equip $C_\alpha(\mathbb{S}^1_+)$ with the equivalent metric $\varphi_\alpha^*(dt^2 + t^{2\alpha} g_{\mathbb{S}^1_+})$ and give U the pull-back metric $\Phi_\alpha^* g_{W_\alpha}$.

Let $\mathfrak{g}$ be a Riemannian metric for M such that $\mathfrak{g} = \Phi_\alpha^* g_{W_\alpha}$ on U. Then U is a semi-circular (α,Γ)-end of $(M,\mathfrak{g})$. In Sect. 8 it is shown that $\Phi_\alpha^* g_{W_\alpha} \sim g/\delta_\Gamma^{2\alpha}$ on U. Thus, if we fix any $\rho \in C^\infty(M,(0,1])$ with $\rho \sim \delta_\Gamma^{2\alpha}$ on U and $\rho \sim \mathbf{1}$ on $M \backslash U$, it follows from Theorem 1.6 that $(M,\ g/\rho^2)$ is uniformly regular.

These considerations and Corollary 1.7 show that the Zaremba problem on $\mathscr{M}$ for (1.3), in which Dirichlet boundary conditions are assigned on one half of the boundary of the ellipsoid $\mathscr{M}$ and Neumann conditions on the other half, is well-posed provided $\mathscr{A}$ is regularly uniformly strongly ρ-elliptic where $\rho \sim \delta_\Gamma^{2\alpha}$ near Γ and $\rho \sim \mathbf{1}$ away from Γ. They also show that (M,g) can be visualized as a manifold with a cuspidal end of type (α,Γ). This is illustrated by Fig. 5 for the case where $\alpha = 1$.

The arguments used in this simple case extend to the general setting. This leads in Sect. 8 to the proof of the following proposition which clarifies our choice of the name for (1.5).

Proposition 1.8. *Let (M,g) have smooth cuspidal singularities of type $[\mathbf{V},\boldsymbol{\alpha},\boldsymbol{\Gamma},\boldsymbol{\beta}]$ and let $\beta = \beta_\Gamma$ be the cuspidal weight for $\Gamma \in \boldsymbol{\Gamma}$. Then there exists an open neighborhood $\mathscr{U}$ of Γ in $\mathscr{M}$ such that $U := \mathscr{U} \setminus \Gamma$ is a (β,Γ)-cuspidal end of (M,g).*

The preceding treatment indicates that there are two possible ways of looking at these problems. In the first one we put forward the differential equation setting. Then the singular manifold has an inferior position and it is only the singularity function ρ which comes into play. In the second approach the geometric appearance of the singular manifold is relevant. In this case we start off with a singular manifold (M,g) which may not be obtained from a uniformly regular ambient manifold by cutting out lower-dimensional submanifolds. Instead, (M,g) can have more general singular ends U; namely such that U is isometric to a model (α,Γ)-wedge over (B,g_B), where (B,g_B) is as in (1.1), and B replaces $\mathbb{S}$ in definition (1.7). Theorem 8.1 and Proposition 8.2(i) guarantee then the existence of singularity functions ρ, modeling again the geometric structure of (M,g), such that (M,g) is singular of type ρ. Consequently, we can obtain well-posedness theorems for degenerate parabolic equations on singular manifolds by applying Theorem 1.5.

Up to now we have considered the case in which we introduce a conformal metric g/ρ^2 on M in order to render it uniformly regular. This means that we restrict ourselves to differential operators with isotropic degenerations. However, other choices are possible also. For example, in the setting (1.8) we can endow $(0,1)\times\mathbb{S}_+^1\times\Gamma$ with the metric $t^{-2\alpha}dt^2 + g_{\mathbb{S}_+^1} + g_\Gamma$ instead of $dt^2 + t^{2\alpha}g_{\mathbb{S}_+^1} + g_\Gamma$ as above. This is a consequence of the next theorem which is also proved in Sect. 8. For simplicity, we consider the case where M has only one singular end. The extension to the general case is straightforward. Moreover,

$$(0,1)\times\mathbb{S}\times\Gamma \tag{1.9}$$

is the canonical representation of a tubular neighborhood U of Γ in (M,g) in the sense made precise later in this paper.

Theorem 1.9. *Let (1.5) be satisfied with $\mathbf{V} = \emptyset$ and $\boldsymbol{\Gamma} = \{\Gamma\}$, and fix $\alpha > 0$. Let U be a tubular neighborhood of Γ in (M,g). Suppose $\mathfrak{g}$ is a metric for M which* *coincides on $M\backslash U$ with g and equals near Γ*

$$t^{-2}dt^2+t^{-2\alpha}(g_{\mathbb{S}}+g_\Gamma) \quad \text{if } 0<\alpha\le 1 ,$$

respectively

$$t^{-2(\alpha+1)}dt^2+g_{\mathbb{S}}+g_\Gamma \quad \text{if } \alpha>1 , \tag{1.10}$$

in the canonical representation (1.9) of U. Then $(M,\mathfrak{g})$ is uniformly regular.

Recall that $g_{\mathbb{S}}$, respectively g_Γ, is absent if $\ell=m$, respectively $\ell=1$.

By applying Theorem 1.4 to the setting of Theorem 1.9 we obtain well-posedness results for parabolic problems with anisotropic degeneration. To indicate the inherent potential of such applications we consider the particularly interesting setting in which Γ is a compact connected component of the boundary of $\mathscr{M}$. We also suppose, for simplicity, that $\mathscr{A}$ is the negative Laplace–Beltrami operator $-\Delta$ of (M,g) and assume $\alpha>1$. Then it follows from (1.10) that the (interior) flux vector field satisfies in a collar neighborhood of Γ

$$\text{grad} \sim (\delta^{2\alpha}\partial_\nu, \text{grad}_\Gamma) .$$

Hence it degenerates in the normal direction only and there is no degeneration at all in tangential directions. This is in contrast to the isotropic case in which Corollary 1.7 applies and, in the present setting, gives

$$\text{grad} \sim \delta^{2\alpha}(\partial_\nu, \text{grad}_\Gamma)$$

near Γ.

There has been done an enormous amount of research on *elliptic* equations on singular manifolds. All of it is related, in one way or another, to the seminal paper by V.A. Kondrat′ev [20]. It is virtually impossible to review this work here and to do justice to the many authors who contributed. It may suffice to mention the three most active groups and some of their principal exponents. First, there is the Russian school which builds directly on Kondrat′ev's work and is also strongly application-oriented (for example, see the numerous papers and books by V.G. Maz′ya, S.A. Nazarov, and their coauthors). Second, the group gathering around B.-W. Schulze has constructed an elaborate calculus of pseudo-differential algebras on manifolds with singularities, mainly of conical and cuspidal type. For a lucid presentation of some of its aspects in the simplest setting of manifolds with cuspidal points and wedges we refer to the book of V.E. Nazaikinskii, A.Yu. Savin, B.-W. Schulze, and B.Yu. Sternin [25]. Third, another general approach to pseudo-differential operators on manifolds with singularities has been developed by R. Melrose and his coworkers. A brief explanation, stressing the differences of the techniques used by the latter two groups, is found in the section 'Bibliographical Remarks' of [25]. Henceforth, we call these methods "classical" for easy reference.

To explain to which extent our point of view differs from the classical approach we consider the simplest case, namely, a manifold with one conical singularity. By means of the stretching diffeomorphism the model cone $C_1(\mathbb{S})$ is represented by the "stretched manifold" $(0,1)\times\mathbb{S}$ whose metric is

$$g=\mathrm{d}t^2+t^2g_{\mathbb{S}}=t^2\big((\mathrm{d}t/t)^2+g_{\mathbb{S}}\big)\ .$$

Thus the corresponding Laplace–Beltrami operator is given by $t^{-2}\big((t\partial_t)^2+\Delta_{\mathbb{S}}\big)$. More generally, in the classical theories there are considered differential operators which, on the stretched manifold, are (in the second order case) of the form $t^{-2}L$, where L is a uniformly elliptic operator generated by the vector fields $t\partial_t,\partial_{\theta^1},\dots,\partial_{\theta^{m-1}}$ with $(\theta^1,\dots,\theta^{m-1})$ being local coordinates for $\mathbb{S}$.

Instead, our approach is based on the metric

$$\hat{g}=g/t^2=(\mathrm{d}t/t)^2+g_{\mathbb{S}}$$

whose Laplacian is $(t\partial_t)^2+\Delta_{\mathbb{S}}$. Hence our theory addresses operators of type L. As has been shown in [3], and explained above, this amounts to the study of degenerate differential operators in the original setting. (Let us mention, in passing, that the variable transformation $t=e^{-s}$ carries $\big((0,1)\times\mathbb{S},\ \mathrm{d}t^2/t^2+g_{\mathbb{S}}\big)$ onto $\big((1,\infty)\times\mathbb{S},\ \mathrm{d}s^2+g_{\mathbb{S}}\big)$ whose Laplacian is $\partial_s^2+\Delta_{\mathbb{S}}$. The latter Riemannian manifold is easily seen to be uniformly regular "near infinity," that is, cofinally uniformly regular as defined in Sect. 6. These trivial observations form part of the basis of this paper.)

The factor t^{-2} multiplying L in the classical approach does not play a decisive role for the proof of many results in the elliptic theory since it can be "moved to the right-hand side." However, the situation changes drastically if a spectral parameter is included since $t^{-2}L+\lambda=t^{-2}(L+\lambda t^2)$ is no longer of the same type as L. This is the reason why—at least up to now—there is no general theory of "classical" parabolic equations on singular manifolds.

All singular manifolds discussed so far belong to the class of manifolds with "smooth singularities." By this we mean that the bases of the cusps themselves do not have singularities. If they are also singular, we model manifolds with cuspidal corners and more complicated higher order singularities. For the sake of simplicity we do not consider such cases in this paper. However, all definitions and theorems presented below have been "localized" so that an extension to 'corner manifolds' can be built directly on the present work.

In the next section, besides fixing our basic notation, we give precise (localized) definitions of Riemannian manifolds which are uniformly regular, respectively singular of type ρ. All subsequent considerations are given for the latter class. Corresponding assertions for uniformly regular manifolds are obtained by setting $\rho=\mathbf{1}$.

Section 3 contains preliminary technical results and, in particular, the proof of (an extended version of) Example 1.1(e). As a first application of these investigations we present, in Sect. 4, some easy examples of uniformly regular Riemannian manifolds.

In Sect. 5 we introduce a general class of 'cusp characteristics' which provides us with ample families of singularity functions ρ. It is a consequence of Example 5.1(b) that our results do not only apply to manifolds with cuspidal singularities, but also to manifolds with "exponential" cusps and wedges, or in more general situations (see Example 5.1(b) and Lemma 8.4).

In the proximate section we introduce model wedges and explore their singularity behavior under various Riemannian metrics. The case of the 'natural' metric, induced by the embedding in the ambient Euclidean space, is treated in Sect. 7. The last section contains the main results and the proofs left out in the introduction.

2 Notations and Definitions

By a *manifold* we always mean a smooth, that is, C^∞ manifold with (possibly empty) boundary such that its underlying topological space is separable and metrizable. Thus we work in the smooth category. A manifold does not need to be connected, but all connected components are of the same dimension.

Let M be a submanifold of some manifold N. Then $\iota : M \hookrightarrow N$, or simply $M \hookrightarrow N$, denotes the natural embedding $p \mapsto p$, for which we also write ι_M. (The meaning of N will always be clear from the context.) This embedding induces the natural (fiber-wise linear) embedding $\iota : TM \hookrightarrow TN$ of the tangent bundle of M into the one of N.

Let (N,h) be a Riemannian manifold. Then ι^*h denotes the restriction of h to $M \hookrightarrow N$, that is, $(\iota^*h)(p)(X,Y) = h(p)(X,Y)$ for $p \in M$ and $X,Y \in T_pM \hookrightarrow T_pN$. If g is a Riemannian metric for M, then (M,g) is a *Riemannian submanifold* of (N,h), in symbols: $(M,g) \hookrightarrow (N,h)$, if $g = \iota^*h$. If M has codimension 0, then we write again h for ι^*h.

The Euclidean metric

$$|dx|^2 = (dx^1)^2 + \cdots + (dx^m)^2$$

of $\mathbb{R}^m$ is also denoted by g_m. Unless explicitly stated otherwise, we identify $\mathbb{R}^m$ with $(\mathbb{R}^m, g_m)$.

Given a finite-dimensional normed vector space $E = (E, |\cdot|)$ and an open subset V of $\mathbb{R}^m$ or $\mathbb{H}^m$, we write $\|\cdot\|_{k,\infty}$ for the usual norm of $BC^k(V,E)$, the Banach space of all $v \in C^k(V,E)$ such that $|\partial^\alpha v|$ is uniformly bounded for $\alpha \in \mathbb{N}^m$ with $|\alpha| \le k$. (We use standard multi-index notation.) As usual, $C^k(V) = C^k(V,\mathbb{R})$ etc., and $\|\cdot\|_\infty = \|\cdot\|_{\infty,0}$.

Suppose M and N are manifolds and $\varphi : M \to N$ is a diffeomorphism. By φ^* we denote the pull-back by φ (of general tensor fields) and $\varphi_* := (\varphi^{-1})^*$ is the corresponding push-forward. Thus $\varphi^* v = v \circ \varphi$ for a function v on N. Recall that the pull-back $\varphi^* h$ of a Riemannian metric h on N is given by

$$(\varphi^* h)(X,Y) = \varphi^*\big(h(\varphi_* X, \varphi_* Y)\big) \tag{2.1}$$

for all vector fields X and Y on M.

As usual, $(\partial/\partial x^1,\dots,\partial/\partial x^m)$ is the coordinate frame for $T_{U_\kappa}M$ associated with the local coordinates $\kappa=(x^1,\dots,x^m)$ on $U_\kappa:=\mathrm{dom}(\kappa)$. Here $T_{U_\kappa}M$ denotes the restriction of TM to $U_\kappa\hookrightarrow M$. Thus $\kappa_*(\partial/\partial x^i)=e_i$, where $(e_1,\dots,e_m)$ is the standard basis for $\mathbb{R}^m$. The basis for $T^*_{U_\kappa}M$, dual to $(\partial/\partial x^1,\dots,\partial/\partial x^m)$, is $(\mathrm{d}x^1,\dots,\mathrm{d}x^m)$ with $\mathrm{d}x^i$ being the differential of the coordinate function x^i.

Let g be a Riemannian metric on M. For a local chart $\kappa=(x^1,\dots,x^m)$ the local representation for g with respect to these coordinates is given by

$$g=g_{ij}\mathrm{d}x^i\,\mathrm{d}x^j\ ,\qquad g_{ij}:=g\Big(\frac{\partial}{\partial x^i},\frac{\partial}{\partial x^j}\Big)\ .$$

Here and below, we employ the standard summation convention. Then, given vector fields $\xi=\xi^ie_i$ and $\eta=\eta^je_j$ on $\kappa(U_\kappa)$, it follows from (2.1) that

$$\begin{aligned}\kappa_*g(\xi,\eta)&=\kappa_*\big(g(\kappa^*\xi,\kappa^*\eta)\big)=\kappa_*\big(g(\xi^i\partial/\partial x^i,\eta^j\partial/\partial x^j)\big)\\&=\kappa_*(g_{ij}\xi^i\eta^j)=\kappa_*g_{ij}\xi^i\eta^j=(g_{ij}\circ\kappa^{-1})\xi^i\eta^j\ .\end{aligned}$$

Thus $\kappa_*g(x)$ is for each $x\in\kappa(U_\kappa)$ a positive definite symmetric bilinear form. Hence there exists $c(x)\geq 1$ such that

$$|\xi|^2/c(x)\leq\kappa_*g(x)(\xi,\xi)\leq c(x)\,|\xi|^2\ ,\qquad \xi\in\mathbb{R}^m\ ,\quad x\in\kappa(U_\kappa)\ ,\tag{2.2}$$

where $|\xi|:=\sqrt{g_m(\xi,\xi)}=\sqrt{(\xi\,|\,\xi)}$ is the Euclidean norm of $\xi\in\mathbb{R}^m$. In other words,

$$g_m/c(x)\leq\kappa_*g(x)\leq c(x)g_m\ ,\qquad x\in\kappa(U_\kappa)\ .$$

We set $Q:=(-1,1)\subset\mathbb{R}$. If κ is a local chart for an m-dimensional manifold M, then it is *normalized* (at p) if $\kappa(U_\kappa)=Q^m$ whenever $U_\kappa\subset\mathring{M}$, the interior of M, whereas $\kappa(U_\kappa)=Q^m\cap\mathbb{H}^m$ if U_κ has a nonempty intersection with the boundary ∂M of M (and $\kappa(p)=0$). We put $Q^m_\kappa:=\kappa(U_\kappa)$ if κ is normalized. (We find it convenient to use normalization by cubes. Of course, we could equally well normalize by employing Euclidean balls.)

Let M be an m-dimensional manifold and S a (nonempty) subset thereof. Given an atlas $\mathfrak{K}$ for M, we set

$$\mathfrak{K}_S:=\{\,\kappa\in\mathfrak{K}\ ;\ U_\kappa\cap S\neq\emptyset\,\}\ .$$

Then $\mathfrak{K}_S$ has *finite multiplicity* or: $\mathfrak{K}$ has *finite multiplicity on* S, if there exists $k\in\mathbb{N}$ such that any intersection of more than k coordinate patches U_κ with $\kappa\in\mathfrak{K}_S$ is empty. The least such k is then the multiplicity, $\mathrm{mult}(\mathfrak{K}_S)$, of $\mathfrak{K}_S$. The atlas $\mathfrak{K}$ is *shrinkable on* S, or: $\mathfrak{K}_S$ is *shrinkable*, if $\mathfrak{K}_S$ consists of normalized charts and there exists $r\in(0,1)$ such that

$$\{\,\kappa^{-1}(rQ^m_\kappa)\ ;\ \kappa\in\mathfrak{K}_S\,\}\tag{2.3}$$

is a cover of S. It is *shrinkable on* S *to* $r_0\in(0,1)$ if (2.3) holds for each $r\in(r_0,1)$.

An atlas $\mathfrak{K}$ for M is *uniformly regular on S* if

$$\begin{aligned} &\text{(i)} \quad \mathfrak{K}_S \text{ is shrinkable and has finite multiplicity ;} \\ &\text{(ii)} \quad \|\tilde{\kappa}\circ\kappa^{-1}\|_{k,\infty} \le c(k)\ , \quad \kappa,\tilde{\kappa}\in\mathfrak{K}_S\ , \quad k\in\mathbb{N}\ . \end{aligned} \tag{2.4}$$

In (ii) and in similar situations it is understood that only $\kappa,\tilde{\kappa}\in\mathfrak{K}_S$ with $U_\kappa\cap U_{\tilde{\kappa}}\neq\emptyset$ are being considered. Two atlases $\mathfrak{K}$ and $\tilde{\mathfrak{K}}$ for M, which are uniformly regular on S, are *equivalent on S*, in symbols: $\mathfrak{K}\underset{S}{\approx}\tilde{\mathfrak{K}}$, if

$$\begin{aligned} &\text{(i)} \quad \operatorname{card}\{\,\tilde{\kappa}\in\tilde{\mathfrak{K}}_S\ ,\ U_{\tilde{\kappa}}\cap U_\kappa\neq\emptyset\,\}\le c\ , \quad \kappa\in\mathfrak{K}_S\ ; \\ &\text{(ii)} \quad \|\tilde{\kappa}\circ\kappa^{-1}\|_{k,\infty}+\|\kappa\circ\tilde{\kappa}^{-1}\|_{k,\infty}\le c(k)\ , \quad \kappa\in\mathfrak{K}_S\ , \quad \tilde{\kappa}\in\tilde{\mathfrak{K}}_S\ , \quad k\in\mathbb{N}\ . \end{aligned} \tag{2.5}$$

This defines an equivalence relation on the class of all atlases for M which are uniformly regular on S. Each equivalence class is a *structure of uniform regularity on S*. We write $[\![\mathfrak{K}]\!]_S$ for it to indicate that it is *generated* by $\mathfrak{K}$, that is, contains $\mathfrak{K}$ as a representative. If M is endowed with a structure $[\![\mathfrak{K}]\!]_S$ of uniform regularity on S, then $\bigl(M,[\![\mathfrak{K}]\!]_S\bigr)$ is a *uniformly regular manifold on S*.

Let $\bigl(M,[\![\mathfrak{K}]\!]_S\bigr)$ be a uniformly regular manifold on S and let g be a Riemannian metric for M. Suppose

$$\begin{aligned} &\text{(i)} \quad \kappa_*g\sim g_m\ , \quad \kappa\in\mathfrak{K}_S\ . \\ &\text{(ii)} \quad \|\kappa_*g\|_{k,\infty}\le c(k)\ , \quad \kappa\in\mathfrak{K}_S\ , \quad k\in\mathbb{N}\ . \end{aligned} \tag{2.6}$$

It follows from (2.5) that (2.6) prevails if $\mathfrak{K}_S$ is replaced by any $\tilde{\mathfrak{K}}_S$ with $\tilde{\mathfrak{K}}\underset{S}{\approx}\mathfrak{K}$. Thus it is meaningful to say that g is a *Riemannian metric for* $\bigl(M,[\![\mathfrak{K}]\!]_S\bigr)$ which is *uniformly regular on S* if (2.6) applies to some, hence every, representative of $[\![\mathfrak{K}]\!]_S$. We also say that two such metrics g and $\bar{g}$ are *equivalent on S*, $g\underset{S}{\sim}\bar{g}$, if $g\,|\,S\sim\bar{g}\,|\,S$. This defines an equivalence relation on the class of all Riemannian metrics for $\bigl(M,[\![\mathfrak{K}]\!]_S\bigr)$ which are uniformly regular on S. Similarly as above, $[\![g]\!]_S$ is the equivalence class containing the representative g.

By a **uniformly regular Riemannian manifold** *on S*, written as $\bigl(M,[\![\mathfrak{K}]\!]_S,[\![g]\!]_S\bigr)$, we mean a uniformly regular manifold $\bigl(M,[\![\mathfrak{K}]\!]_S\bigr)$ on S equipped with an equivalence class of uniformly regular Riemannian metrics on S. It is a convenient abuse of language to say instead that $(M,\mathfrak{K},g)$ is a Riemannian manifold which is uniformly regular on S. Even more loosely, (M,g) is (a manifold which is) **uniformly regular** *on S*, if there exists an atlas $\mathfrak{K}$ which is uniformly regular on S such that $\bigl(M,[\![\mathfrak{K}]\!]_S,[\![g]\!]_S\bigr)$ is a uniformly regular Riemannian manifold on S.

Suppose $\rho\in C^\infty\bigl(M,(0,\infty)\bigr)$ and let g be a Riemannian metric for M. Then ρ is a *singularity function for* (M,g) *on S*, if there exists an atlas $\mathfrak{K}$ which is uniformly regular on S such that $(M,\mathfrak{K},g/\rho^2)$ is a Riemannian manifold which is uniformly regular on S. Two singularity functions are *equivalent on S*, $\rho\underset{S}{\approx}\tilde{\rho}$, if $\mathfrak{K}\underset{S}{\approx}\tilde{\mathfrak{K}}$ and $g/\rho^2\underset{S}{\sim}g/\tilde{\rho}^2$. We denote by $[\![\rho]\!]_S$ the equivalence class of singularity functions containing the representative ρ, the *singularity type* of (M,g) on S. Finally, the **Riemannian manifold** (M,g) **is singular of type** $[\![\rho]\!]_S$—more loosely: **of type** ρ *on S*—if

$(M,g/\rho^2)$ is uniformly regular on S. Clearly, (M,g) is singular of type $[\![\mathbf{1}]\!]_S$ iff it is uniformly regular on S.

A pair $(\rho,\mathfrak{K})$ is a *singularity datum* for (M,g) on S if

$$\begin{array}{rl} \text{(i)} & \rho \in C^\infty\big((M,(0,\infty))\big)\ . \\ \text{(ii)} & \mathfrak{K} \text{ is an atlas which is uniformly regular on } S\ . \\ \text{(iii)} & \|\kappa_*\rho\|_{k,\infty} \le c(k)\rho_\kappa\ ,\quad \kappa \in \mathfrak{K}_S\ ,\quad k \in \mathbb{N}\ , \\ & \text{where } \rho_\kappa := \kappa_*\rho(0) = \rho\big(\kappa^{-1}(0)\big)\ . \\ \text{(iv)} & \rho\,|\,U_\kappa \sim \rho_\kappa\ ,\quad \kappa \in \mathfrak{K}_S\ . \\ \text{(v)} & \kappa_*g \sim \rho_\kappa^2 g_m\ ,\quad \kappa \in \mathfrak{K}_S\ . \\ \text{(vi)} & \|\kappa_*g\|_{k,\infty} \le c(k)\rho_\kappa^2\ ,\quad \kappa \in \mathfrak{K}_S\ ,\quad k \ge 0\ . \end{array} \tag{2.7}$$

It is easily verified that $(M,\mathfrak{K},g/\rho^2)$ is uniformly regular on S if $(\rho,\mathfrak{K})$ is a singularity datum for (M,g) on S. Thus ρ is a singularity function for (M,g) if $(\rho,\mathfrak{K})$ is a singularity datum for it.

The "localization" of all these quantities "to S" is introduced for technical reasons. Our principal interest concerns the choice $S=M$. In this case the qualifiers "on S" and the symbol S are omitted, of course.

3 Preliminaries

Let (M,g) be a Riemannian manifold and $X \subset M$. For $p,q \in X$ we denote by $d_X(p,q) = d_{g,X}(p,q)$ the distance between p and q *in* X. Thus $d_X(p,q)$ is the infimum of the lengths of all piece-wise smooth paths of M joining p to q within X. If p and q lie in different connected components, then $d_X(p,q) := \infty$.

We suppose $\mathbb{X} \in \{\mathbb{R}^m, \mathbb{H}^m\}$, X is open in $\mathbb{X}$, and $S \subset X$. We denote by δ_S the distance in X from S to $\mathbb{X}\backslash X$, that is, $\delta_S := \inf_{p\in S} d_X(p,\mathbb{X}\backslash X)$, where $d_X(p,\emptyset) := \infty$. Then we assume $0 < \delta \le \delta_S/\sqrt{m}$ and set

$$Z_{\delta,X} := \big\{ z \in \mathbb{Z}^m \cap \mathbb{X}\ ;\ \delta(z+Q_z)\cap X \ne \emptyset \big\}\ ,$$

where $Q_z := Q^m$ if $z \in \mathring{X}$ and $Q_z := Q^m \cap \mathbb{H}^m$ otherwise. Given $z \in Z_{\delta,X}$,

$$\lambda_{\delta,z}(x) := -z + x/\delta\ , \qquad x \in \delta(z+Q_z)\cap X\ . \tag{3.1}$$

Then

$$\mathfrak{L} = \mathfrak{L}(\delta,X) := \{\, \lambda_{\delta,z}\ ;\ z \in Z_{\delta,X}\,\}$$

is an atlas for X of multiplicity 2^m. Since $\operatorname{diam}\big(\delta(z+Q_z)\big) = \sqrt{m}\delta \le \delta_S$ we see that $\mathfrak{L}_S$ is normalized and shrinkable to $1/2$. Given $\lambda, \tilde{\lambda} \in \mathfrak{L}_S$ with $\lambda = \lambda_{\delta,z}$ and $\tilde{\lambda} = \lambda_{\delta,\tilde{z}}$,

$$\tilde{\lambda}\circ\lambda^{-1}(y) = z - \tilde{z} + y\ , \qquad y \in \lambda(U_\lambda \cap U_{\tilde{\lambda}})\ .$$

This shows that $\mathfrak{L}$ is uniformly regular on S. Furthermore, denoting by ∂ the Fréchet derivative,

$$\partial\lambda^{-1} = \delta 1_m\,, \quad \lambda_* g_X = \delta^2 g_m\,, \qquad \lambda \in \mathfrak{L}_S\,, \tag{3.2}$$

where $g_X = \iota_X^* g_m$ and 1_m is the identity in $\mathbb{R}^{m\times m}$. In particular, setting $X := \mathbb{X}$ it follows that

$$\mathbb{R}^m \text{ and } \mathbb{H}^m \text{ are uniformly regular Riemannian manifolds}\,. \tag{3.3}$$

Let M be an m-dimensional manifold and $S \subset M$. Suppose $\mathfrak{K}$ is an atlas for M which is uniformly regular on S. Then there exists $r \in (0,1)$ such that (2.3) is a cover of S. Given $\kappa \in \mathfrak{K}_S$, we fix $\delta \in \bigl(0,(1-r)/\sqrt{m}\bigr)$ and put $\mathfrak{L}_\kappa := \mathfrak{L}(\delta, Q_\kappa^m)$. By the above $\mathfrak{L}_\kappa$ is an atlas for Q_κ^m of multiplicity 2^m which is uniformly regular on rQ_κ^m and shrinkable to $1/2$ on rQ_κ^m. Hence

$$\mathfrak{M} = \mathfrak{M}(\delta,\mathfrak{K}) := \{\,\lambda\circ\kappa\,;\ \kappa\in\mathfrak{K}_S,\ \lambda\in\mathfrak{L}_\kappa\} \cup (\mathfrak{K}\backslash\mathfrak{K}_S) \tag{3.4}$$

is an atlas for M such that

$$U_{\lambda\circ\kappa} = \kappa^{-1}(U_\lambda) \subset U_\kappa\,, \qquad \kappa\in\mathfrak{K}_S\,, \quad \lambda\in\mathfrak{L}_\kappa\,. \tag{3.5}$$

It has multiplicity at most $2^m\mathrm{mult}(\mathfrak{K}_S)$ on S and is shrinkable to $1/2$ on S. For $\mu,\tilde\mu \in \mathfrak{M}_S$ with $\mu = \lambda\circ\kappa$ and $\tilde\mu = \tilde\lambda\circ\tilde\kappa$ we get from (3.1) and (3.2)

$$\|\partial^\alpha(\tilde\mu\circ\mu^{-1}\|_\infty \le \delta^{-1}\delta^{|\alpha|}\,\|\partial^\alpha(\tilde\kappa\circ\kappa^{-1})\|_\infty\,, \qquad \alpha\in\mathbb{N}^m\backslash\{0\}\,. \tag{3.6}$$

Note that $\lambda\circ\kappa \in \mathfrak{M}_S$ implies $\kappa\in\mathfrak{K}_S$. Thus, since $\mathfrak{K}_S$ is uniformly regular and $\delta\le 1$,

$$\|\partial^\alpha(\tilde\mu\circ\mu^{-1})\|_\infty \le c(\alpha)$$

for $\mu,\tilde\mu\in\mathfrak{M}_S$ with $\mu=\lambda\circ\kappa$ and $\tilde\mu=\tilde\lambda\circ\tilde\kappa$ and $\alpha\in\mathbb{N}^m\backslash\{0\}$. Hence

$$\mathfrak{M} \text{ is uniformly regular on } S\,. \tag{3.7}$$

Let g be a Riemannian metric for M. Then (3.2) implies

$$\mu_* g = \lambda_*\kappa_* g = \delta^2\kappa_* g\,, \qquad \mu = \lambda\circ\kappa\in\mathfrak{M}_S\,. \tag{3.8}$$

Consequently,

$$\|\partial^\alpha(\mu_* g)\|_\infty \le c(\alpha)\delta^2\,\|\partial^\alpha(\kappa_* g)\|_\infty\,, \qquad \mu\in\mathfrak{M}_S\,, \quad \alpha\in\mathbb{N}^m\,. \tag{3.9}$$

Suppose $(\rho,\mathfrak{K})$ is a singularity datum for M on S. Then we infer from (2.7)(iii) and (iv) and from (3.5)

$$\mu_*\rho = (\kappa_*\rho)\circ\lambda^{-1} \sim (\kappa_*\rho)(0) = \rho_\kappa \sim \rho_\mu \tag{3.10}$$

and, using $\delta \leq 1$ once more,

$$\|\partial^\alpha(\mu_*\rho)\|_\infty \leq \delta^{|\alpha|}\,\|\partial^\alpha(\mu_*\rho)\|_\infty \leq c(\alpha)\rho_\kappa \leq c(\alpha)\rho_\mu \tag{3.11}$$

for $\mu = \lambda \circ \kappa \in \mathfrak{M}_S$ and $\alpha \in \mathbb{N}^m$.

These considerations show, in particular, that a uniformly regular Riemannian manifold possesses a uniformly regular atlas consisting of arbitrarily small charts; also see Lemma 3.2.

Let $(\tilde{M}, \tilde{g})$ be a Riemannian manifold without boundary. Then we endow the product manifold $M \times \tilde{M}$ with the product metric, denoted (slightly loosely) by $g + \tilde{g}$.

Theorem 3.1. *Suppose ρ is a bounded singularity function for (M,g) on $S \subset M$ and $\tilde{\rho}$ is one for $(\tilde{M},\tilde{g})$ on $\tilde{S} \subset \tilde{M}$. Then $\rho \otimes \tilde{\rho}$ is a singularity function for $(M \times \tilde{M}, g + \tilde{g})$ on $S \times \tilde{S}$.*

Proof. (1) We choose $0 < \bar{r} < r < 1$ and an atlas $\mathfrak{K}$ for M, respectively $\tilde{\mathfrak{K}}$ for $\tilde{M}$, such that $\mathfrak{K}$, respectively $\tilde{\mathfrak{K}}$, is shrinkable to $\bar{r}$ on S, respectively $\tilde{S}$, and $(\rho, \mathfrak{K})$, respectively $(\tilde{\rho}, \tilde{\mathfrak{K}})$, is a singularity datum for (M,g) on S, respectively $(\tilde{M}, \tilde{g})$ on $\tilde{S}$. Denoting by m, respectively $\tilde{m}$, the dimension of M, respectively $\tilde{M}$, we set $\delta := (1-r)/\sqrt{m+\tilde{m}}$. Given $\kappa \in \mathfrak{K}_S$ and $\tilde{\kappa} \in \mathfrak{K}_{\tilde{S}}$, we put

$$\delta_{\tilde{\kappa}} := \min\{\tilde{\rho}_{\tilde{\kappa}}, \delta\}\,, \quad \tilde{\delta}_\kappa := \min\{\rho_\kappa, \delta\}\,. \tag{3.12}$$

We set

$$\mathfrak{M}' := \big\{ (\lambda \circ \kappa) \times (\tilde{\lambda} \circ \tilde{\kappa}) \;;\; \kappa \in \mathfrak{K}_S,\ \tilde{\kappa} \in \tilde{\mathfrak{K}}_{\tilde{S}},\ \lambda \in \mathfrak{L}(\delta_{\tilde{\kappa}}, Q^m_\kappa),\ \tilde{\lambda} \in \mathfrak{L}(\tilde{\delta}_\kappa, Q^{\tilde{m}}) \big\}$$

and

$$\mathfrak{M}'' := \big\{ \kappa \times \tilde{\kappa} \;;\; \text{either } \kappa \in \mathfrak{K} \backslash \mathfrak{K}_S \text{ or } \tilde{\kappa} \in \tilde{\mathfrak{K}} \backslash \tilde{\mathfrak{K}}_{\tilde{S}} \big\}\,.$$

Then $\mathfrak{M} := \mathfrak{M}' \cup \mathfrak{M}''$ is an atlas for $M \times \tilde{M}$ and a refinement of the product atlas $\mathfrak{K} \otimes \tilde{\mathfrak{K}}$ in the sense that for each $\mu \in \mathfrak{M}$ there exists $\kappa \times \tilde{\kappa} \in \mathfrak{K} \otimes \tilde{\mathfrak{K}}$ such that $U_\mu \subset U_{\kappa \times \tilde{\kappa}}$. Moreover,

$$\mathfrak{M}_{S \times \tilde{S}} \subset \mathfrak{M}'\,. \tag{3.13}$$

Note that $\mathfrak{M}$ is normalized on $S \times \tilde{S}$ and has finite multiplicity thereon.

Suppose $\mu_i = (\lambda_i \circ \kappa_i) \times (\tilde{\lambda}_i \circ \tilde{\kappa}_i) \in \mathfrak{M}'$ for $i = 1,2$, and $U_{\mu_1} \cap U_{\mu_2} \neq \emptyset$. Then both $U_{\kappa_1} \cap U_{\kappa_2}$ and $U_{\tilde{\kappa}_1} \cap U_{\tilde{\kappa}_2}$ are nonempty. Hence $\tilde{\rho}_{\tilde{\kappa}_1} \sim \tilde{\rho}_{\tilde{\kappa}_2}$ and $\rho_{\kappa_1} \sim \rho_{\kappa_2}$. From this, $\delta_{\tilde{\kappa}_i} \leq \tilde{\rho}_{\tilde{\kappa}_i}$, and the boundedness of $\tilde{\rho}$ we infer $\delta_{\tilde{\kappa}_1}/\delta_{\tilde{\kappa}_2} \leq c$ and, analogously, $\tilde{\delta}_{\kappa_1}/\tilde{\delta}_{\kappa_2} \leq c$. Thus, using (3.1), (3.2), the finite multiplicity of $\mathfrak{M}_{S \times \tilde{S}}$ and the fact that κ_i and $\tilde{\kappa}_i$ are normalized, we obtain (cf. (3.6))

$$\|\mu_1 \circ \mu_2^{-1}\|_{k,\infty} \leq c\big(\|\kappa_1 \circ \kappa_2^{-1}\|_{k,\infty} + \|\tilde{\kappa}_1 \circ \tilde{\kappa}_2^{-1}\|_{k,\infty}\big) \leq c(k)$$

for $\mu_1, \mu_2 \in \mathfrak{M}_{S \times \tilde{S}}$ and $k \in \mathbb{N}$. This proves that $\mathfrak{M}$ is uniformly regular on $S \times \tilde{S}$.

(2) By adapting (3.10) and (3.11) to the present setting we find, due to (3.13),

$$\mu_*(\rho\otimes\tilde{\rho}) \sim (\rho\otimes\tilde{\rho})_\mu \sim \rho_\kappa\tilde{\rho}_{\tilde{\kappa}} \tag{3.14}$$

for $\mu = (\lambda\circ\kappa)\times(\tilde{\lambda}\circ\tilde{\kappa}) \in \mathfrak{M}_{S\times\tilde{S}}$ and

$$\|\mu_*(\rho\otimes\tilde{\rho})\|_{k,\infty} \le c(k)(\rho\otimes\tilde{\rho})_\mu\,, \qquad \mu\in\mathfrak{M}_{S\times\tilde{S}}\,, \quad k\in\mathbb{N}\,.$$

(3) For $\mu = (\lambda\circ\kappa)\times(\tilde{\lambda}\circ\tilde{\kappa}) \in \mathfrak{M}'$ we find by (3.8)

$$\begin{aligned}\mu_*(g+\tilde{g}) &= (\lambda\circ\kappa)_*g + (\tilde{\lambda}\circ\tilde{\kappa})_*\tilde{g}\\ &\sim \delta_{\tilde{\kappa}}^2\kappa_*g + \tilde{\delta}_\kappa^2\tilde{\kappa}_*\tilde{g} \sim \delta_{\tilde{\kappa}}^2\rho_\kappa^2 g_m + \tilde{\delta}_\kappa^2\tilde{\rho}_{\tilde{\kappa}}^2 g_{\tilde{m}}\,,\end{aligned} \tag{3.15}$$

uniformly with respect to $\mu\in\mathfrak{M}_{S\times\tilde{S}}$. Definition (3.12) and the boundedness of ρ and $\tilde{\rho}$ imply $\delta_{\tilde{\kappa}} \sim \tilde{\rho}_{\tilde{\kappa}}$ and $\tilde{\delta}_\kappa \sim \rho_\kappa$. Using this and (3.14) we get from (3.15)

$$\mu_*(g+\tilde{g}) \sim \rho_\kappa^2\tilde{\rho}_{\tilde{\kappa}}^2(g_m+g_{\tilde{m}}) \sim (\rho\otimes\tilde{\rho})_\mu^2 g_{m+\tilde{m}}\,, \qquad \mu\in\mathfrak{M}_{S\times\tilde{S}}\,.$$

Lastly, we infer from (3.9) and (3.14)

$$\begin{aligned}\|\mu_*(g+\tilde{g})\|_{k,\infty} &\le c(k)\big(\delta_{\tilde{\kappa}}^2\,\|\kappa_*g\|_{k,\infty} + \tilde{\delta}_\kappa^2\,\|\tilde{\kappa}_*\tilde{g}\|_{k,\infty}\big)\\ &\le c(k)\rho_\kappa^2\tilde{\rho}_{\tilde{\kappa}}^2 \le c(k)(\rho\otimes\tilde{\rho})_\mu^2\end{aligned}$$

for $\mu\in\mathfrak{M}_{S\times\tilde{S}}$ and $k\in\mathbb{N}$. This proves the assertion. □

Our next considerations exploit the "localization to S."

Lemma 3.2. *Let (M,g) be uniformly regular on $S\subset M$. Suppose V is open in M and*

$$d_{\bar{V}}(S, M\backslash V) > 0\,. \tag{3.16}$$

Then there exists an atlas $\mathfrak{M}$ for M belonging to the structure of uniform regularity on S such that $U_\mu\subset V$ for $\mu\in\mathfrak{M}_S$.

Proof. Let $\mathfrak{K}$ be an atlas belonging to the structure of uniform regularity on S. Choose $r\in(0,1)$ such that (2.3) is a cover of S. Fix $\delta\in\big(0,(1-r)/\sqrt{m}\big)$ and set $\mathfrak{M} := \mathfrak{M}(\delta,\mathfrak{K})$. Then $\mathfrak{M}_S$ is uniformly regular by (3.7).

It follows from $\kappa_*g \sim g_m$ for $\kappa\in\mathfrak{K}_S$, (3.8), and (3.16) that we can choose δ so small that $\mathrm{diam}(U_\mu) < d_{\bar{V}}(S, M\backslash V)$ for $\mu\in\mathfrak{M}_S$.

Lastly, we infer from (2.4), (3.1), and (3.2) that

$$\|\kappa\circ\mu^{-1}\|_{k,\infty} + \|\mu\circ\kappa^{-1}\|_{k,\infty} \le c(k)\,, \qquad \kappa\in\mathfrak{K}_S\,, \quad \mu\in\mathfrak{M}_S\,.$$

Thus $\mathfrak{M} \underset{S}{\approx} \mathfrak{K}$, which proves the claim. □

The following lemma will be fundamental for the construction of singular Riemannian manifolds by "patching together simpler pieces."

Lemma 3.3. *Suppose:*

(i) $\{V_\alpha \; ; \; \alpha \in \mathsf{A}\}$ *is a finite family of open subsets of* M.
(ii) $S_\alpha \subset V_\alpha$ *and* $\{S_\alpha \; ; \; \alpha \in \mathsf{A}\}$ *is a covering of* M.
(iii) $(\rho_\alpha, \mathfrak{K}_\alpha)$ *is a singularity datum for* (V_α, g) *on* S_α.
(iv) $\rho_\alpha \,|\, V_\alpha \cap V_{\tilde\alpha} \sim \rho_{\tilde\alpha} \,|\, V_\alpha \cap V_{\tilde\alpha}$, $\alpha, \tilde\alpha \in \mathsf{A}$.
(v) $\|\kappa_{\tilde\alpha} \circ \kappa_\alpha^{-1}\|_{k,\infty} + \|\kappa_\alpha \circ \kappa_{\tilde\alpha}^{-1}\|_{k,\infty} \le c(k)$ *for* $(\kappa_\alpha, \kappa_{\tilde\alpha}) \in \mathfrak{K}_{\alpha,S_\alpha} \times \mathfrak{K}_{\tilde\alpha,S_{\tilde\alpha}}$, $\alpha, \tilde\alpha \in \mathsf{A}$, $\alpha \ne \tilde\alpha$, $k \in \mathbb{N}$.

Then $\mathfrak{K} := \bigcup_\alpha \mathfrak{K}_{\alpha,S_\alpha}$ *is a uniformly regular atlas for* M *and there exists* ρ *belonging to* $C^\infty\big(M,(0,\infty)\big)$ *and satisfying*

$$\rho \,|\, S_\alpha \sim \rho_\alpha \,, \qquad \alpha \in \mathsf{A}\,, \tag{3.17}$$

such that $(\rho, \mathfrak{K})$ *is a singularity datum for* (M,g).

Proof. (1) It is a consequence of (i)–(iii) and (v) that $\mathfrak{K}$ is a uniformly regular atlas for M.

(2) Since M is locally compact, separable, and metrizable the same applies to V_α. Thus V_α is paracompact. Hence there exists a smooth partition of unity $\{\chi_{\alpha,\beta} \; ; \; \beta \in \mathfrak{K}_\alpha\}$ on V_α subordinate to $\{U_\beta \; ; \; \beta \in \mathfrak{K}_\alpha\}$ (e.g., [7]). We extend each $\chi_{\alpha,\beta}$ over M by setting it equal to 0 outside V_α and set $\psi_\alpha := \sum_{\beta \in \mathfrak{K}_{\alpha,S_\alpha}} \chi_{\alpha,\beta}$. Then $\psi_\alpha \in C^\infty\big(M,[0,1]\big)$ with $\psi_\alpha \,|\, S_\alpha = \mathbf{1}$. We put $\varphi_\alpha := \psi_\alpha \big/ \sum_{\tilde\alpha \in \mathsf{A}} \psi_{\tilde\alpha}$. Assumptions (i) and (ii) guarantee that $\{\varphi_\alpha \; ; \; \alpha \in \mathsf{A}\}$ is a smooth partition of unity on M subordinate to the open cover $\{V_\alpha \; ; \; \alpha \in \mathsf{A}\}$ of M.

We put $\rho := \sum_\alpha \varphi_\alpha \rho_\alpha$. Then, given $\alpha \in \mathsf{A}$ and $x \in V_\alpha$, we infer from (iv)

$$\begin{aligned}\rho(x) = \sum_{V_\beta \cap V_\alpha \ne \emptyset} \varphi_\beta(x)\rho_\beta(x) &\sim \rho_\alpha(x) \sum_{V_\beta \cap V_\alpha \ne \emptyset} \varphi_\beta(x) \\ &= \rho_\alpha(x) \sum_\beta \varphi_\beta(x) = \rho_\alpha(x)\,.\end{aligned} \tag{3.18}$$

This proves (3.17).

(3) By (iii)

$$\kappa_*(g/\rho_\alpha^2) \sim g_m\,, \quad \|\kappa_*(g/\rho_\alpha^2)\|_{k,\infty} \le c(k)$$

for $\kappa \in \mathfrak{K}_{\alpha,S_\alpha}$, $\alpha \in \mathsf{A}$, and $k \in \mathbb{N}$. We deduce from (3.18)

$$\kappa_*(g/\rho^2) = \kappa_* g / \kappa_* \rho^2 \sim \kappa_* g / \kappa_* \rho_\alpha^2 = \kappa_*(g/\rho_\alpha^2) \sim g_m \tag{3.19}$$

for $\kappa \in \mathfrak{K}_{\alpha,S_\alpha}$ and $\alpha \in \mathsf{A}$, that is, for $\kappa \in \mathfrak{K}$. The definition of ρ implies

$$\kappa_* \rho = \sum_\alpha (\kappa_* \varphi_\alpha)(\kappa \circ \kappa_\alpha^{-1})_* \kappa_{\alpha *} \rho_\alpha\,, \qquad \kappa \in \mathfrak{K}\,.$$

From this, (iii), the chain rule, and the uniform regularity of $\mathfrak{K}$ we deduce the estimate $\|\kappa_* \rho\|_{k,\infty} \le c(k)$ for $\kappa \in \mathfrak{K}$ and $k \in \mathbb{N}$. Consequently, we infer from the chain rule and (3.19)

$$\|\kappa_*(g/\rho^2)\|_{k,\infty} \le c(k)\,, \qquad \kappa\in\mathfrak{K}\,, \quad k\in\mathbb{N}\,.$$

This proves the last part of the assertion. □

The following (almost trivial) lemma shows that the class of singular manifolds is invariant under Riemannian isometries.

Lemma 3.4. *Let $f\colon \tilde M\to M$ be a diffeomorphism of manifolds. Suppose g is a Riemannian metric for M and ρ is a singularity function for (M,g) on $S\subset M$. Then $f^*\rho$ is a singularity function for $(\tilde M, f^*g)$ on $f^{-1}(S)$.*

Proof. Let $\mathfrak{K}$ be an atlas which is uniformly regular on S. It is easily verified that $f^*\mathfrak{K} := \{\,f^*\kappa\ ;\ \kappa\in\mathfrak{K}\}$ is an atlas for $\tilde M$ which is uniformly regular on $f^{-1}(S)$. Note

$$(f^*\kappa)_*f^*\rho = (\rho\circ f)\circ(\kappa\circ f)^{-1} = \rho\circ\kappa = \kappa_*\rho$$

and

$$(f^*\kappa)_*(f^*g) = (\kappa\circ f)_*(f^{-1})_*g = \bigl((\kappa\circ f)\circ f^{-1}\bigr)_*g = \kappa_*g$$

for $\kappa\in\mathfrak{K}$. From this it is obvious that conditions (2.7) carry over from ρ, $\mathfrak{K}$, and g to $f^*\rho$, $f^*\mathfrak{K}$, and f^*g. □

Suppose $\partial M\ne\emptyset$ and let $\dot\iota\colon \partial M\hookrightarrow M$ be the natural embedding. Let g be a Riemannian metric for M. Then $\dot g := \dot\iota^*g$ is the Riemannian metric for ∂M induced by g. Given a local chart κ for M with $\partial U_\kappa = U_\kappa\cap\partial M\ne\emptyset$, we set $U_{\dot\kappa} := \partial U_\kappa$ and $\dot\kappa := \iota_0\circ(\iota^*\kappa)\colon U_{\dot\kappa}\to\mathbb{R}^{m-1}$, where $\iota_0\colon \{0\}\times\mathbb{R}^{m-1}\to\mathbb{R}^{m-1}$, $(0,x')\mapsto x'$. Moreover, $\dot\rho := \iota^*\rho = \rho\,|\,\partial M$ for $\rho\colon M\to\mathbb{R}$.

Lemma 3.5. *Let $\mathfrak{K}$ be an atlas for M which is uniformly regular on S. Then*

$$\dot{\mathfrak{K}} := \{\,\dot\kappa\ ;\ \kappa\in\mathfrak{K}_{\partial M}\}$$

is one for ∂M and it is uniformly regular on $\partial M\cap S$. If $(\rho,\mathfrak{K})$ is a singularity datum for (M,g) on S, then $(\dot\rho,\dot{\mathfrak{K}})$ is one for $(\partial M,\dot g)$ on $\partial M\cap S$.

Proof. Obvious. □

In this lemma it is implicitly assumed that $m\ge 2$. However, calling—in abuse of language—every 0-dimensional manifold uniformly regular, Lemma 3.5 holds for $m=1$ also, employing obvious interpretations and adaptions.

4 Uniformly Regular Riemannian Manifolds

On the basis of the preceding considerations we now provide proofs for some of the claims made in Examples 1.1.

Let (M,g) be a Riemannian manifold. It has *bounded geometry* if it has an empty boundary, is complete, has a positive injectivity radius, and all covariant derivatives of the curvature tensor are bounded.

Theorem 4.1. *If (M,g) has bounded geometry, then it is uniformly regular.*

Proof. This follows from Th. Aubin [6, Lemma 2.2.6] and J. Eichhorn [11] (also see M.A. Shubin [28]). □

A uniformly regular Riemannian manifold without boundary is complete (cf. M. Disconzi, Y. Shao, and G. Simonett [10]). It has been shown by R.E. Greene [12] that every manifold M without boundary admits a Riemannian metric g such that (M,g) has bounded geometry. However, in view of applications to differential equations which we have in mind, this result is of restricted interest, in general. Indeed, the metric is then given a priori and is closely related to the differential operators under consideration.

Although Theorem 4.1 is very general it has the disadvantage that it applies only to manifolds without boundary. The following results do not require ∂M to be empty.

Lemma 4.2. *Let (M,g) be a Riemannian manifold and suppose $S \subset M$ is compact. Then there exists a unique uniformly regular structure for M on S, and (M,g) is uniformly regular on S.*

Proof. (1) For each $p \in M$ there exists a local chart $\tilde{\kappa}_p$ of M with $p \in U_{\tilde{\kappa}}$. We set $W_p := Q^m$ if $p \in \mathring{M}$, and $W_p := Q^m \cap \mathbb{H}^m$ for $p \in \partial M$. Then we can fix $\delta_p > 0$ such that $\overline{\tilde{\kappa}_p(p) + \delta_p W_p} \subset \kappa_p(U_{\kappa_p})$. From this it follows that, by translation and dilation, we find for each pair $p,q \in M$ local charts κ_p and κ_q, normalized at p and q, respectively, such that $\| \kappa_p \circ \kappa_q^{-1} \|_{k,\infty} \leq c(p,q,k)$ for $k \in \mathbb{N}$.

By the compactness of S we can determine a finite subset Σ of S such that $\{ \kappa_p^{-1}(2^{-1}Q^m_{\kappa_p}) \; ; \; p \in \Sigma \}$ is an open cover of S. Let $\mathfrak{N}$ be an atlas for the open submanifold $M \backslash S$ of M. Then

$$\mathfrak{K} := \{ \kappa_p \; ; \; p \in \Sigma \} \cup \mathfrak{N}$$

is an atlas for M, and $\mathfrak{K}_S = \{ \kappa_p \; ; \; p \in \Sigma \}$. Since Σ is finite $\mathfrak{K}$ is uniformly regular on S and (cf. (2.2)) condition (2.6) is satisfied.

(2) Let $\mathfrak{L}$ be an atlas for M which is uniformly regular on S. By the compactness of S we find a subatlas $\mathfrak{M}$ of $\mathfrak{L}$ such that $\mathfrak{M}_S$ is a finite subset of $\mathfrak{L}_S$. It is obvious that $\mathfrak{M}$ can be chosen such that $\mathfrak{M} \underset{S}{\approx} \mathfrak{L}$. Since $\mathfrak{K}_S$ and $\mathfrak{M}_S$ are both finite, $\mathfrak{M} \underset{S}{\approx} \mathfrak{K}$. Consequently, $\mathfrak{L} \underset{S}{\approx} \mathfrak{K}$. This proves the uniqueness assertion. □

Corollary 4.3. *Every compact Riemannian manifold is uniformly regular.*

The next theorem concerns submanifolds of codimension 0 of uniformly regular Riemannian manifolds.

Theorem 4.4. *Let (N,g) be an m-dimensional uniformly regular Riemannian manifold and (M,g) an m-dimensional Riemannian submanifold with compact boundary. Then (M,g) is uniformly regular.*

Proof. By the preceding corollary we can assume $\partial M \neq \emptyset$.

Since M is locally compact and ∂M is compact there exist relatively compact open neighborhoods W_1 and W_2 of ∂M in M with $W_1 \subset \bar{W}_1 \subset W_2$. We set $V_1 := W_2$ and $S_1 := \bar{W}_1$ as well as $V_2 := \mathring{M}$ and $S_2 := M \backslash W_1$. Then V_i is open in M, $S_i \subset V_i$, and $S_1 \cup S_2 = M$.

The compactness of S_1 in M and $d_M(S_1, M\backslash W_2) > 0$ imply, due to Lemmas 3.2 and 4.2, that there exists an atlas $\mathfrak{K}_1$ for M such that $(\mathbf{1}, \mathfrak{K}_1)$ is a singularity datum for V_1 on S_1.

Note that $d_M(S_2, \partial M) > 0$. Hence Lemma 3.2 and the uniform regularity of (N,g) imply the existence of an atlas $\mathfrak{K}_2$ for $\mathring{M}$ such that $(\mathbf{1}, \mathfrak{K}_2)$ is a singularity datum for V_2 on S_2.

Since $S := S_1 \cap S_2 = \bar{W}_1 \backslash W_1$ is compact we can assume that $\mathfrak{K}_{1,S}$ and $\mathfrak{K}_{2,S}$ are finite. Hence it is obvious that condition (v) of Lemma 3.3 is satisfied. Thus that lemma guarantees the validity of the claim. □

Corollary 4.5. *Let M be an m-dimensional Euclidean submanifold of $\mathbb{R}^m$ with compact boundary. Then M is a uniformly regular Riemannian manifold.*

Proof. Set $N := \mathbb{R}^m$ and recall (3.3). □

5 Characteristics

We write $J_0 := (0,1]$, $J_\infty := [1,\infty)$, and assume throughout that $J \in \{J_0, J_\infty\}$. A subinterval I of J is *cofinal* if $1 \notin I$ and $J \backslash \mathring{I}$ is a compact interval.

We denote by $\mathscr{R}(J)$ the set of all $R \in C^\infty\big(J,(0,\infty)\big)$ satisfying $R(1) = 1$, such that $R(\omega) := \lim_{t\to\omega} R(t)$ exists in $[0,\infty]$ if $J = J_\omega$. Then we write $R \in \mathscr{C}(J)$ if

$$
\begin{aligned}
&\text{(i)} \quad R \in \mathscr{R}(J) \text{ and } R(\infty) = 0 \text{ if } J = J_\infty\,;\\
&\text{(ii)} \quad \int_J dt/R(t) = \infty\,;\\
&\text{(iii)} \quad \|\partial^k R\|_\infty < \infty\,,\ \ k \geq 1\,.
\end{aligned}
\tag{5.1}
$$

The elements of $\mathscr{C}(J)$ are called *cusp characteristics* on J.

On J_∞ we introduce, in addition, the set $\mathscr{F}(J_\infty)$ of *funnel characteristics* by: $R \in \mathscr{F}(J_\infty)$ if

$$
\begin{aligned}
&\text{(i)} \quad R \in \mathscr{R}(J_\infty) \text{ and } R(\infty) > 0\,;\\
&\text{(ii)} \quad \|\partial^k R\|_\infty < \infty\,,\ \ k \geq 1\,.
\end{aligned}
\tag{5.2}
$$

Examples 5.1. *(a) We set $\mathsf{R}_\alpha(t) := t^\alpha$ for $\alpha \in \mathbb{R}$. Then*
$\mathsf{R}_\alpha \in \mathscr{C}(J_0)$ *if* $\alpha \geq 1$, $\mathsf{R}_\alpha \in \mathscr{C}(J_\infty)$ *if* $\alpha < 0$, *and* $\mathsf{R}_\alpha \in \mathscr{F}(J_\infty)$ *if* $0 \leq \alpha \leq 1$.

(b) Suppose $\beta > 0$ *and* $\gamma \in \mathbb{R}$. *Put* $R(t) := \mathrm{e}^{\beta(1-t^\gamma)}$. *Then* $R \in \mathscr{C}(J_0)$ *if* $\gamma < 0$, *whereas* $R \in \mathscr{C}(J_\infty)$ *for* $\gamma > 0$.

(c) For $\alpha \geq -2/\pi$ *and* $\beta > 0$ *we put* $R_{\mathrm{arctan},\alpha,\beta}(t) := 1 + \alpha \arctan\big(\beta(t-1)\big)$. *Then* $R_{\mathrm{arctan},-2/\pi,\beta} \in \mathscr{C}(J_\infty)$ *and* $R_{\mathrm{arctan},\alpha,\beta} \in \mathscr{F}(J_\infty)$ *if* $\alpha > -2/\pi$. □

Let $R \in \mathscr{C}(J)$, respectively $R \in \mathscr{F}(J_\infty)$. Then the *R-gauge diffeomorphism*

$$\sigma = \sigma[R] : J \to \mathbb{R}^+$$

is defined by

$$\sigma(t) := \begin{cases} \operatorname{sign}(t-1) \displaystyle\int_1^t \mathrm{d}\tau / R & \text{if} \quad R \in \mathscr{C}(J)\,, \\ \displaystyle\int_1^t \sqrt{1+\dot{R}^2}\,\mathrm{d}\tau & \text{if}\ R \in \mathscr{F}(J_\infty)\,. \end{cases}$$

Note that $\sigma(J) = \mathbb{R}^+$ and $\dot{\sigma}(t) \neq 0$ for $t \in J$. Hence σ is indeed a diffeomorphism whose inverse is written $\tau = \tau[R] := \sigma^{-1} : \mathbb{R}^+ \to J$. We define the *R-sequence* (t_j) by $t_j = t_j[R] := \tau(j)$ for $j \in \mathbb{N}$. Then (t_j) is strictly increasing to ∞ if $J = J_\infty$, whereas it strictly decreases to 0 otherwise. For $k \geq 1$ we put

$$I_k = I_k[R] := \begin{cases} (0, t_k] & \text{if } J = J_0\,, \\ [k, \infty) & \text{if } J = J_\infty\,. \end{cases}$$

Thus I_k is a cofinal interval of J.

Lemma 5.2. *Suppose* $R \in \mathscr{C}(J)$ *or*[3] $R \in \mathscr{F}(J_\infty)$. *Set*

$$r = r[R] := \begin{cases} R & \textit{if } R \in \mathscr{C}(J)\,, \\ \mathbf{1} & \textit{if } R \in \mathscr{F}(J_\infty)\,. \end{cases} \tag{5.3}$$

Then r *is a singularity function for* $(\mathring{J}, \mathrm{d}t^2)$ *on* I_2.

Proof. (1) We set

$$J_j = J_j[R] := \begin{cases} (t_{j+1}, t_{j-1}) & \text{if } J = J_0\,, \\ (t_{j-1}, t_{j+1}) & \text{if } J = J_\infty\,. \end{cases}$$

Then J_j is a nonempty open subinterval of $\mathring{J}$ for $j \geq 1$, and $\{J_j\,;\ j \geq 1\}$ is a covering of $\mathring{J}$ of multiplicity 2. We let

$$\sigma_j := \sigma \,|\, J_j - j\,, \qquad j \geq 1\,. \tag{5.4}$$

[3] More precisely: $J = J_\infty$ and $R \in \mathscr{F}(J)$.

Then $\mathfrak{S} = \mathfrak{S}[R] := \{\, \sigma_j \; ; \; j \geq 1 \,\}$ is a normalized atlas, the *R-atlas*, for $\mathring{J}$ of multiplicity 2 which is shrinkable to $1/2$. Note that $\tau_j = \tau_j[R] := \sigma_j^{-1}$ satisfies

$$\tau_j(s) = \tau(s+j) \,, \qquad s \in Q \,, \quad j \geq 1 \,. \tag{5.5}$$

By (5.4) and (5.5) we see that $\sigma_j \circ \tau_k(s) = s + k - j \in Q$ if $s \in Q$ and $\tau_k(s) \in J_j$. This proves that $\mathfrak{S}$ is uniformly regular on I.

(2) We set $\rho := R \circ \tau = \tau^* R$. Then

$$\dot{\rho} = (\tau^* \dot{R}) \dot{\tau} \,. \tag{5.6}$$

Furthermore, $\sigma \circ \tau = \mathrm{id}$ implies

$$\dot{\tau} = 1/\tau^* \dot{\sigma} \,. \tag{5.7}$$

(3) Assume $R \in \mathscr{C}(J)$. If $J = J_0$, then $R(0) = 0$ by (5.1)(ii). Thus, for each choice of J,

$$0 < \rho \leq c \,. \tag{5.8}$$

Since $\dot{\sigma}(t) = \mathrm{sign}(t-1)/R(t)$ we get from (5.7)

$$\dot{\tau} = \mathrm{sign}(\tau - 1) \rho \,. \tag{5.9}$$

Hence, by (5.6) and setting $\varepsilon := \mathrm{sign}(\tau - 1)$,

$$\dot{\rho} = b_1 \rho \,, \qquad b_1 := \varepsilon \tau^* \dot{R} \in BC(\mathbb{R}^+) \,. \tag{5.10}$$

Furthermore,

$$\dot{b}_1 = \varepsilon (\tau^* \ddot{R}) \dot{\tau} = (\tau^* \ddot{R}) \rho \in BC(\mathbb{R}^+) \,, \tag{5.11}$$

due to (5.9) and (5.1)(iii). Consequently, we obtain from (5.10)

$$\ddot{\rho} = b_2 \rho \,, \qquad b_2 := \dot{b}_1 + b_1^2 \in BC(\mathbb{R}) \,.$$

By induction

$$\partial^k \rho = b_k \rho \,, \qquad b_k := \dot{b}_{k-1} + b_{k-1} b_1 \,, \quad k \geq 2 \,. \tag{5.12}$$

Thus b_k is a polynomial function in the variables $b_1, \dot{b}_1, \ldots, \partial^{k-1} b_1$ with coefficients in $\mathbb{Z}$.

From (5.9)–(5.11) we get

$$\ddot{b}_1 = \varepsilon (\tau^* \partial^3 R) \rho^2 + \tau^* (\ddot{R}) (\tau^* \dot{R}) \rho \,.$$

Hence we find, once more inductively, that $\partial^\ell b_1$ is a polynomial function in the variables $\rho, \tau^* \partial R, \ldots, \tau^* \partial^{\ell+1} R$ with coefficients in $\mathbb{Z}$. Consequently, b_k is a polynomial function in the variables $\rho, \tau^* \partial R, \ldots, \tau^* \partial^{k+1} R$. Hence $b_k \in BC(\mathbb{R})$ by (5.8) and

(5.1)(iii). Thus we obtain from (5.12)

$$|\partial^k \rho| \le c(k)\rho\;, \qquad k \ge 1\;. \tag{5.13}$$

It follows from $\partial \log\rho = \dot\rho/\rho$ and the last estimate that $\beta := \|\partial \log\rho\|_\infty < \infty$. Hence, by the mean-value theorem,

$$\bigl|\log\bigl(\rho(s)/\rho(t)\bigr)\bigr| = |\log\rho(s) - \log\rho(t)| \le \beta|s-t|\;, \qquad s,t \ge 0\;.$$

This implies $\mathrm{e}^{-\beta} \le \rho(s)/\rho(t) \le \mathrm{e}^{\beta}$ for $|s-t| \le 1$, that is,

$$\rho(s) \sim \rho(t)\;, \qquad s,t \in \mathbb{R}^+\;, \quad |s-t| \le 1\;. \tag{5.14}$$

Since $\rho_j := \tau_j^* R = \rho(\cdot + j)$ we deduce from (5.14)

$$\rho_j \sim \rho_j(0)\;, \qquad j \ge 1\;. \tag{5.15}$$

Furthermore, since $\partial\rho_j = (\partial\rho)(\cdot + j)$, we obtain from (5.13) and (5.15)

$$\|\partial^k \rho_j\|_\infty \le c(k)\rho_j(0)\;, \qquad j \ge 1\;, \quad k \ge 0\;. \tag{5.16}$$

Due to $R = r$ and $\tau_j^* r = \kappa_{j*} r$ we see from (5.15) and (5.16) that $r \in C\bigl(\mathring{J}, (0,\infty)\bigr)$ satisfies (2.7)(iii), (iv) with $\mathfrak{K} = \mathfrak{S}$ and $S = I_2$.

(4) Suppose $R \in \mathscr{F}(J_\infty)$. Since $\dot\sigma = (1 + \dot R^2)^{1/2}$ we get from (5.7)

$$\dot\tau = \bigl(1 + (\tau^* \dot R)^2\bigr)^{-1/2}\;. \tag{5.17}$$

Using this and $\|\dot R\|_\infty < \infty$ we obtain

$$1/c \le \dot\tau \le 1\;. \tag{5.18}$$

From (5.17), (5.18), and (5.2)(ii) we deduce inductively

$$\|\partial^k \tau\|_\infty < \infty\;, \qquad k \ge 1\;. \tag{5.19}$$

Hence (5.5) implies

$$\dot\tau_j \sim \mathbf{1}\;, \qquad \|\partial^k \tau_j\|_\infty \le c(k)\;, \qquad j \ge 1\;, \quad k \ge 1\;. \tag{5.20}$$

Thus $r = \mathbf{1}$ satisfies (2.7)(iv) with $\mathfrak{K} = \mathfrak{S}$ and $S = I_2$, and (2.7)(iii) is trivially true.

(5) Again we assume $R \in \mathscr{C}(J)$ or $R \in \mathscr{F}(J_\infty)$. Then

$$\sigma_{j*}\,\mathrm{d}t^2 = \tau_j^*\,\mathrm{d}t^2 = \mathrm{d}\tau_j^2 = \dot\tau_j^2\,\mathrm{d}s^2\;. \tag{5.21}$$

If $R \in \mathscr{C}(J)$, then we get $\dot\tau_j^2 = \rho_j^2$ from (5.9). Hence

$$\sigma_{j*}\,\mathrm{d}t^2 = \rho_j^2\,\mathrm{d}s^2 = (r_{\sigma_j})^2\,\mathrm{d}s^2\;, \qquad j \ge 1\;.$$

If $R \in \mathscr{F}(J_\infty)$, then we obtain from (5.20) and (5.21)

$$\sigma_{j*}\, dt^2 \sim ds^2 = (r_{\sigma_j})^2\, ds^2 , \qquad j \geq 1 .$$

Hence (2.7)(v) applies to r and $g = dt^2$ with $\mathfrak{K} = \mathfrak{S}$ and $S = I_2$ as well.

(6) Using (5.21), we infer from (5.9) and (5.16) if $R \in \mathscr{C}(J)$, respectively from (5.9) and (5.19) if $R \in \mathscr{F}(J_\infty)$, that

$$\|\partial^k(\sigma_*\, dt^2)\|_\infty \leq c(k) r_\sigma^2 , \qquad \sigma \in \mathfrak{S}_I , \quad k \geq 0 .$$

Thus (2.7)(vi) is also satisfied. This proves the assertion. □

6 Model Cusps and Funnels

We suppose $R \in \mathscr{R}(J)$, $0 \leq d \leq \bar{d}$, and B is a d-dimensional submanifold of $\mathbb{R}^{\bar{d}}$. Let I be an open cofinal subinterval of $\mathring{J}$. We set

$$P(R,B;I) := \bigl\{ \bigl(t, R(t)y\bigr) \; ; \; t \in \mathring{J},\ y \in B \bigr\} \subset \mathbb{R} \times \mathbb{R}^{\bar{d}} = \mathbb{R}^{1+\bar{d}} .$$

Then $P = P(R,B) = P(R,B;\mathring{J})$ is a $(1+d)$-dimensional submanifold of $\mathbb{R}^{1+\bar{d}}$, the *(model)* (R,B)*-pipe on* J, also called *(model) R-pipe over (the basis) B on J*. Note

$$\partial P(R,B) = P(R,\partial B) ,$$

where $P(R,\emptyset) := \emptyset$. An R-pipe is an *R-cusp* if $R(\omega) = 0$, where $\omega \in \{0,\infty\}$ and $J = J_\omega$, and an *R-funnel* otherwise. The map

$$\varphi = \varphi[R] \colon P \to \mathring{J} \times B , \quad \bigl(t, R(t)y\bigr) \mapsto (t,y) \tag{6.1}$$

is a diffeomorphism, the *canonical stretching diffeomorphism* of P.

If $d = 0$, then B is a countable discrete subset of $\mathbb{R}^{\bar{d}}$. In abuse of language and for a unified presentation we call it uniformly regular Riemannian manifold as well and write formally (B, g_B) for B, although g_B has no proper meaning. In this case g_B has to be replaced by 0 in the formulas below.

Suppose $p \in C^\infty\bigl(P,(0,\infty)\bigr)$ and g_P is a Riemannian metric for P. Then p is a *cofinal singularity function for* (P, g_P) *on* $S \subset B$ if there exists a cofinal subinterval I of J such that p is a singularity function for (P, g_P) on $\varphi^{-1}(I \times S)$. It follows from Lemma 4.2 that this is then true for every cofinal subinterval of J. In related situations the qualifier "cofinal" has similar (obvious) meanings.

We consider the following assumption:

$$\begin{aligned} &g_B \text{ is a Riemannian metric for } B,\ S \subset B, \text{ and} \\ &b \text{ is a bounded singularity function for } (B, g_B) \text{ on } S. \end{aligned} \tag{6.2}$$

Lemma 6.1. *Let condition (6.2) apply. Suppose* $a \in C^\infty\big(J,(0,\infty)\big)$ *and* r *is a bounded singularity function for* $(\mathring{J}, a\,dt^2)$ *on some cofinal subinterval of* J*. Let* $R \in \mathscr{R}(J)$ *and set*

$$g := \varphi^*(a\,dt^2 + g_B)\ , \quad p := \varphi^*(r \otimes b)\ . \tag{6.3}$$

Then p *is a cofinal singularity function for* (P,g) *on* S*.*

Proof. By Theorem 3.1 $r \otimes b$ is a cofinal singularity function for $(\mathring{J} \times B,\ a\,dt^2 + g_B)$ on S. Hence the assertion follows from (6.1) and Lemma 3.4. □

Corollary 6.2. *Put*

$$\hat{g} := \varphi^*\big((a\,dt^2 + g_B)/(r \otimes b)^2\big)\ . \tag{6.4}$$

Then $(P,\hat{g})$ *is cofinally uniformly regular on* S*.*

The following two propositions are cornerstones for the construction of wide classes of singular manifolds.

Proposition 6.3. *Let (6.2) be satisfied and suppose* $R \in \mathscr{C}(J)$ *or* $R \in \mathscr{F}(J_\infty)$*. Set* $a := \mathbf{1}$*. Define* r *by (5.3) and* g *by (6.3). Then* p *is a cofinal singularity function for* (P,g) *on* S*.*

Proof. Lemmas 5.2 and 6.1. □

We write $\psi \in \mathscr{R}_0(J,\hat{J})$ if $\psi \in \mathscr{R}(J)$ with $\psi(J) = \hat{J} \in \{J_0, J_\infty\}$ and $\dot{\psi}(t) \neq 0$ for $t \in J$. Thus ψ is a diffeomorphism from J onto $\hat{J}$.

Proposition 6.4. *Suppose (6.2) applies,* $\psi \in \mathscr{R}_0(J,\hat{J})$*, and* $\hat{R} \in \mathscr{C}(\hat{J})$*, or* $\hat{J} = J_\infty$ *and* $\hat{R} \in \mathscr{F}(J_\infty)$*. Set*

$$R := \psi^*\hat{R}\ , \quad \varphi := \varphi[R]\ , \quad g := \varphi^*(\dot{\psi}^2\,dt^2 + g_B)\ ,$$

and

$$r = r[R] := \begin{cases} R & \text{if } \hat{R} \in \mathscr{C}(\hat{J})\ , \\ \mathbf{1} & \text{if } \hat{R} \in \mathscr{F}(J_\infty)\ . \end{cases}$$

Then $p := \varphi^*(r \otimes b)$ *is a cofinal singularity function for* (P,g) *on* S*.*

Proof. We write $\hat{P} := P(\hat{R},B)$, $\hat{\varphi} := \varphi[\hat{R}]$, and $\hat{g} := \hat{\varphi}^*(ds^2 + g_B)$. Then

$$\Phi := \hat{\varphi}^{-1} \circ (\psi \times \mathrm{id}_B) \circ \varphi \colon P \to \hat{P} \tag{6.5}$$

is a diffeomorphism and

$$\Phi^*\hat{g} = \varphi^*(\psi^* \times \mathrm{id}_B)\hat{\varphi}_*\hat{g} = \varphi^*(\psi^* \times \mathrm{id}_B)(ds^2 + g_B) = \varphi^*(\dot{\psi}^2\,dt^2 + g_B) = g\ . \tag{6.6}$$

Furthermore, setting $\hat{r} := r[\hat{R}]$ and $\hat{p} := \hat{\varphi}^*(\hat{r} \otimes b)$,

$$\Phi^*\hat{p} = \varphi^*(\psi^* \times \mathrm{id}_B)(\hat{r} \otimes b) = \varphi^*(r \otimes b) = p\ . \tag{6.7}$$

Proposition 6.3 guarantees that $\hat{p}$ is a cofinal singularity function for $(\hat{P}, \hat{g})$ on S. Hence the assertion follows from (6.5)–(6.7) and Lemma 3.4. □

Now we provide some examples. The most important ones concern *α-pipes*, that is, R_α-pipes over B on J. We write $P_\alpha = P_\alpha(B) := P(\mathsf{R}_\alpha, B)$ and $\varphi_\alpha := \varphi[\mathsf{R}_\alpha]$ for $\alpha \in \mathbb{R}$.

Examples 6.5. *Let (6.2) be satisfied.*

(a) Set $g_\alpha := \varphi_\alpha^*(dt^2 + g_B)$,

$$p_\alpha := \varphi_\alpha^*(\mathsf{R}_\alpha \otimes b) \textit{ if either } J = J_0 \textit{ and } \alpha \geq 1\,,\ \textit{or } J = J_\infty \textit{ and } \alpha < 0\,,$$

and

$$p_\alpha := \varphi_\alpha^*(\mathbf{1} \otimes b) \textit{ if } J = J_\infty \textit{ and } 0 \leq \alpha \leq 1\,.$$

Then p_α *is a cofinal singularity function for* (P_α, g_α) *on* S.

Proof. Example 5.1(a) and Proposition 6.3. □

(b) We put

$$p_\alpha := \varphi_\alpha^*(\mathsf{R}_\alpha \otimes b) \textit{ if } J = J_0 \textit{ and } 0 < \alpha \leq 1\,,$$

and

$$p_\alpha := \varphi_\alpha^*(\mathbf{1} \otimes b) \textit{ if either } J = J_0 \textit{ and } \alpha \leq 0\,,\ \textit{or } J = J_\infty \textit{ and } \alpha \geq 1\,.$$

We also fix $\beta \neq 0$ *such that* $0 < \beta \leq \alpha$ *if* $J = J_0$ *and* $0 < \alpha \leq 1$, $\beta \geq \alpha$ *if* $J = J_\infty$ *and* $\alpha > 1$, *and* $\beta \leq \alpha$ *if* $J = J_0$ *and* $\alpha \leq 0$. *Then* p_α *is a cofinal singularity function for* $(P_\alpha, g_{\alpha,\beta})$ *on* S, *where* $g_{\alpha,\beta} := \varphi_\alpha^*(t^{2(\beta-1)}\,dt^2 + g_B)$.

Proof. Note that $\mathsf{R}_\beta \in \mathscr{R}_0(J, \hat{J})$ with $\hat{J} = J$ if $\beta > 0$, and $\hat{J} = J_\infty$ if $J = J_0$ and $\beta < 0$. Moreover, $\mathsf{R}_\beta^* \mathsf{R}_\gamma = \mathsf{R}_{\beta\gamma}$ for $\gamma \in \mathbb{R}$.

We put $\psi := \mathsf{R}_\beta$ and $\hat{R} := \mathsf{R}_{\alpha/\beta}$ so that $\psi^* \hat{R} = \mathsf{R}_\alpha$. It follows from Example 5.1(a) that $\hat{R} \in \mathscr{C}(J_0)$ if $J = J_0$ and $0 < \alpha \leq 1$, and $\hat{R} \in \mathscr{F}(J_\infty)$ otherwise. Moreover, $\dot{\psi} = \dot{\mathsf{R}}_\beta \sim \mathsf{R}_{\beta-1}$. Now the claim follows from Proposition 6.4. □

(c) Suppose $J = J_0$ *and* $R(t) := 1 - \alpha \arctan(1 - 1/t)$ *with* $\alpha \geq -2/\pi$. *Set*

$$p := \varphi^*(\mathbf{1} \otimes b) \textit{ if } \alpha > -2/\pi\,, \quad p := \varphi^*(R \otimes b) \textit{ if } \alpha = -2/\pi\,,$$

and $g := \varphi^*(t^{-4}\,dt^2 + g_B)$. *Then* p *is a cofinal singularity function for* (P, g) *on* S.

Proof. Put $\hat{R} := R_{\mathrm{arctan},\alpha,1}$ (see Example 5.1(c)) and $\psi := \mathsf{R}_{-1}$. Then $R = \psi^* \hat{R}$ and $\dot{\psi} \sim \mathsf{R}_{-2}$. Hence Example 5.1(c) and Proposition 6.4 imply the assertion. □

7 Submanifolds of Euclidean Spaces

Now we consider the case where (M,g) is a Riemannian submanifold of $(\mathbb{R}^n, g_n)$ for some $n \in \mathbb{N}^\times$. In other words, we assume

$$(M,g) \hookrightarrow (\mathbb{R}^n, g_n)\,.$$

By Nash's theorem this is no restriction of generality. It is now natural and convenient to describe M by local parametrizations. Hereby, given a local chart κ for M, the map

$$i_\kappa := \iota_M \circ \kappa^{-1} \in C^\infty\big(\kappa(U_\kappa), \mathbb{R}^n\big)$$

is the *local parametrization associated with* κ. The following lemma provides a useful tool for establishing that a given function ρ on M is a singularity function for (M,g).

By a parametrization-regular (p-r) singularity datum for (M,g) *on* $S \subset M$ we mean a pair $(\rho, \mathfrak{K})$ with the following properties:

$$\begin{aligned}
&\text{(i)} && \mathfrak{K} \text{ is an atlas for } M \text{ such that } \mathfrak{K}_S \text{ is shrinkable}\\
& && \text{and has finite multiplicity}\,.\\
&\text{(ii)} && \rho \in C^\infty\big((M,(0,\infty)\big) \text{ satisfies (2.7)(iii) and (iv)}\,.\\
&\text{(iii)} && \kappa_* g \geq \rho_\kappa^2 g_m / c\,, \quad \kappa \in \mathfrak{K}_S\,.\\
&\text{(iv)} && \|\partial^k i_\kappa\|_\infty \leq c(k)\rho_\kappa\,, \quad \kappa \in \mathfrak{K}_S\,, \quad k \geq 1\,,
\end{aligned} \tag{7.1}$$

where ∂ denotes the Fréchet derivative. Clearly, ρ is a *p-r singularity function for* (M,g) *on* S if there exists an atlas $\mathfrak{K}$ such that $(\rho, \mathfrak{K})$ is a p-r singularity datum for (M,g) on S.

Lemma 7.1. *Suppose* $(\rho, \mathfrak{K})$ *is a p-r singularity datum for* (M,g) *on* S. *Then it is a singularity datum for* (M,g) *on* S.

Proof. (1) In the following, we identify a linear map $a\colon \mathbb{R}^m \to \mathbb{R}^n$ with its representation matrix $[a] \in \mathbb{R}^{n \times m}$ with respect to the standard bases. Then

$$\kappa_* g = \kappa_*(\iota_M^* g_n) = i_\kappa^* g_n = (\partial i_\kappa)^\top \partial i_\kappa\,, \qquad \kappa \in \mathfrak{K}\,. \tag{7.2}$$

From this and (7.1)(iii) and (iv) it follows

$$\kappa_* g \sim \rho_\kappa^2 g_m\,, \quad \|\kappa_* g\|_{k,\infty} \leq c(\kappa)\rho_\kappa^2\,, \qquad \kappa \in \mathfrak{K}_S\,, \quad k \in \mathbb{N}\,. \tag{7.3}$$

Hence $[\kappa_* g]$ has its spectrum in $[\rho_\kappa^2/c,\ c\rho_\kappa^2] \subset \mathbb{R}$ for $\kappa \in \mathfrak{K}_S$. Consequently, the spectrum of $[\kappa_* g]^{-1}$ is contained in $[\rho_\kappa^{-2}/c,\ c\rho_\kappa^{-2}]$ for $\kappa \in \mathfrak{K}_S$. This implies

$$\|[\kappa_* g]^{-1}\|_\infty \leq c/\rho_\kappa^2\,, \qquad \kappa \in \mathfrak{K}_S\,.$$

Thus, by the chain rule and (7.3), it follows

$$\|[\kappa_* g]^{-1}\|_{k,\infty} \leq c(k)\rho_\kappa^{-2}\,, \qquad \kappa \in \mathfrak{K}_S\,, \quad k \in \mathbb{N}\,. \tag{7.4}$$

(2) We set

$$\Lambda_\kappa(x) := [\kappa_* g]^{-1}(\partial i_\kappa)^\top \in \mathbb{R}^{m\times n}\,, \qquad x \in Q_\kappa^m\,, \quad \kappa \in \mathfrak{K}_S\,.$$

Then (7.1)(iv) and (7.4) imply

$$\|\Lambda_\kappa\|_{k,\infty} \le c(k)\rho_\kappa^{-1}\,, \qquad \kappa \in \mathfrak{K}_S\,, \quad k \in \mathbb{N}\,. \tag{7.5}$$

Given $\kappa \in \mathfrak{K}_S$ and $p \in U_\kappa$,

$$T_pM = \{p\} \times \partial i_\kappa\big((\kappa(p))(\mathbb{R}^m) \hookrightarrow \{p\} \times \mathbb{R}^n = T_p\mathbb{R}^n\,. \tag{7.6}$$

We read off (7.2) that Λ_κ is a left inverse for ∂i_κ. Furthermore,

$$\ker(\Lambda_\kappa) = \ker\big((\partial i_\kappa)^\top\big) = \big(\mathrm{im}(\partial i_\kappa)\big)^\perp\,.$$

It follows from this and (7.6) that $T_p\kappa$, the tangential of κ at p, is given by

$$T_p\kappa\colon T_pM \to T_{\kappa(p)}\mathbb{R}^m\,, \quad (p,\xi) \mapsto \big(\kappa(p), \Lambda_\kappa(\kappa(p))\xi\big)$$

(cf. [5, Remark 10.3(d)]). Thus we find $\partial(\tilde\kappa \circ \kappa^{-1}) = \Lambda_{\tilde\kappa}\partial i_\kappa$ for $\kappa, \tilde\kappa \in \mathfrak{K}_S$ with $U_\kappa \cap U_{\tilde\kappa} \ne \emptyset$. Hence (7.1)(iv) and (7.5) imply

$$\|\tilde\kappa \circ \kappa^{-1}\|_{k,\infty} \le c(k)\,, \qquad \kappa, \tilde\kappa \in \mathfrak{K}_S\,, \quad k \in \mathbb{N}\,,$$

due to $\mathrm{im}(\tilde\kappa \circ \kappa^{-1}) \subset Q^m$. Thus, recalling (7.1)(i), we see that $\mathfrak{K}$ is uniformly regular on S. This proves the claim. □

In the next lemma we consider a particularly simple, but important, p-r regular singularity datum. In this special situation it is the converse of the preceding lemma.

Lemma 7.2. *Suppose $S \subset M$ and S is compact in $\mathbb{R}^n$. If $(M, \mathfrak{K}, g)$ is uniformly regular on S, then $(\mathbf{1}, \mathfrak{K})$ is a p-r singularity datum for (M, g) on S.*

Proof. Due to the hypotheses, conditions (7.1)(i)–(iii) are trivially satisfied with $\rho = \mathbf{1}$.

For each $p \in S$ there is a normalized local chart φ_p for $\mathbb{R}^n$ such that $\varphi_p(p) = 0$,

$$\|\varphi_p^{-1}\|_{k,\infty} \le c(k,p)\,, \qquad k \in \mathbb{N}\,, \tag{7.7}$$

and $\kappa_p := \varphi_p \,|\, (M \cap U_{\varphi_p})$ is a normalized local chart for M with $\kappa_p(p) = 0 \in \mathbb{R}^m$. By the compactness of S in $\mathbb{R}^n$ there exists a finite subset P of S such that $\{\, U_p := \mathrm{dom}(\kappa_p)\ ;\ p \in P\,\}$ is an open covering of S in M. We set $\hat{\mathfrak{K}} := \{\, \kappa_p\ ;\ p \in P\,\}$ and $\tilde{\mathfrak{K}} := \hat{\mathfrak{K}} \cup (\mathfrak{K} \backslash \mathfrak{K}_S)$. Then $\tilde{\mathfrak{K}}$ is an atlas for M and $\tilde{\mathfrak{K}}_S = \hat{\mathfrak{K}}$. For $p \in P$ we define $f_p := \varphi_p^{-1}\colon Q_{\varphi_p}^n \to \mathbb{R}^n$. Then $i_{\kappa_p} = f_p \,|\, Q_{\kappa_p}^m$, where $\mathbb{R}^m$ is identified with the subspace $\mathbb{R}^m \times \{0\}$ of $\mathbb{R}^n$, of course. Since $\tilde{\mathfrak{K}}_S$ is finite, it is obvious from (7.7) that

$$\|\partial^k i_\kappa\|_\infty \le c(k)\,, \qquad \kappa \in \tilde{\mathfrak{K}}_S\,, \quad k \ge 1\,. \tag{7.8}$$

By the same reason, and since $\mathfrak{K}$ has finite multiplicity on S, we see that $\tilde{\mathfrak{K}} \underset{S}{\approx} \mathfrak{K}$. Hence (7.8) holds for $\mathfrak{K}_S$ as well, that is, condition (7.1)(iv) is valid also. □

Now we return to the setting of the preceding section. It follows from Corollary 6.2 and Proposition 6.3 that, given $R \in \mathscr{C}(J)$ or $R \in \mathscr{F}(J_\infty)$, the R-pipe $P = P(R,B)$ can be equipped with countably many nonequivalent metrics which make it into a cofinally uniformly regular Riemannian manifold. However, since $\iota_P \colon P \hookrightarrow \mathbb{R}^{1+\bar{d}}$, it is most natural to endow P with the metric $g_P := \iota_P^* g_{1+\bar{d}}$. The following proposition gives sufficient conditions guaranteeing that g in Proposition 6.3 can be replaced by g_P.

Proposition 7.3. *Suppose (B, g_B) is a d-dimensional bounded Riemannian submanifold of $(\mathbb{R}^{\bar{d}}, g_{\bar{d}})$ and b is a p-r singularity function for (B, g_B) on $S \subset B$. Also suppose $R \in \mathscr{C}(J)$ or $R \in \mathscr{F}(J_\infty)$ and define r by (5.3). Then $p = \varphi^*(R \otimes b)$ is a cofinal p-r singularity function for (P, g_P) on S.*

Proof. (1) Let $\mathfrak{B}$ be an atlas for B such that $(b, \mathfrak{B})$ is a p-r singularity datum for (B, g) on S, and let $\mathfrak{S}$ be the R-atlas for $\mathring{J}$. We write

$$f := \iota_P \circ \varphi^{-1} \colon \mathring{J} \times B \to \mathbb{R}^{1+\bar{d}}\,, \quad Y_\beta := i_\beta\,,$$

and use the notations of Sect. 5. Then

$$f_{j\beta} := f \circ (\tau_j \times \beta^{-1}) = (\tau_j, \rho_j Y_\beta) \colon Q \times \beta(U_\beta) \to \mathbb{R}^{1+\bar{d}}$$

is a diffeomorphism onto an open subset $U_{j\beta}$ of P. We denote by ϖ the permutation $\mathbb{R}^{1+d} \to \mathbb{R}^{d+1}$, $(t, y) \mapsto (y, t)$ (which is only needed if $\partial B = \emptyset$). Then, see (6.1),

$$\kappa_{j\beta} := \varpi \circ f_{j\beta}^{-1}(\beta, \tau_j) \colon U_{j\beta} \to \beta(U_\beta) \times Q$$

is a local chart for P and $f_{j\beta} = i_{\kappa_{j\beta}}$. We set

$$\mathfrak{K} := \{\, \kappa_{j\beta}\ ;\ j \ge 1,\ \beta \in \mathfrak{B} \,\} = \varphi^*\big(\varpi \circ (\mathfrak{B} \otimes \mathfrak{S})\big)$$

where $\mathfrak{B} \otimes \mathfrak{S}$ is the product atlas on $B \times \mathring{J}$ and

$$\varpi \circ (\mathfrak{B} \otimes \mathfrak{S}) := \{\, \sigma \times \beta\ ;\ \beta \in \mathfrak{B},\ \sigma \in \mathfrak{S} \,\}\,.$$

By Lemma 5.2 we know that $\mathfrak{S}$ is uniformly regular on $I := I_2[R]$. Hence $\mathfrak{B} \otimes \mathfrak{S}$ is uniformly regular on $S \times I$ by Theorem 3.1. From this and Lemma 3.4 it follows that $\mathfrak{K}$ is a uniformly regular atlas for P on $V := \varphi^{-1}(I \times S)$.

(2) Given $\kappa = \kappa_{j\beta} \in \mathfrak{K}$,

$$\begin{aligned}\kappa_* g_P &= \kappa_* \iota_P^* g_{1+\bar{d}} = (\iota_P \circ \kappa^{-1})^* g_{1+\bar{d}} \\ &= f_{j\beta}^*(dt^2 + |dy|^2) = d\tau_j^2 + |d(\rho_j Y_\beta)|^2\,.\end{aligned}$$

Hence $\mathrm{d}(\rho_j Y_\beta) = \dot\rho_j\, \mathrm{d}s\, Y_\beta + \rho_j\, \mathrm{d}Y_\beta$ implies

$$\kappa_* g_P = (\dot\tau_j^2 + \dot\rho_j^2 |Y_\beta|^2)\, \mathrm{d}s^2 + 2\rho_j \dot\rho_j\, \mathrm{d}s\, (Y_\beta \,|\, \mathrm{d}Y_\beta) + \rho_j^2 |\mathrm{d}Y_\beta|^2 \, .$$

Using $|\mathrm{d}Y_\beta|^2 = \beta_* g_B$ and estimating the next to the last term by the Cauchy-Schwarz inequality gives

$$\kappa_* g_P \geq (\dot\tau_j^2 + (1-1/\varepsilon)\dot\rho_j^2 |Y_\beta|^2)\, \mathrm{d}s^2 + (1-\varepsilon)\rho_j^2 \beta_* g_B$$

for $0 < \varepsilon < 1$, $j \geq 1$, and $\beta \in \mathfrak{B}$.

(3) Suppose $R \in \mathscr{C}(J)$. Then

$$\dot\tau_j^2 = \rho_j^2 \tag{7.9}$$

by (5.9), and $\dot\rho_j^2 \leq c\rho_j^2$ by (5.15) and (5.16). Thus the boundedness of B in $(\mathbb{R}^{\bar d}, g_{\bar d})$ implies that we can choose ε sufficiently close to 1 such that

$$\kappa_* g_B \geq \rho_j^2 (\mathrm{d}s^2 + \beta_* g_B)/c \geq \rho_j^2 (\mathrm{d}s^2 + b_\beta^2 g_d)/c \tag{7.10}$$

for $j \geq 1$, $\beta \in \mathfrak{B}_S$, and $\kappa = \kappa_{j\beta}$, where the last inequality holds since $(b, \mathfrak{K})$ is a p-r singularity datum for (B, g_B) on S.

(4) Assume $R \in \mathscr{F}(J_\infty)$. Then (5.18) implies

$$\dot\tau_j \sim \mathbf{1} \, , \qquad j \geq 1 \, .$$

From this and

$$\dot\rho_j = (\dot R \circ \tau_j)\dot\tau_j \, , \qquad \|\dot R\|_{k,\infty} < \infty \, , \quad k \in \mathbb{N} \, ,$$

we get

$$\|\dot\rho_j\|_{k,\infty} \leq c(k) \, , \qquad j \geq 1 \, , \quad k \in \mathbb{N} \, . \tag{7.11}$$

Thus, similarly as above,

$$\kappa_* g_B \geq \big(\mathrm{d}s^2 + \rho_j^2(0) b_\beta^2 g_d\big)/c \, , \qquad j \geq 1 \, , \quad \beta \in \mathfrak{B}_S \, , \quad \kappa = \kappa_{j\beta} \, . \tag{7.12}$$

(5) Now we proceed analogously to the proof of Theorem 3.1. Recalling that $\mathfrak{S}$ is shrinkable to $1/2$ on I we fix $r \in (1/2, 1)$ such that $\{\, \kappa^{-1}(rQ_\kappa^{d+1}) \, ; \, \kappa \in \mathfrak{K}_V \,\}$ is a covering of V. Then we set $\delta := (1-r)\big/\sqrt{d+1}$,

$$\delta_\beta := \min\{b_\beta, \delta\} \, , \quad \delta_j := \min\{1/R(t_j), \, \delta\} \, ,$$

and

$$\mathfrak{L}_\beta := \mathfrak{L}(\delta_\beta, Q) \, , \quad \mathfrak{L}_j := \mathfrak{L}(\delta_j, Q_\beta^d)$$

for $\beta \in \mathfrak{B}_S$ and $j \geq 1$. Note that the boundedness of b implies

$$\delta_\beta \sim b_\beta \, , \qquad \beta \in \mathfrak{K}_S \, . \tag{7.13}$$

Furthermore,

$$\delta_j \sim 1/\rho_j(0)\,, \quad j \geq 1\,, \quad \text{if } r \in \mathscr{F}(J_\infty)\,, \tag{7.14}$$

since $R(t_j) = \rho_j(0)$ and $1/R \leq c$ in this case.

Given $\kappa = \kappa_{j\beta} \in \mathfrak{K}_V$, we define

$$\mathfrak{N}_\kappa := \begin{cases} \{\mu \times \lambda\;;\; \lambda \in \mathfrak{L}_\beta\,,\; \mu := \mathrm{id}_{Q^d_\beta}\} & \text{if } R \in \mathscr{C}(J)\,, \\ \{\mu \times \lambda\;;\; \lambda \in \mathfrak{L}_\beta\,,\; \mu \in \mathfrak{L}_j\} & \text{if } R \in \mathscr{F}(J_\infty)\,. \end{cases}$$

Then $\mathfrak{N}_\kappa$ is an atlas for Q^{d+1}_κ which is uniformly regular on rQ^{d+1}_κ. Consequently, cf. (3.4),

$$\mathfrak{P} := \{\nu \circ \kappa\;;\; \kappa \in \mathfrak{K}_V\,,\; \nu \in \mathfrak{N}_\kappa\} \cup (\mathfrak{K} \backslash \mathfrak{K}_V)$$

is an atlas for P which is uniformly regular on V. Observe

$$\mathfrak{P}_V \subset \{\nu \circ \kappa\;;\; \kappa \in \mathfrak{K}_V\,,\; \nu \in \mathfrak{N}_\kappa\}\,.$$

Hence condition (7.1)(i) is satisfied.

(6) By the assumption on $(b, \mathfrak{B})$

$$\|\beta_* b\|_{k,\infty} \leq c(k) b_\beta\,, \quad b\,|\,U_\beta \sim b_\beta\,, \qquad \beta \in \mathfrak{B}_S\,, \quad k \geq 0\,. \tag{7.15}$$

Furthermore,

$$\|\rho_j\|_{k,\infty} \leq c(k)\rho_j(0)\,, \quad \rho_j\,|\,J_j \sim \rho_j(0)\,, \qquad j \geq 1\,, \quad k \geq 0\,. \tag{7.16}$$

Indeed, if $R \in \mathscr{C}(J)$, then this is a consequence of (5.16) and (5.15), respectively. If $R \in \mathscr{F}(J_\infty)$, then $\rho_j(0) = R(t_j) \geq 1/c$ for $j \geq 1$. Hence (7.16) follows from (7.11).

We deduce from (3.10) that

$$b_\beta = \beta_* b(0) = \kappa_* b(0) \sim (\nu \circ \kappa)_* b(0) = \pi_* b(0) \tag{7.17}$$

and

$$\rho_j(0) = (\sigma_j)_* R(0) = \kappa_* R(0) \sim (\nu \circ \kappa)_* R(0) = \pi_* R(0) \tag{7.18}$$

for $\pi = \nu \circ \kappa \in \mathfrak{P}_V$ with $\kappa = \kappa_{j\beta} \in \mathfrak{K}_V$ and $\nu \in \mathfrak{L}_\kappa$. From (7.15)–(7.18) we derive

$$\|\pi_* p\|_{k,\infty} \leq c(k) p_\pi\,, \quad p\,|\,U_\pi \sim p_\pi\,, \qquad \pi \in \mathfrak{P}_V\,, \quad k \geq 0\,.$$

Thus condition (7.1)(ii) applies.

(7) From (7.10), (7.12), (7.17), (7.18), and (3.8) we get

$$\pi_* g_P \geq \rho_j^2(0)(\delta_\beta^2\, ds^2 + b_\beta^2 g_d)/c \quad \text{if } R \in \mathscr{C}(J)\,,$$

respectively

$$\pi_* g_P \geq \big(\delta_\beta^2\, ds^2 + \rho_j^2(0)\delta_j^2 b_\beta^2 g_d\big)/c \quad \text{if } R \in \mathscr{F}(J_\infty)\,,$$

for $\pi = \nu \circ \kappa \in \mathfrak{P}_V$ with $\kappa = \kappa_{j\beta}$ and $\nu \in \mathfrak{N}_\kappa$. From this, (7.13), and (7.14) we obtain in either case $\pi_* g_P \geq p_\pi^2 g_{1+d}/c$ for $\pi \in \mathfrak{P}_V$. Thus condition (7.1)(iii) is fulfilled.

(8) By the assumption on $(p, \mathfrak{B})$

$$\|\partial^\alpha Y_\beta\|_\infty \leq c(\alpha) b_\beta \ , \qquad \beta \in \mathfrak{B}_S \ , \quad \alpha \in \mathbb{N}^d \backslash \{0\} \ . \tag{7.19}$$

Given $\pi = \nu \circ \kappa \in \mathfrak{P}_V$ with $\kappa = \kappa_{j\beta} \in \mathfrak{K}_V$ and $\nu = \mu \times \lambda \in \mathfrak{N}_{\kappa_{j\beta}}$,

$$i_\pi = i_\kappa \circ \nu^{-1} = \big(\lambda_* \tau_j, (\lambda_* \rho_j) \mu_* Y_\beta\big) \ . \tag{7.20}$$

Suppose $R \in \mathscr{C}(J)$. Then we get from (7.9) and (3.2)

$$\|\partial^k(\lambda_* \tau_j)\|_\infty = \delta_\beta \|\partial^{k-1}(\lambda_* \rho_j)\|_\infty = \delta_\beta^k \|\lambda_*(\partial^{k-1}\rho_j)\|_\infty \leq c(k) b_\beta \rho_j(0) \tag{7.21}$$

for $j,k \geq 1$ and $\beta \in \mathfrak{B}_S$, due to $0 < \delta_\beta \leq 1$ and (7.13). By means of (7.17)–(7.21) and $\mu = \mathrm{id}$ we deduce

$$\|\partial^\alpha i_\pi\|_\infty \leq c(\alpha) p_\pi \ , \qquad \pi \in \mathfrak{P}_V \ , \quad \alpha \in \mathbb{N}^{1+d} \backslash \{0\} \ , \tag{7.22}$$

if $R \in \mathscr{C}(J)$.

Assume $R \in \mathscr{F}(J_\infty)$. Then (5.20) and the definition of r imply, similarly as above,

$$\|\partial^k(\lambda_* \tau_j)\|_\infty \leq c(k) b_\beta \leq c(k) p_\pi$$

for $\pi = \nu \circ \kappa \in \mathfrak{P}_V$, $\kappa = \kappa_{jp}$, and $\nu \in \mathfrak{N}_\kappa$. Analogously, we get from (7.11)

$$\|\partial^k(\lambda_* \rho_j)\|_\infty \leq c(k) p_\pi \ , \qquad \pi = \nu \circ \kappa \in \mathfrak{P}_V \ , \quad \kappa = \kappa_{j\beta} \ , \quad \nu \in \mathfrak{N}_\kappa \ , \tag{7.23}$$

for $k \geq 1$. Finally, similar arguments invoking (7.19) lead to

$$\|\partial^\alpha(\mu_* Y_\beta)\|_\infty \leq c(\alpha) \delta_j p_\pi \ , \qquad \alpha \in \mathbb{N}^{1+d} \backslash \{0\} \ , \tag{7.24}$$

for $\pi = \nu \circ \kappa \in \mathfrak{P}_V$ with $\kappa = \kappa_{j\beta}$ and $\nu \in \mathfrak{N}_\kappa$. By (7.11) $|\rho_j(s) - \rho_j(0)| \leq c$ for $s \in Q$ and $j \geq 1$. Hence

$$1 - c/\rho_j(0) \leq \rho_j(s)/\rho_j(0) \leq 1 + c/\rho_j(0) \ , \qquad s \in Q \ , \quad j \geq 1 \ . \tag{7.25}$$

Assume $R(\infty) < \infty$. Then $1/c \leq \rho_j(s) \leq c$ for $s \in Q$ and $j \geq 1$. In this case it is obvious that

$$\rho_j \sim \rho_j(0) \ , \qquad j \geq 1 \ . \tag{7.26}$$

If, however, $R(\infty) = \infty$, then we see from (7.25) that there exists j_0 such that (7.26) holds for $j \geq j_0$. As above, we observe that (7.26) applies for $1 \leq j \leq j_0$ also. Thus (7.26) is true in general. Using this we infer from (7.14) and (7.24) that

$$\|(\lambda_* \rho_j) \partial^\alpha(\mu_* Y_\beta)\|_\infty \leq c(\alpha) p_\pi \ , \qquad \alpha \in \mathbb{N}^d \backslash \{0\} \ ,$$

for $\pi = \nu \circ \kappa \in \mathfrak{P}_V$ with $\kappa = \kappa_{j\beta}$ and $\nu \in \mathfrak{N}_\kappa$. Moreover, (7.23), (7.24), $0 < \delta_j \le 1$, and the boundedness of b guarantee

$$\|\partial^k(\lambda_* \rho_j)\partial^\alpha(\mu_* Y_\beta)\|_\infty \le c(k,\alpha) p_\pi , \qquad k \ge 1 , \quad \alpha \in \mathbb{N}^d ,$$

for $\pi = \nu \circ \kappa \in \mathfrak{P}_V$ with $\kappa = \kappa_{j\beta}$ and $\nu \in \mathfrak{N}_\kappa$. Here we also use the boundedness of B in $\mathbb{R}^{\bar d}$ if $\alpha = 0$. This implies that estimate (7.22) holds in this case as well. Hence condition (7.1)(iv) is also satisfied. This proves the assertion. □

Remark 7.4. *Let the hypotheses of Proposition 7.3 be satisfied with $R \in \mathscr{C}(J_0)$. Set $(B_1, g_{B_1}) := (P, g_P)$, $\bar d_1 := 1 + \bar d$, $b_1 := p$, and $S_1 := V = \varphi^{-1}(I \times S)$. Then (B_1, g_{B_1}) is a bounded Riemannian submanifold of $(\mathbb{R}^{\bar d_1}, g_{\bar d_1})$ and b_1 is a bounded p-r singularity function for (B_1, g_{B_1}) on S_1.*

Fix $J_1 \in \{J_0, J_\infty\}$ and $R_1 \in \mathscr{C}(J_1)$, respectively $R_1 \in \mathscr{F}(J_\infty)$. Set $r_1 := R_1$ if $R_1 \in \mathscr{C}(J_1)$, respectively $r_1 := \mathbf{1}$ if $R_1 \in \mathscr{F}(J_1)$. Denote by $\varphi_1 : P_1 = P(R_1, B_1) \to \mathring{J}_1 \times B_1$ the canonical stretching diffeomorphism of P_1 and set $g_{P_1} := \iota_{P_1}^* g_{1+\bar d_1}$. Then Proposition 7.3 applies to guarantee that $p_1 := \varphi_1^*(r_1 \otimes b_1)$ is a cofinal singularity function for (P_1, g_{P_1}) on S_1. In particular, $(P_1, g_{P_1}/p_1^2)$ is cofinally uniformly regular and, given cofinal subintervals I_1 of J_1 and I of J, respectively,

$$\varphi_{1*}(g_{P_1}/p_1^2) \underset{I_1 \times I \times S}{\sim} (r_1 \otimes R \otimes b_1)^{-2}(ds_1^2 + ds^2 + g_B)$$

and $\varphi_1(P_1) = \mathring{J}_1 \times \mathring{J} \times B$. □

This remark shows that we can iterate Proposition 7.3 to handle ‘higher order’ singularities, e.g. cuspidal corners or funnels with edges.

8 Singular Ends

Throughout this section, (M, g) is an m-dimensional Riemannian manifold and $J \in \{J_0, J_\infty\}$.

Suppose:

(i) $R \in \mathscr{C}(J)$, $\ell \in \{1, \dots, m\}$, $\bar\ell \ge \ell$.
(ii) (B, g_B) is a compact $(\ell - 1)$-dimensional Riemannian submanifold of $\mathbb{R}^{\bar\ell}$.
(iii) (Γ, g_Γ) is a compact connected $(m - \ell)$-dimensional Riemannian manifold without boundary .

Then

$$W = W(R, B, \Gamma) := P(R, B) \times \Gamma$$

is the *smooth model Γ-wedge over the (R,B)-pipe $P = P(R,B)$*. It is a submanifold of $\mathbb{R}^{1+\bar\ell} \times \Gamma$ of dimension m, and $\partial W = \partial P \times \Gamma$. If $\ell = m$, then Γ is a one-point space and W is naturally identified with P (equivalently: there is no (Γ, g_Γ)). Thus every pipe is also a wedge. This convention allows for a uniform language by speaking, in what follows, of wedges only. Given a cofinal subinterval I of J, we set

$$W[I] := P(R,B;I) \times \Gamma .$$

We fix a Riemannian metric h_P for P and set $g_W := h_P + g_\Gamma$.

Let V be open in M. Then (V,g), more loosely: V, is a *smooth wedge of type* (W,g_W) in (M,g) if it is isometric to (W,g_W). More precisely, (Vg) is said to be *modeled* by $[\Phi,W,g_W]$ if Φ is an isometry from (V,g) onto (W,g_W), a *modeling isometry* for (V,g).

Assume

$$\begin{aligned} &\{V_0,V_1,\dots,V_k\} \text{ is a finite open covering of } M \text{ such that}\\ &\quad \text{(i)}\quad V_i \cap V_j = \emptyset\,,\ 1 \le i < j \le k\,;\\ &\quad \text{(ii)}\quad V_0 \cap V_i \text{ is relatively compact for } 1 \le i \le k\,;\\ &\quad \text{(iii)}\quad (V_i,g) \text{ is a smooth wedge in } (M,g) \text{ for } 1 \le i \le k\,. \end{aligned} \tag{8.1}$$

Then (M,g) is a Riemannian *manifold with* (finitely many) *smooth singularities.*

The following theorem is the main result of this paper. It is shown thereafter that we can derive from it all results stated in the introduction—and many more—by appropriate choices of the modeling data.

Theorem 8.1. *Suppose (M,g) is a Riemannian manifold with smooth singularities. Let ρ_0 be a singularity function for (M,g) on V_0 and assume that ρ_i is a cofinal singularity function for (V_i,g), $1 \le i \le k$. Then there exists a singularity function ρ for (M,g) such that $\rho \sim \rho_j$ on V_j for $0 \le j \le k$. Thus $(M,\ g/\rho^2)$ is uniformly regular.*

Proof. Suppose (V_i,g) is modeled by $[\Phi_i,W_i,g_i]$ for $1 \le i \le k$, where we write W_i for $W(R_i,B_i,\Gamma_i)$ with $R_i \in \mathscr{R}(J_i)$ and $g_i := g_{W_i}$. Given a cofinal subinterval I_i of J_i, we set $S_i := \Phi_i^{-1}\big(W_i[I_i]\big)$. By the relative compactness of $V_0 \cap V_i$ we can find a closed subset S_0 of V_0 such that $S_0 \supset V_0 \setminus \bigcup_{i=1}^k V_i$ and $\operatorname{dist}(S_0 \cap V_i,\ V_i \backslash V_0) > 0$ as well as closed cofinal subintervals I_i of J_i, $1 \le i \le k$, such that $\{S_0,S_1,\dots,S_k\}$ is a covering of M. By the assumptions on ρ_j, $0 \le j \le k$, we can find atlases $\mathfrak{K}_j$, $0 \le j \le k$, such that $(\rho_j,\mathfrak{K}_j)$ is a singularity datum for (V_j,g_j) on S_j. Since $V_0 \cap V_i$ is relatively compact it follows that $\rho_0 \sim \mathbf{1}$ and $\rho_i \sim \mathbf{1}$ on $V_0 \cap V_i$ for $1 \le i \le k$. Thus $\rho_i \,|\, (V_i \cap V_j) \sim \rho_j \,|\, (V_i \cap V_j)$ for $0 \le i < j \le k$, due to (8.1)(i). Note that $S_0 \cap S_i$ is relatively compact in $V_0 \cap V_i$. Hence we can assume that $\mathfrak{K}_{i,\ S_0 \cap S_i}$ is finite for $1 \le i \le k$. From this and (8.1)(i) it is clear that condition (v) of Lemma 3.3 is satisfied. Hence that lemma guarantees the validity of the assertion. □

Let (V,g) be a smooth wedge in (M,g) modeled by $[\Phi,W,g_W]$. Then $W = P \times \Gamma$ with $P = P[R,B]$, and $\varphi = \varphi[R]$ is the canonical stretching isometry from (P,h_P) onto $(J \times B,\ \varphi_* h_P)$. Hence

$$\Psi := (\varphi \times \mathrm{id}_\Gamma) \circ \Phi \colon (V,g) \mapsto (J \times B \times \Gamma,\ \varphi_* h_P + g_\Gamma) \tag{8.2}$$

is a modeling isometry for (V,g). Since B and Γ are compact, $\mathbf{1}_B$ and $\mathbf{1}_\Gamma$ are singularity functions for B and Γ, respectively. Suppose $r \in C^\infty\big(J,(0,\infty)\big)$. Then $r \otimes \mathbf{1}_B \otimes \mathbf{1}_\Gamma$ is the 'constant extension' of r over $J \times B \times \Gamma$. It satisfies

$$(\varphi \times \mathrm{id}_\Gamma)^*(r \otimes \mathbf{1}_B \otimes \mathbf{1}_\Gamma)(t,y,z) = r(t) \ , \qquad (t,y,z) \in J \times B \times \Gamma \ .$$

Thus, in abuse of notation, we set

$$\Phi^* r := \Psi^*(r \otimes \mathbf{1}_B \otimes \mathbf{1}_\Gamma) \tag{8.3}$$

without fearing confusion. In other words: we identify r with its point-wise extension over $P \times \Gamma$.

Proposition 8.2. *Let (V,g) be a smooth wedge in (M,g) modeled by $[\Phi,W,g_W]$. Assume that one of the following conditions is satisfied:*

(i) *$R \in \mathscr{C}(J)$ or $R \in \mathscr{F}(J_\infty)$, $h_P = g_P$, and $r := R$ if $R \in \mathscr{C}(J)$, whereas $r = \mathbf{1}$ otherwise.*

(ii) (α) *$J = J_0$, $\alpha \in (-\infty,1]$, and $R = \mathsf{R}_\alpha$.*
(β) *$\beta \neq 0$ and satisfies $\beta \leq \alpha$ with $\beta > 0$ if $\alpha > 0$.*
(γ) *$h_P = \varphi_\alpha^*(t^{2(\beta-1)}dt^2 + g_B)$.*
(δ) *$r := \mathsf{R}_\alpha$ if $0 < \alpha \leq 1$ and $r := \mathbf{1}$ otherwise.*

Then $\rho := \Phi^* r$ is a cofinal singularity function for (V,g).

Proof. Suppose p is a cofinal singularity function for (P,h_P). Then $p \otimes \mathbf{1}_\Gamma$ is one for $W = P \times \Gamma$, due to Theorem 3.1.

If (i) is satisfied, then Lemma 7.1 and Proposition 7.3 guarantee that $\varphi^*(R \otimes \mathbf{1}_B)$ is a cofinal singularity function for (P,g_P).

Let (ii) apply. Then it follows from Example 6.5(b) that $\varphi_\alpha^*(r \otimes \mathbf{1}_B)$ is a cofinal singularity function for (P_α,h_P). Now the considerations preceding the proposition imply the claims. □

For the next lemma we recall definition (1.6) where now $\mathscr{M}$ is replaced by M.

Lemma 8.3. *Suppose $R \in \mathscr{C}(J_\infty) \cup \mathscr{F}(J_\infty)$. Let (V,g) be a smooth wedge in (M,g) modeled by $\big[\Phi, P(R,B), g_{P(R,B)}\big]$. If $R \in \mathscr{C}(J_\infty)$, then there exists a cofinal singularity function ρ for (V,g) satisfying $\rho \sim R \circ \delta_V$. If $R \in \mathscr{F}(J_\infty)$, then (V,g) is cofinally uniformly regular.*

Proof. Suppose $R \in \mathscr{C}(J_\infty)$. Then $\Phi^* R$ is a cofinal singularity function for (V,g) by Proposition 8.2(i). Since Φ is an isometry it follows $\Phi^* R \sim R \circ \delta_V$. This implies the assertion in the present case. If $R \in \mathscr{F}(J_\infty)$, then the claim follows also from the cited proposition. □

Proof of Theorem 1.2. The foregoing lemma shows that a tame end is cofinally uniformly regular. Let $\{V_0, V_1, \dots, V_k\}$ be an open covering of (M,g) as in the definition preceding Theorem 1.2. Then we can shrink V_0 slightly to $\tilde V_0$ such that $\{\tilde V_0, V_1, \dots, V_k\}$ is still an open covering and Lemma 4.2 applies to guarantee that (M,g) is uniformly regular on $\tilde V_0$. Now the assertion follows from Theorem 8.1. □

With the help of Theorem 8.1 and Proposition 8.2 it is easy to construct uniformly regular Riemannian metrics in a great variety of geometric constellations. We leave this to the reader and proceed to study manifolds with smooth cuspidal singularities. For this we suppose:

$$\begin{aligned} &\text{(i)} \quad (\mathcal{M},g) \text{ is an } m\text{-dimensional Riemannian manifold}\,. \\ &\text{(ii)} \quad (\Gamma,g_\Gamma) \text{ is a compact connected Riemannian submanifold} \\ &\qquad\ \text{of } (\mathcal{M},g) \text{ without boundary and codimension } \ell\geq 1\,. \\ &\text{(iii)} \quad \Gamma\subset\partial\mathcal{M} \text{ if } \Gamma\cap\partial\mathcal{M}\neq\emptyset\,. \end{aligned} \tag{8.4}$$

In the following, we use the notation preceding definition (1.7). First we assume $\Gamma\subset\mathring{\mathcal{M}}$. Then there exists a uniform open tubular neighborhood $\mathcal{U}$ of Γ in $\mathring{\mathcal{M}}$ (e.g. M.W. Hirsch [19] or A.A. Kosinski [21]). More precisely, there exist $\varepsilon\in(0,1)$, an open subset $\mathcal{U}=\mathcal{U}_\varepsilon$ of $\mathcal{M}$ with $\mathcal{U}\cap\Gamma=\Gamma$, and a "tubular" diffeomorphism $\tau\colon\mathcal{U}\to\mathbb{B}^\ell\times\Gamma$ such that $\tau(\Gamma)=\{0\}\times\Gamma$, the tangential $T\tau$ of τ equals on $T\Gamma$ the identity multiplied with the factor ε, and

$$\tau_* g\sim g_{\mathbb{B}^\ell}+g_\Gamma\,. \tag{8.5}$$

Let $T^\perp\Gamma$ be the normal bundle of Γ. For $\xi\in\mathbb{S}^{\ell-1}$ and $q\in\Gamma$ there exists a unique $\nu_\xi(q)\in T_q^\perp\Gamma$ satisfying

$$(T_q\tau)\nu_\xi(q)=\big((0,\xi),q\big)\in T_0\mathbb{R}^\ell\times\Gamma\,.$$

Let $\gamma_{\nu,q}\colon[0,\varepsilon]\to\mathcal{M}$ be the geodesic emanating from q in direction $\nu\in T_q^\perp\Gamma$. Then

$$p=p(t,\xi,q):=\tau^{-1}(t,\xi,q)=\gamma_{\varepsilon\nu_\xi(q),q}(t)\,,\qquad (t,\xi,q)\in[0,1)\times\mathbb{S}^{\ell-1}\times\Gamma\,.$$

From this we infer

$$t\sim\delta_{\mathcal{U}}\big(p(t,\xi,q),\Gamma\big)\,,\qquad (t,\xi,q)\in[0,1)\times\mathbb{S}^{\ell-1}\times\Gamma\,. \tag{8.6}$$

Next we suppose $\Gamma\subset\partial\mathcal{M}$. Let $\mathcal{U}^\bullet=\mathcal{U}_\varepsilon^\bullet$ be an open tubular neighborhood of Γ in $\partial\mathcal{M}$ with associated tubular diffeomorphism

$$\tau^\bullet\colon\mathcal{U}^\bullet\to\mathbb{B}^{\ell-1}\times\Gamma\,. \tag{8.7}$$

Furthermore, there exists a uniform collar $\mathcal{V}=\mathcal{V}_\varepsilon$ for $\partial\mathcal{M}$ over $\mathcal{U}^\bullet$. That is to say: by making ε smaller, if necessary, we can assume that $\mathcal{V}$ is an open subset of $\mathcal{M}$ such that $\mathcal{V}\cap\partial\mathcal{M}=\mathcal{U}^\bullet$ and there exists a diffeomorphism $\tau^+\colon\mathcal{V}\to[0,1)\times\mathcal{U}^\bullet$ with $\tau^+(\mathcal{U}^\bullet)=\{0\}\times\mathcal{U}^\bullet$, $T\tau^+$ equals the identity in $T_\Gamma\partial\mathcal{M}$ multiplied by ε, and

$$\tau_*^+ g\sim dt^2+g_{\partial\mathcal{M}}\,. \tag{8.8}$$

Note that $\mathbb{B}_+^\ell \subset [0,1) \times \mathbb{B}^{\ell-1}$. Hence it follows from (8.7) that there exists an open subset $\mathscr{U} = \mathscr{U}_\varepsilon$ of $\mathscr{W}$ such that $\mathscr{U} \cap \partial\mathscr{M} = \mathscr{U}^\bullet$ and

$$\tau := (\mathrm{id}_{[0,1)} \times \tau^\bullet) \circ \tau^+ : \mathscr{U} \to \mathbb{B}_+^\ell \times \Gamma \tag{8.9}$$

is a diffeomorphism satisfying

$$\tau(\mathscr{U}^\bullet) = \{0\} \times \mathbb{B}^{\ell-1} \times \Gamma , \quad \tau(\Gamma) = \{0\} \times \Gamma .$$

By (8.5) and (8.8) we find

$$\tau_* g \sim dt^2 + g_{\mathbb{B}^{\ell-1}} \times g_\Gamma \sim g_{\mathbb{B}_+^\ell} \times g_\Gamma .$$

We let $\gamma^\bullet_{\nu^\bullet,q}$ be the geodesic in $\partial\mathscr{M}$ emanating from $q \in \Gamma$ in direction $\nu^\bullet \in T^\perp_{\partial\mathscr{M}}\Gamma$, where $T^\perp_{\partial\mathscr{M}}\Gamma$ is the orthogonal complement of $T_q\Gamma$ in $T_q\partial\mathscr{M}$. Suppose $\xi = (s,\eta)$ belongs to $\mathbb{S}_+^{\ell-1}$ with $s \in [0,1)$ and $\eta \in \mathbb{R}^{\ell-2}$, $0 \le t \le 1$, and $q \in \Gamma$. Define $\nu^\bullet_\eta(q)$ in $T^\perp_{\partial\mathscr{M},q}\Gamma$ by $(T^\bullet{}_q\tau^\bullet)\nu^\bullet_\eta(q) = \big((0,\eta),q\big) \in T_0\mathbb{R}^{\ell-2} \times \Gamma$, where $T^\bullet\tau^\bullet$ is the tangential of $\tau^\bullet$ in $\partial\mathscr{M}$. Set $r = r(t,\eta,q) := \gamma^\bullet_{\varepsilon\nu^\bullet_\eta(q),q}(t) \in \mathscr{U}^\bullet$. Analogously, let $\mu_s(r)$ in $T_r^\perp\partial\mathscr{M}$ be given by $(T_r\tau^+)\mu_s(r) = \big((0,s),r\big) \in T_0\mathbb{R} \times \mathscr{U}^\bullet$. Then

$$p = p(t,\xi,q) := \tau^{-1}(t\xi,q) = \gamma_{\varepsilon\mu_s(r(t,\eta,q)),r(t,\eta,q)}(t) \in \mathscr{U} .$$

This means that we reach p from $q \in \Gamma$ in two steps. First we go from q to $r \in \mathscr{U}^\bullet$ by following during the time interval $[0,t]$ the geodesic in $\mathscr{U}^\bullet$ which emanates from q in direction $\varepsilon\nu^\bullet_\eta(q)$. Second, we follow during the time interval $[0,t]$ the geodesic in $\mathscr{U}$ emanating from r in direction $\varepsilon\mu_s(r)$ to arrive at p. Observe

$$\mathrm{dist}_{\mathscr{U}}(p,r) = \mathrm{dist}_{\mathscr{U}}(p,\mathscr{U}^\bullet) = \delta_{\mathscr{U}}(p,\mathscr{U}^\bullet) .$$

Hence

$$t \sim \mathrm{dist}_{\mathscr{U}}(p,r) \le \mathrm{dist}_{\mathscr{U}}(p,q) \le \mathrm{dist}_{\mathscr{U}}(p,r) + \mathrm{dist}_{\mathscr{U}^\bullet}(r,q) \le 2t .$$

From this we infer

$$t \sim \delta_{\mathscr{U}}\big(p(t,\xi,q),\Gamma\big) , \qquad (t,\xi,q) \in [0,1) \times \mathbb{S}^{\ell-1} \times \Gamma . \tag{8.10}$$

Henceforth, $\mathbb{B} := \mathbb{B}^\ell$ and $\mathbb{S} := \mathbb{S}^{\ell-1}$ if $\Gamma \in \mathring{\mathscr{M}}$, whereas $\mathbb{B} := \mathbb{B}_+^\ell$ and $\mathbb{S} := \mathbb{S}_+^{\ell-1}$ otherwise. Then $U = U_\Gamma := \mathscr{U} \backslash \Gamma$ is, in either case, a *tubular neighborhood of* Γ *in* (M,g) and $\tau = \tau | U : U \to \dot{\mathbb{B}} \times \Gamma$ is the (*associated*) *tubular diffeomorphism*, defined by (8.9) if $\Gamma \in \partial\mathscr{M}$. By δ_Γ we denote the restriction of $\mathrm{dist}_{\mathscr{U}}(\cdot,\Gamma)$ to U.

Let $R \in \mathscr{C}(J_0)$ and $\varphi = \varphi[R]$. With the (ℓ-dimensional) polar coordinate diffeomorphism π the composition

$$U \xrightarrow{\tau} \dot{\mathbb{B}} \times \Gamma \xrightarrow{\pi \times \mathrm{id}_\Gamma} (0,1) \times \mathbb{S} \times \Gamma \xrightarrow{\varphi^{-1} \times \mathrm{id}_\Gamma} W(R,\mathbb{S},\Gamma) \tag{8.11}$$

defines a diffeomorphism Φ from U onto the model Γ-wedge $W = W(R,\mathbb{S},\Gamma)$ over the spherical, respectively semi-spherical, R-cusp $P = P(R,\mathbb{S})$. We call U *smooth singular end of* (M,g) *of type* (R,Γ) if Φ is an isometry from (U,g) onto (W,g_W), where $h_P := g_P$.

Lemma 8.4. *Let U be a smooth singular end of (M,g) of type (R,Γ). Then there exists a cofinal singularity function ρ for (U,g) satisfying $\rho \sim R\circ\delta_\Gamma$.*

Proof. It is a consequence of Proposition 7.3, Lemma 7.1, and Lemma 3.4 that $\rho := \Phi^* R$ is a cofinal singularity function for (U,g). From (8.11) and (8.2) we deduce $\Psi = (\pi \otimes \mathrm{id}_\Gamma)\circ\tau$. Moreover, $\Psi\big(p(t,\xi,q)\big) = (t,\xi,q)$ for (t,ξ,q) belonging to $(0,1)\times\mathbb{S}\times\Gamma$. Hence $(\Phi^*R)\big(p(t,\xi,q)\big) = R(t)$ by (8.3). Now the claim is implied by (8.6), respectively (8.10). □

It is clear that the assertion of this lemma is independent of the particular choice of U, that is, of ε.

Proof of Theorem 1.6 and Proposition 1.8. The statements follow directly from Lemma 8.4 with $R = \mathsf{R}_\alpha$, Lemma 8.3, and Theorem 8.1. □

Proof of Theorem 1.9. We set $R := \mathsf{R}_\alpha$ if $0 < \alpha \leq 1$, and $R := \mathsf{R}_{-\alpha}$ for $\alpha > 1$. It follows from Example 6.5(b) (setting $\beta := \alpha$ if $\alpha \leq 1$ and $\beta := -\alpha$ otherwise) and Theorem 3.1 that

$$g_W := \varphi^*\big(t^{-2\alpha}(t^{2(\alpha-1)}dt^2 + g_{\mathbb{S}})\big) + t^{-2\alpha}g_\Gamma = \varphi^*(t^{-2}dt^2 + t^{-2\alpha}g_{\mathbb{S}}) + t^{-2\alpha}g_\Gamma$$

is a cofinally uniformly regular metric for $W = W(R,\mathbb{S},\Gamma)$ if $0 < \alpha \leq 1$, whereas

$$g_W := \varphi^*(t^{-2(\alpha+1)}dt^2 + g_{\mathbb{S}}) + g_\Gamma$$

is one if $\alpha > 1$. Thus Φ, defined by (8.11), is an isometry from $(U,\mathfrak{g})$ onto (W,g_W). Hence the claim follows once more from Lemma 3.4. □

Lastly, we mention that there occur interesting and important singular manifolds if assumption (8.4)(iii) is dropped, that is, if Γ intersects $\mathring{\mathscr{M}}$ as well as $\partial\mathscr{M}$. Then Γ is no longer a smooth singular end but has cuspidal corners, for example. Such cases are not considered here although the technical means for their study have been provided in the preceding sections.

References

1. H. Amann, Anisotropic function spaces on singular manifolds (2012). arXiv:1204.0606
2. H. Amann, Function spaces on singular manifolds. Math. Nachr. **286**, 436–475 (2012)
3. H. Amann, Parabolic equations on uniformly regular Riemannian manifolds and degenerate initial boundary value problems (2014). arXiv:1403.2418
4. H. Amann. Parabolic equations on noncompact Riemannian manifolds. (in preparation)
5. H. Amann, J. Escher, *Analysis II* (English translation) (Birkhäuser, Basel, 2006)

6. Th. Aubin, *Nonlinear Analysis on Manifolds. Monge-Ampère Equations* (Springer, New York, 1982)
7. L. Conlon, *Differentiable Manifolds* (Birkhäuser, Boston, 2001)
8. E.B. Davies, *Heat Kernels and Spectral Theory* (Cambridge University Press, Cambridge, 1989)
9. R. Denk, M. Hieber, J. Prüss, $\mathscr{R}$-boundedness, Fourier multipliers and problems of elliptic and parabolic type. Mem. Am. Math. Soc., **166**(788) (2003)
10. M. Disconzi, Y. Shao, G. Simonett, Remarks on uniformly regular Riemannian manifolds (2014). arXiv:1410.8627v2
11. J. Eichhorn, The boundedness of connection coefficients and their derivatives. Math. Nachr. **152**, 145–158 (1991)
12. R.E. Greene, Complete metrics of bounded curvature on noncompact manifolds. Arch. Math. (Basel) **31**(1), 89–95 (1978/1979)
13. A. Grigor'yan, *Heat kernel and analysis on manifolds* (Am. Math. Soc. 2009)
14. A. Grigor'yan, L. Saloff-Coste. Heat kernel on manifolds with ends. Ann. Inst. Fourier (Grenoble) **59**(5), 1917–1997 (2009).
15. G. Grubb. Parameter-elliptic and parabolic pseudodifferential boundary problems in global L_p Sobolev spaces. Math. Z. **218** (1995), 43–90.
16. G. Grubb. *Functional Calculus of Pseudodifferential Boundary Problems* (Birkhäuser, Boston, 1996)
17. G. Grubb, N.J. Kokholm. A global calculus of parameter-dependent pseudodifferential boundary problems in L_p Sobolev spaces. Acta Math. **171** (1993), 165–229.
18. M. Hieber, J. Prüss. Heat kernels and maximal L^p-L^q estimates for parabolic evolution equations. Comm. Partial Differ. Equ. **22**(9–10), 1647–1669 (1997)
19. M.W. Hirsch, in *Differential Topology*. Graduate Texts in Mathematics, vol. 33 (Springer, New York, 1994)
20. V.A. Kondrat′ev, Boundary value problems for elliptic equations in domains with conical or angular points. Trudy Moskov. Mat. Obšč. **16**, 209–292 (1967)
21. A.A. Kosinski, *Differential Manifolds*. (Academic, Boston, 1993)
22. Th. Krainer, Maximal L^p-L^q regularity for parabolic partial differential equations on manifolds with cylindrical ends. Integral Equ. Oper. Theory. **63**(4), 521–531 (2009)
23. O.A. Ladyzhenskaya, V.A. Solonnikov, N.N. Ural'ceva. *Linear and Quasilinear Equations of Parabolic Type* (Transl. Math. Monographs). (American Mathematical Society, Providence, 1968)
24. A.L. Mazzucato, V. Nistor. Mapping properties of heat kernels, maximal regularity, and semi-linear parabolic equations on noncompact manifolds. J. Hyperbolic Differ. Equ. **3**(4), 599–629 (2006)
25. V.E. Nazaikinskii, A.Yu. Savin, B.-W. Schulze, B.Yu. Sternin, *Elliptic Theory on Singular Manifolds*. (Chapman & Hall/CRC, Boca Raton, 2006)
26. L. Saloff-Coste, in *The Heat Kernel and its Estimates*. Probabilistic Approach to Geometry. Adv. Stud. Pure Math., vol. 57 (Mathematical Society of Japan, Tokyo, 2010), pp. 405–436
27. E. Schrohe, in *Spaces of Weighted Symbols and Weighted Sobolev Spaces on Manifolds*. Pseudodifferential operators, Oberwolfach, 1986. Lecture Notes in Math., vol. 1256 (Springer, Berlin, 1987), pp. 360–377
28. M.A. Shubin, Spectral theory of elliptic operators on noncompact manifolds. Astérisque 207(5), 35–108 (1992). (Méthodes semi-classiques, vol. 1, Nantes, 1991)

Eigenvalue Estimates on Bakry–Émery Manifolds

Nelia Charalambous, Zhiqin Lu and Julie Rowlett

Abstract We demonstrate lower bounds for the eigenvalues of compact Bakry–Émery manifolds with and without boundary. The lower bounds for the first eigenvalue rely on a generalized maximum principle which allows gradient estimates in the Riemannian setting to be directly applied to the Bakry–Émery setting. Lower bounds for all eigenvalues are demonstrated using heat kernel estimates and a suitable Sobolev inequality.

1 Introduction

Let (M, g) be a Riemannian manifold and $\phi \in \mathcal{C}^2(M)$. A *Bakry–Émery* manifold is a triple (M, g, ϕ), where the measure on M is the weighted measure $e^{-\phi} dV_g$. The naturally associated Bakry–Émery Laplacian is

$$\Delta_\phi = \Delta - \nabla\phi \cdot \nabla, \tag{1}$$

where

$$\Delta = \frac{1}{\sqrt{\det(g)}} \sum_{i,j} \partial_i g^{ij} \sqrt{\det(g)} \partial_j .$$

The operator can be extended as a self-adjoint operator with respect to the weighted measure $e^{-\phi} dV_g$. It is also known as a "drifting" or "drift" Laplacian. Bakry and Émery [3] observed that this generalized notion of Laplace operator has analogous

N. Charalambous (✉)
Department of Mathematics and Statistics, University of Cyprus, 1678 Nicosia, Cyprus
e-mail: nelia@ucy.ac.cy

Z. Lu
Department of Mathematics, University of California, Irvine, CA 92697, USA
e-mail: zlu@uci.edu

J. Rowlett
Mathematical Sciences, Chalmers University of Technology
and the University of Gothenburg, SE-412 96 Göteborg, Sweden
e-mail: julie.rowlett@chalmers.se

J. Escher et al., (eds.), *Elliptic and Parabolic Equations*, Springer Proceedings in Mathematics & Statistics 119, DOI 10.1007/978-3-319-12547-3_2

properties to the standard Laplacian (which is none other than a Bakry–Émery Laplacian with $\phi \equiv 0$) and can be used to study a much larger class of diffusion equations and relations between energy and entropy. For example, a Bakry–Émery Laplacian appears in the Ornstein-Uhlenbeck equation, and Bakry–Émery manifolds play a key role in the log-Sobolev inequality of Gross [26] and Federbush [20] as well as the hypercontractivity inequality of Nelson [35].

The Bakry–Émery Laplacian has a canonically associated heat operator,

$$\partial_t - \Delta_\phi.$$

The fundamental solution is known as the Bakry–Émery heat kernel. The naturally associated curvature tensor for (M, g, ϕ) is the Bakry–Émery Ricci curvature defined by[1]

$$\mathrm{Ric}_\phi = \mathrm{Ric} + \mathrm{Hess}(\phi).$$

Above Ric and Δ are, respectively, the Ricci curvature and Laplacian with respect to the Riemannian metric g.

A collection of geometric results for Bakry–Émery manifolds is contained in [43]. We are interested in the analysis of the Bakry–Émery Laplacian and associated heat kernel. It turns out that some results can be extracted from the analysis of the Laplacian and heat kernel on an appropriately defined Riemannian manifold. In this article we are interested in obtaining new lower bounds for the eigenvalues of the Bakry–Émery Laplacian which are presented in Sect. 2. Upper bounds are also known to hold for compact Bakry–Émery manifolds. We include here a brief summary of some recent results in this area in order to compare them to the lower bounds that we obtain. Although our survey is certainly not comprehensive, it gives a flavor of the type of estimates that one can show.

2 Eigenvalue Estimates

2.1 One-dimensional Collapse

We discovered in [33] that the eigenvalues of a Bakry–Émery Laplacian on a compact n-dimensional manifold are the limit under one-dimensional collapse of Neumann eigenvalues for the classical Laplacian on a related $(n+1)$-dimensional manifold.

Theorem 1 *(L.-R.) Let (M, g, ϕ) be a compact Bakry–Émery manifold. Let*

$$M_\varepsilon := \{(x, y) \mid x \in M, \quad 0 \le y \le \varepsilon e^{-\phi(x)}\} \subset M \times \mathbb{R}^+,$$

with $\phi \in \mathcal{C}^2(M)$ and $e^{-\phi} \in \mathcal{C}(M \cup \partial M)$. Let $\{\mu_k\}_{k=0}^\infty$ be the eigenvalues of the Bakry–Émery Laplacian on M. If $\partial M \neq \emptyset$, assume the Neumann boundary condition. Let $\mu_k(\varepsilon)$ be the Neumann eigenvalues of M_ε for $\tilde{\Delta} := \Delta + \partial_y^2$. Then

$$\mu_k(\varepsilon) = \mu_k + O(\varepsilon^2), \quad \forall k \ge 0.$$

[1] In the notation of [29], this is the ∞ Bakry–Émery Ricci curvature.

2.2 Maximum Principle and Gradient Estimates

One of the classical methods for obtaining eigenvalue estimates is via gradient estimates, which was first used by Li-Yau [31]. The papers [1, 4, 28, 40, 45–48] appear to be the most influential. These estimates, which are often quite complicated and tricky, are based on the following maximum principle.

Let M be a compact Riemannian manifold. Let u be a smooth function on M. Assume that

$$H = \frac{1}{2}|\nabla u|^2 + F(u),$$

where F is a smooth function of one variable, and let x_0 be an interior point of M at which H reaches its maximum. Then at x_0

$$0 \geq |\nabla^2 u|^2 + \nabla u \nabla(\Delta u) + \mathrm{Ric}(\nabla u, \nabla u) + F'(u)\Delta u + F''(u)|\nabla u|^2.$$

The above inequality is useful for obtaining lower bounds on the first eigenvalue of a Laplace or Schrödinger operator; see [38]. This together with the eigenvalue convergence under one-dimensional collapse in Theorem 1 would indicate that similar estimates could be obtained for M using M_ε. However, there are two major problems with this naive approach:

1. M_ε need not be convex, even if M is. As we know, if M is convex, the maximum of H must be reached in the interior of M. In general, we do not have such a property for M_ε.
2. The natural Ricci curvature attached to the problem is Ric_ϕ, not the Ricci curvature of M_ε, which is essentially Ric.

Nonetheless, carefully estimating the eigenfunctions and their derivatives on M_ε, we proved the following maximum principle for Bakry–Émery manifolds. For a smooth function u on M_ε, we define the following function on M

$$\psi(x) := u(x, 0), \quad x \in M.$$

Theorem 2 *(Maximum Principle (L.-R.)) Assume that* $(x_0, 0)$ *is the maximum point of* H *on* $M \times \{0\} \subset M_\varepsilon$*. Then*

$$o(1) \geq |\nabla^2 \psi|^2 + \nabla\psi\nabla(\Delta_\phi u) + \mathrm{Ric}_\phi(\nabla\psi, \nabla\psi) + F'(\psi)\Delta_\phi u + F''(\psi)|\nabla\psi|^2$$

as $\varepsilon \to 0$*, where* $o(1)$ *depends on a certain weighted Hölder norm of* u *(see [33, Theorem 5] for details). In particular, if* u *is an eigenfunction of unit* L^2 *norm, then* $o(1) \to 0$ *as* $\varepsilon \to 0$.

It therefore follows from this maximum principle that one may apply all the proofs of gradient estimates directly to Bakry–Émery geometry which we summarize as follows.

Bakry–Émery Gradient Estimate Principle. *There is a one-one correspondence between the gradient estimate on a Riemannian manifold and on a Bakry–Émery manifold. More precisely, the eigenvalue estimate on the Bakry–Émery manifold* (M, g, ϕ) *is equivalent to that on the Riemannian manifold* $(M_\varepsilon, g + dy^2)$ *for* ε *small enough.*

Remark 1 It is known that the Bakry–Émery Laplacian is unitarily equivalent the Schrödinger operator

$$\Delta + \frac{1}{2}\Delta\phi + \frac{1}{4}|\nabla\phi|^2,$$

(see [18, 19, 39]). Using this observation, we are able to prove several eigenvalue inequalities in the Bakry–Émery setting virtually effortlessly, as long as the analogous results have been obtained in the Riemannian case. However, to obtain results involving gradient estimates, the equivalent estimates for the eigenfunctions are also required; these were demonstrated in [33].

Remark 2 Similar estimates can also be obtained by taking the warped product with the unit ball [9], the advantage of that treatment being to avoid the boundary estimates. However, further work is necessary in that case to eliminate the extra eigenvalues which are created in that process.

2.3 *Lower Bounds for the First Eigenvalue*

Using the maximum principle and Theorem 1, we are able to provide the Bakry–Émery version of the first eigenvalue estimates. Throughout this subsection, let (M, g, ϕ) be a compact n-dimensional Bakry–Émery manifold either without boundary or with convex boundary, in which case we assume the Neumann boundary condition. We first consider the case in which the (Bakry–Émery) Ricci curvature has a nonpositive lower bound.

Theorem 3 *Assume* $\mathrm{Ric}_\phi \geq -(n-1)k$ *for some* $k \geq 0$. *Then the first (positive) eigenvalue of the Bakry–Émery Laplacian* μ_1 *satisfies*

$$\mu_1 \geq \frac{\pi^2}{d^2}\exp(-c_n\sqrt{kd^2}),$$

where d *is the diameter of* M *with respect to* g, *and* c_n *is a constant depending only on* n.

In the Riemannian case ($\phi \equiv 0$), the result is due to Yang [44] following a similar idea of Zhong and Yang [48].

Proof Let f be the first eigenfunction and assume without loss of generality

$$\max f = 1;$$
$$\min f = -\beta$$

for some $0 < \beta \leq 1$. The following gradient estimate was demonstrated in [44, Lemma 2],

$$\frac{|\nabla f|}{\sqrt{1-f^2}} \leq \sqrt{\mu_1} + \frac{1}{2}\max(\sqrt{n-1}, \sqrt{2})\sqrt{(n-1)k}\,\sqrt{1-f^2} \tag{2}$$

for the Riemannian case. By our principle, the same estimate is true in the Bakry–Émery case. As a result, we have

$$\mu_1 \geq \frac{\pi^2}{16} \cdot \frac{\max(n-1,2)(n-1)k}{(\exp(1/2\max(\sqrt{n-1}, \sqrt{2})\sqrt{(n-1)kd^2}) - 1)^2} \tag{3}$$

which is obtained by integrating (2) over the geodesic connecting the maximum and minimum points of f.

Define the *normalized* eigenfunction

$$\varphi := \frac{f - (1-\beta)/2}{(1+\beta)/2}$$

so that $\max \varphi = 1$ and $\min \varphi = -1$. Similarly, the following gradient estimate in [44, Lemma 5],

$$|\nabla\varphi|^2 \leq \mu_1 + (n-1)k + \mu_1 \xi(\varphi),$$

is also true in the Bakry–Émery case, where $\mu_1 + (n-1)k + \mu_1\xi(\varphi)$ satisfies an ordinary differential equation in [44, Eq. (40)]. Consequently, we have

$$\mu_1 \geq \frac{\pi^2}{d^2} \cdot \frac{1}{1 + (n-1)k/\mu_1}. \tag{4}$$

Combining the above inequality with (3) proves the theorem; for further details we refer to [44]. □

When the (Bakry–Émery) Ricci curvature has a positive lower bound, we obtain a result of Futaki-Sano [22] by our maximum principle and the corresponding Riemannian case proven by Ling [32].

Theorem 4 *Assume that* $\mathrm{Ric}_\phi \geq (n-1)k$ *for some positive constant* $k > 0$. *Then the first (positive) eigenvalue of the Bakry–Émery Laplacian satisfies*

$$\mu_1 \geq \frac{\pi^2}{d^2} + \frac{3}{8}(n-1)k, \qquad \text{for } n = 2;$$

$$\mu_1 \geq \frac{\pi^2}{d^2} + \frac{31}{100}(n-1)k, \qquad \text{for } n > 2,$$

where d is the diameter of the manifold.

A slightly stronger estimate was shown by Andrews and Ni [2] for convex domains in a Bakry–Émery manifold.

Theorem 5 *(Andrews-Ni) Assume that for the Bakry–Émery manifold* (M, g, ϕ) *the associated curvature* $\mathrm{Ric}_\phi \geq (n-1)k > 0$. *Then for any convex domain* Ω, *the first positive eigenvalue of the Bakry–Émery Laplacian with Neumann boundary condition satisfies*

$$\mu_1 \geq \frac{\pi^2}{d^2} + \frac{(n-1)k}{2},$$

where d *is the diameter of* Ω.

2.4 Lower Bounds of Higher Eigenvalues

We will also demonstrate a lower bound for *all* the eigenvalues which holds whenever the manifold satisfies an appropriate Sobolev inequality (S) as in Sect. 3 below.

Theorem 6 *Let* M *be a compact manifold without boundary, on which the Sobolev inequality* (S) *holds. Then the* k*th eigenvalue of the Bakry–Émery Laplacian satisfies the lower bound*

$$\lambda_k^{\frac{\nu}{2}} \geq c(\nu)\, k V_\phi^{-1}\, C_1^{\frac{\nu}{2}}$$

where V_ϕ *is the weighted volume of* M *and* $c(\nu) > 0$ *is a uniform constant that only depends on* ν *and* C_1 *is as in Lemma 2. The same inequality holds when* $\partial M \neq \emptyset$ *for the positive Neumann eigenvalues.*

Remark 3 We note that in the case $\partial M \neq \emptyset$, if the Sobolev inequality (S) holds for all $u|_{\partial M} = 0$ with constant C_2, then the kth Dirichlet eigenvalue of the Bakry–Émery Laplacian satisfies the same inequality with constant C_2.

Assuming the Sobolev inequalities on M_ε, then the result follows from Cheng-Li [12] and Theorem 1. This however, would entail a uniform Sobolev constant for all the M_ε. Instead, we shall prove this result using the Bakry–Émery heat kernel estimates which we will present in the following sections.

2.5 Upper Bounds

In this section we provide some recent upper bound estimates for eigenvalues of Bakry–Émery manifolds. We first recall that in the case of a Riemannian manifold, Cheng proved the following (see [38, Theorem III.2])

Theorem 7 *(Cheng) Let* M *be a compact Riemannian manifold without boundary or with Neumann boundary condition. Let* d *be the diameter of* M. *Then for* $j \geq 1$,

1. *If* $\mathrm{Ric} \geq 0$, *then* $\mu_j \leq 8n(n+4)j^2/d^2$;
2. *If* $\mathrm{Ric} \geq n-1$, *then* $\mu_j \leq 4nj^2/d^2$;
3. *If* $\mathrm{Ric} \geq -(n-1)k$ *for* $k \geq 0$, *then* $\mu_j \leq \frac{1}{4}k + 8n(n+4)j^2/d^2$

Using [9], the above inequalities are true under slightly stronger assumptions in the Bakry–Émery case.

Theorem 8 *Let M be a compact Bakry–Émery manifold without boundary or with Neumann boundary condition. Let d be the diameter of M. Let $\varepsilon > 0$. If*

$$\mathrm{Ric}_\phi - \varepsilon \nabla\phi \otimes \nabla\phi \geq -(n-1)k, \quad \textit{for} \quad k \geq 0,$$

then

$$\mu_j \leq C(n, \varepsilon)(k + j^2/d^2), \quad \forall j \in \mathbb{N},$$

where $C(n, \varepsilon)$ is a constant depending on n and ε.

Using this result we are able to prove the following which is essentially due to [18, 19, 39].

Theorem 9 *Assume that* $\mathrm{Ric}_\phi - \varepsilon \nabla\phi \otimes \nabla\phi \geq 0$. *Then we have*

$$\mu_j \leq C(n, \varepsilon)\mu_1.$$

Proof By Theorem 3, we have

$$\frac{\pi^2}{d^2} \leq \mu_1.$$

The result therefore follows from Theorem 8. □

Recently, Funano and Shioya proved [21] the following stronger and somewhat surprising result.

Theorem 10 *(Funano-Shioya) Let (M, g, ϕ) be a compact Bakry–Émery manifold with nonnegative Bakry–Émery Ricci curvature. Then there exists a positive constant C_j which depends only on j (and not even on the dimension!) and in particular is independent of (M, g, ϕ) such that*

$$\mu_j \leq C_j \mu_1.$$

Moreover, this result also holds if the $\partial M \neq \emptyset$ is C^2 under the Neumann boundary condition.

Using an example, they showed that the nonnegativity of curvature is a necessary condition. The proof relies on a geometric theory of concentration of metric measure spaces due to Gromov [25].

Hassannezhad demonstrated upper bounds for the eigenvalues without curvature assumptions [27].

Theorem 11 *(Hassannezhad) There exist constants A_n and B_n depending only on n such that for every n-dimensional compact Bakry–Émery manifold (M, g, ϕ) with*

$|\nabla\phi| \le \sigma$ for some $\sigma \ge 0$, and for every $j \in \mathbb{N}$ we have

$$\mu_j \le A_n \max\{\sigma^2, 1\} \left(\frac{V_\phi([g])}{V(M,g)}\right)^{2/n} + B_n \left(\frac{j}{V(M,g)}\right)^{2/n}.$$

Above $V(M,g)$ denotes the volume of M with respect to g, and $V([g])$ denotes the min-conformal volume,

$$V([g]) = \inf\{V(M,g_0), \textit{such that } g_0 \in [g], \textit{and } \mathrm{Ric}(g_0) \ge -(n-1)\}.$$

This theorem was proven by first demonstrating an analogous estimate for the Schrödinger operator

$$\Delta + \frac{1}{2}\Delta\phi + \frac{1}{4}|\nabla\phi|^2,$$

which is unitarily equivalent to the Bakry–Émery Laplace operator; see [39, p. 28].

The proof of the following theorem is based on constructing a family of test functions supported on a suitable family of balls and is known as a Buser type upper bound, since this idea goes back to Buser [6], and has also been used by Cheng [11] as well as Li and Yau [30].

Theorem 12 *(Hassannezhad) There are positive constants A_n and B_n which depend only on the dimension n such that for every compact Bakry–Émery manifold (M,g,ϕ) with $\mathrm{Ric}_\phi \ge -k^2(n-1)$ and $|\nabla\phi| \le \sigma$, for some constants k, $\sigma \ge 0$, such that for every $j \in \mathbb{N}$ we have*

$$\mu_j \le A_n \max\{\sigma^2, 1\}k^2 + B_n \left(\frac{j}{V_\phi(M)}\right)^{2/n},$$

where

$$V_\phi(M) := \int_M e^{-\phi} dV_g$$

is the weighted volume of M.

3 Sobolev Inequalities

A classical way to obtain lower bounds on the eigenvalues of the Laplacian on a compact set is via the trace of the heat kernel as in [16]. Cheng and Li demonstrated in [12] that one can also find such lower bounds with respect to the Sobolev constant since a Sobolev inequality always holds in the compact case. Their method also ultimately relies on demonstrating upper bounds for the heat trace as in [16].

Definition 1 We say that the Bakry–Émery manifold (M^n, g, ϕ) satisfies the property (S), if there exist constants $\nu = \nu(n) > 2$, $\alpha = \alpha(n)$, and C_o depending only on M such that for all $u \in \mathcal{H}^1(M)$

$$\left(\int_M |u|^{\frac{2\nu}{\nu-2}} e^{-\phi}\right)^{\frac{\nu-2}{\nu}} \le C_o\, V_\phi^{-\frac{2}{\nu}} \int_M (|\nabla u|^2 + \alpha|u|^2)\, e^{-\phi} \qquad (S)$$

where V_ϕ denotes the weighted volume of M.

A global Sobolev inequality as above is known to hold on compact Riemannian manifolds. In a recent article, the first two authors found sufficient conditions for a local Sobolev inequality to hold on a weighted manifold [10]. The local Sobolev inequality points to the geometric features upon which C_o would depend in the case of a weighted manifold. In particular, the authors showed that a volume form comparison assumption is sufficient to ensure a local Sobolev inequality. For any point $x \in M$ we denote the Riemannian volume form in geodesic coordinates at x by

$$dv(\exp_x(r\xi)) = J(x,r,\xi)\,dr\,d\xi$$

for $r > 0$ and ξ any unit tangent vector at x. Then the ϕ-volume form in geodesic coordinates is given by

$$J_\phi(x,r,\xi) = e^{-\phi} J(x,r,\xi).$$

If $y = \exp_x(r\xi)$ is a point that does not belong to the cut-locus of x, then

$$\Delta r(x,y) = \frac{J'(x,r,\xi)}{J(x,r,\xi)} \quad \text{and} \quad \Delta_\phi r(x,y) = \frac{J'_\phi(x,r,\xi)}{J_\phi(x,r,\xi)}$$

where $r(x,y) = d(x,y)$, and the derivatives are taken in the radial direction. The first equality gives Bishop's volume comparison theorem under the assumption of a uniform Laplacian upper bound. On weighted manifolds, the second equality provides us with weighted volume comparison results whenever we have a uniform Bakry–Émery Laplacian upper bound.

Definition 2 We say that the Bakry Émery manifold (M^n, g, ϕ) satisfies the property V_R, if there exists a positive and nondecreasing function $A(R)$ and a uniform constant a (independent of R) such that for all $x \in B_{x_o}(R)$ and $0 < r_1 < r_2 < R$

$$\frac{J_\phi(x,r_2,\xi)}{J_\phi(x,r_1,\xi)} \le \left(\frac{r_2}{r_1}\right)^a e^{A(R)}. \tag{V_R}$$

The above inequality is assumed for all points $\exp_x(r_i\xi)$ that do not belong to the cut locus of x.

We denote by $B_x(r)$ the geodesic ball of radius r at x and by $V_\phi(x,r)$ its weighted volume. The following result was proven in [10].

Lemma 1 *Let (M^n, g, ϕ) be a Bakry–Émery manifold that satisfies the property V_R for all $x \in B_{x_o}(R)$. Then for any $x \in B_{x_o}(R)$, $0 < r < R$ and $u \in C_0^\infty(B_x(r))$ there exist constants $\nu = \nu(n) > 2$, $C_1(n,a)$ and $C_2(n)$ such that*

$$\left(\int_{B_x(r)} |u|^{\frac{2\nu}{\nu-2}} e^{-\phi}\right)^{\frac{\nu-2}{\nu}} \le C_1 \frac{e^{C_2 A(R)}\, r^2}{V_\phi(x,r)^{\frac{2}{\nu}}} \int_{B_x(r)} (\,|\nabla u|^2 + r^{-2}|u|^2\,)\, e^{-\phi}. \tag{5}$$

Previously, similar local Sobolev inequalities were proven in the case of a uniform upper bound on $\Delta_\phi r$. Assumption V_R, however, only requires that the integral of $\Delta_\phi r$

on a geodesic ball be bounded and is thus more general. We refer the interested reader to [10] for specific conditions on Ric_ϕ and ϕ that would guarantee such a uniform upper bound. On a compact manifold they all certainly hold. An interesting question we intend to investigate in future work is to determine the optimal C_o and α of (S). We would also like to remark that in the case $\phi \equiv 0$ one can use the existence of a local Sobolev inequality (5) to find lower bounds for the Neumann eigenvalues of the Laplacian over a geodesic ball. In [7] such lower estimates were also obtained for the Bochner Laplacian on forms. It would also be interesting to consider the analogous problem on weighted manifolds.

The Sobolev inequality (S) allows us to prove an L^2 gradient estimate which together with the heat kernel estimates will be sufficient to prove the eigenvalue lower bounds.

Lemma 2 *Suppose that (S) holds on M. Then for all $u \in H^1(M)$ that satisfy $\int_M u = 0$*

$$\int_M |\nabla u|^2 e^{-\phi} \geq C_1 \left(\int_M u^2 e^{-\phi}\right)^{\frac{2+\nu}{\nu}} \left(\int_M |u| e^{-\phi}\right)^{-\frac{4}{\nu}}$$

for a uniform constant $C_1 = \frac{\lambda_1}{C_o(\lambda_1+\alpha)} V_\phi^{\frac{2}{\nu}}$, where λ_1 is the first nonzero eigenvalue of M.

Proof The Sobolev inequality (S) implies

$$\int_M |\nabla u|^2 \geq C_o^{-1} V_\phi^{\frac{2}{\nu}} \left(\int_M |u|^{\frac{2\nu}{\nu-2}} e^{-\phi}\right)^{\frac{\nu-2}{\nu}} - \alpha \int_M |u|^2 e^{-\phi}. \tag{6}$$

Moreover, whenever $\int_M u = 0$, the definition of λ_1 gives

$$\int_M u^2 e^{-\phi} \geq \frac{1}{\lambda_1} \int_M |\nabla u|^2 e^{-\phi}.$$

Substituting the above inequality to the right side of (6) and solving for $\int_M |\nabla u|^2$ we get

$$\int_M |\nabla u|^2 e^{-\phi} \geq \frac{\lambda_1}{C_o(\lambda_1+\alpha)} V_\phi^{\frac{2}{\nu}} \left(\int_M |u|^{\frac{2\nu}{\nu-2}} e^{-\phi}\right)^{\frac{\nu-2}{\nu}}. \tag{7}$$

By writing $u^2 = |u|^{4/(\nu+2)}|u|^{2\nu/(\nu+2)}$ and applying the Hölder inequality with $p = (\nu+2)/4$ and $q = p/(p-1) = (\nu+2)/(\nu-2)$ we get the estimate

$$\left(\int_M u^2 e^{-\phi}\right)^{\frac{2+\nu}{\nu}} \leq \left(\int_M |u| e^{-\phi}\right)^{\frac{4}{\nu}} \left(\int_M |u|^{\frac{2\nu}{\nu-2}} e^{-\phi}\right)^{\frac{\nu-2}{\nu}}.$$

The lemma follows by solving the above inequality for the second term in the right side and substituting into (7). □

4 Heat Kernel Estimates and the Noncompact Case

We let $H_\phi(x, y, t)$ denote the heat kernel of Δ_ϕ corresponding to the Friedrichs extension. This is certainly unique when M is compact, and on a noncompact manifold it is the smallest positive heat kernel among all other heat kernels that correspond to heat semigroups of self-adjoint extensions of Δ_ϕ. Both in the Riemannian and in the weighted case, heat kernel estimates are closely related to eigenvalue estimates. When $\phi \equiv 0$ Li and Yau in [31] proved upper estimates for the heat kernel of Schrödinger operators whenever the Ricci curvature of the manifold is bounded below. One of the key elements in their proof is the Bochner formula and the Cauchy inequality $|\nabla^2 u|^2 \geq (\Delta u)^2/n$.

In the case $\phi \not\equiv 0$, it was shown by Bakry and Émery in [3] that the analogous Bochner formula can be obtained if one takes as the curvature tensor Ric_ϕ, and it is given by

$$\Delta_\phi |\nabla u|^2 = 2|\nabla^2 u|^2 + 2\langle \nabla u, \nabla \Delta_\phi u\rangle + 2\mathrm{Ric}_\phi(\nabla u, \nabla u). \tag{8}$$

Observe that when $\phi \equiv 0$, (8) becomes the Bochner formula in the Riemannian case. The term $\nabla^2 u$ that appears above is the usual Hessian of u; we do not have a notion of ϕ-Hessian. In other words there is no analogous relationship between the Hessian of u and the $\Delta_\phi u$ as in $|\nabla^2 u|^2 \geq (\Delta u)^2/n$. As a result, there is more subtlety in obtaining gradient and heat kernel estimates estimate in the $\phi \not\equiv 0$ case. Such estimates can be obtained under various assumptions on the curvature of the manifold. We will present a few of these options.

Bakry and Émery also demonstrated that the relevant Ricci tensor for obtaining a gradient estimate is the q-Bakry–Émery Ricci tensor [3] which is defined as

$$\mathrm{Ric}_\phi^q = \mathrm{Ric} + \nabla^2\phi - \frac{1}{q}\nabla\phi \otimes \nabla\phi = \mathrm{Ric}_\phi - \frac{1}{q}\nabla\phi \otimes \nabla\phi$$

where q is a positive number. By generalizing the Li-Yau method, Qian was able to prove a Harnack inequality and heat kernel estimates for the Bakry–Émery Laplacian whenever $\mathrm{Ric}_\phi^q \geq 0$ in [36]. In [9] the first two authors found Gaussian estimates for the Bakry–Émery heat kernel whenever Ric_ϕ^q is bounded below. This was done by associating to the weighted manifold a family of warped product spaces $\tilde{M}_\varepsilon$ and showing that the geometric analysis results on M are closely related to those on $\tilde{M}_\varepsilon$. In particular, the heat kernel estimates on $\tilde{M}_\varepsilon$ implied the Bakry–Émery heat kernel estimates on the weighted manifold.

Theorem 13 *(C.-L.) Let (M^n, g, ϕ) be a Bakry–Émery manifold such that for some positive integer q,*

$$\mathrm{Ric}_\phi^q \geq -K$$

on $B_{x_o}(4R+4) \subset M$. Then for any $x, y \in B_{x_o}(R/4)$, $t < R^2/4$ and $\delta \in (0,1)$

$$C_6 V_\phi^{-1/2}(x, \sqrt{t})\, V_\phi^{-1/2}(y, \sqrt{t}) \cdot \exp[-\, C_7 \frac{d^2(x,y)}{t} - C_8\, K\, t\,]$$

$$\leq H_\phi(x,y,t)$$

$$\leq C_3\, V_\phi^{-1/2}(x,\sqrt{t})\, V_\phi^{-1/2}(y,\sqrt{t}) \cdot \exp\left[-\lambda_{1,\phi}(M)t - \frac{d^2(x,y)}{C_4 t} + C_5\sqrt{Kt}\,\right]$$

for some positive constants C_3, C_4, C_6 and C_7 that only depend on δ and $n+q$ and positive constants C_5, C_8 that only depend on $n+q$, and where $\lambda_{1,\phi}(M)$ is the infimum of the weighted Rayleigh quotient on M.

Whenever $\mathrm{Ric}_\phi^q \geq -K$ on M, then the same bound also holds for all $x, y \in M$ and $t > 0$.

The proof of the above theorem illustrated the strong geometric connection between M and the warped product spaces $\tilde{M}_\varepsilon$ and the fact that the Bakry–Émery Laplacian and the q-Bakry–Émery Ricci tensor are projections (in some sense) of the Laplacian and Ricci tensor of a higher dimensional space. As it was also remarked in [9], one could not get the above estimate by only assuming Ric_ϕ bounded below and ϕ of linear growth at a point (which would be enough for gradient estimates as in [34]). Instead, the Bakry–Émery heat kernel estimate requires an assumption on the *uniform* linear growth of ϕ, which is almost equivalent to assuming that the gradient of ϕ is bounded.

4.1 The Essential Spectrum

On noncompact manifolds, a more interesting part of the spectrum is the essential spectrum of the manifold. In general, the L^2 spectrum of Δ_ϕ, denoted $\sigma(\Delta_\phi)$, consists of all points $\lambda \in \mathbb{C}$ for which $\Delta_\phi - \lambda I$ fails to be invertible on L^2. Since Δ_ϕ is nonnegative definite on L^2, $\sigma(\Delta_\phi)$ is contained in $[0,\infty)$. The essential spectrum of Δ_ϕ on L^2, $\sigma_{\mathrm{ess}}(\Delta_\phi)$, consists of the cluster points in the spectrum and of isolated eigenvalues of infinite multiplicity. One is usually interested in finding sufficient conditions on the manifold so that $\sigma_{\mathrm{ess}}(\Delta) = [0,\infty)$.

In the case $\phi = 0$, it was extremely difficult to directly compute the L^2 spectrum using the classical Weyl's criterion without assuming very strong decay conditions on the curvature of the manifold. By generalizing the Weyl's criterion, the first two authors were able to show that a sufficient condition for such a result is that the Ricci curvature of the manifold is asymptotically nonnegative [8]. The authors were also able to show that in the case of a weighted manifold $\sigma_{\mathrm{ess}}(\Delta_\phi) = [0,\infty)$ whenever the q-Bakry–Émery Ricci tensor is asymptotically nonnegative [9]. In the same article, they also showed an L^p independence result for the essential spectrum of Δ_ϕ on L^p whenever Ric_ϕ^q is bounded below, and the weighted volume of the manifold grows uniformly subexponentially [9]. As in the classical case, the latter result is a consequence of the gaussian estimates for the heat kernel [41]. It would be interesting to find a weighted space analog to the hyperbolic space on which the L^p essential spectrum of Δ_ϕ depends on p, for which the underlying Riemannian manifold is not hyperbolic (see [15] for the classical case).

As mentioned previously, in the noncompact case, Ric_ϕ^q bounded below is not equivalent to Ric_ϕ bounded below, since ϕ and its gradient are not necessarily bounded. As a result, if one would like to assume instead $\mathrm{Ric}_\phi \geq -K$, then some control on ϕ is required. Apart from the Bochner formula, the other main ingredient for obtaining the heat kernel estimates by the Li-Yau method in the Riemannian case is the Laplacian comparison theorem. In particular, the fact that whenever the Ricci curvature of the manifold is bounded below, then there exist uniform constants a, b such that

$$\Delta r(x, y) \leq \frac{a}{r(x, y)} + b$$

where $r(x, y) = d(x, y))$. In the case of weighted manifolds however, $\mathrm{Ric}_\phi(x) \geq -K$ on a ball around x_o does not imply a uniform Bakry–Émery Laplacian estimate $\Delta_\phi r(x, y) \leq C\frac{1}{r(x,y)} + b$ without strong restrictions on ϕ. However, using the techinique of Saloff-Coste as in [37], one can use the local Sobolev inequality [5] to prove a mean value inequality for ϕ-subharmonic functions as well as a mean value inequality for solutions to the Bakry–Émery heat equation. In [9] these were used to prove a Gaussian estimate for the Bakry–Émery heat kernel.

Theorem 14 *(C.-L.) Let (M^n, g, ϕ) be a Bakry–Émery manifold that satisfies the property V_R for all $x \in B_{x_o}(R)$. Let $H_\phi(x, y, t)$ denote the minimal Bakry–Émery heat kernel defined on $M \times M \times (0, \infty)$ Then for any $\varepsilon > 0$ there exist constants $c_1(n, \varepsilon), c_2(n)$ such that*

$$H_\phi(x, y, t) \leq c_1\, V_\phi^{-1/2}(x, \sqrt{t})\, V_\phi^{-1/2}(y, \sqrt{t}) \exp\left[-\lambda_{1,\phi}(M)t - \frac{d^2(x, y)}{4(1+\varepsilon)t} + c_2\, A(R)\right]$$

for any $x, y \in B_{x_o}(R/2)$ and $0 < t < R^2/4$.

For a comprehensive review of heat kernel bounds on noncompact weighted manifolds we refer the interested reader to the extensive summary of results due to Grigor´yan in [24] as well as the book [23], where one can find various equivalence relationships between gaussian heat kernel bounds, Poincaré inequalities and volume doubling, the relative Faber Krahn inequality as well as the Harnack inequality.

4.2 *Proof of Theorem 6*

Proof Let $H_\phi(x, y, t)$ be the heat kernel of the Bakry–Émery Laplacian on M (Neumann kernel in case of boundary). Using the eigenvalues of Δ_ϕ, the Bakry–Émery heat kernel has the following expression

$$H_\phi(x, y, t) = \sum_{i=0}^{\infty} e^{-\lambda_i t} \phi_i(x)\phi_i(y)$$

where $\phi_i(x)$ is the eigenfunction corresponding to λ_i chosen such that $\{\phi_i\}_{i=0}^{\infty}$ are orthonormal in the weighted L^2 norm. Note that $\lambda_0 = 0$ and $\phi_0 = V_\phi^{-1/2}$, and as a result $\int_M \phi_i(y)\, e^{-\phi(y)} = 0$ for all $i \geq 1$. To obtain lower estimates for the eigenvalues, the idea is to find a uniform upper bound for the trace of the Bakry–Émery heat kernel, namely $H_\phi(x,x,t)$. In fact, it will be sufficient to find a uniform upper bound for

$$G(x,y,t) = H_\phi(x,y,t) - \frac{1}{V_\phi} = \sum_{i=1}^{\infty} e^{-\lambda_i t} \phi_i(x)\phi_i(y)$$

from the above remark. We observe that G satisfies

$$\int_M G(x,y,t)\, e^{-\phi(y)} = 0$$

and the semigroup property

$$G(x,z,t+s) = \int_M G(x,y,t)\, G(y,z,s)\, e^{-\phi(z)}$$

for all $x, z \in M$ and $t, s \in [0,\infty)$. The definition of G together with the properties $\int_M H_\phi(x,y,t)\, e^{-\phi(y)} = 1$ and $H_\phi(x,y,t) \geq 0$ imply

$$\int_M |G(x,y,t)|\, e^{-\phi(y)} \leq \int_M |H_\phi(x,y,t)|\, e^{-\phi(y)} + 1 = 2. \tag{9}$$

From the semigroup property

$$G(x,x,t) = \int_M G(x,y,t/2)\, G(x,y,t/2)\, e^{-\phi(y)}.$$

Differentiating both sides with respect to t we obtain

$$\begin{aligned} G'(x,x,t) &= \int_M G'(x,y,t/2)\, G(x,y,t/2)\, e^{-\phi(y)} \\ &= \int_M \Delta_{\phi,y} G(x,y,t/2)\, G(x,y,t/2)\, e^{-\phi(y)} \end{aligned}$$

since G also solves the heat equation. Integration by parts now gives

$$\begin{aligned} -G'(x,x,t) &= \int_M |\nabla_y G(x,y,t/2)|^2\, e^{-\phi(y)} \\ &\geq 2^{-\frac{4}{\nu}} C_1 \left(\int_M |G(x,y,t/2)|^2\, e^{-\phi(y)} \right)^{\frac{2+\nu}{\nu}} \end{aligned}$$

by Lemma 2 and Eq. 9. By the semigroup property of G,

$$-G'(x,x,t)\, (G(x,x,t))^{-(2+\nu)/\nu} \geq 2^{-4/\nu}\, C_1$$

Integrating both sides with respect to t gives

$$\frac{\nu}{2}\,(G(x,x,t))^{-2/\nu} \geq 2^{-4/\nu}\, C_1\, t$$

since G tends to $+\infty$ as t goes to zero. Finally we have the upper estimate

$$G(x,x,t) \leq 4\left(\frac{\nu}{2\,C_1}\right)^{\frac{\nu}{2}}\, t^{-\frac{\nu}{2}}.$$

If we now integrate both sides with respect to x, the eigenvalue expansion for G gives

$$\sum_{i=1}^{\infty} e^{-\lambda_i t} \leq 4\left(\frac{\nu}{2\,C_1}\right)^{\frac{\nu}{2}}\, t^{-\frac{\nu}{2}}\, V_\phi.$$

Setting $t = 1/\lambda_k$ we get

$$4\left(\frac{\nu\,\lambda_k}{2\,C_1}\right)^{\frac{\nu}{2}}\, V_\phi \geq \sum_{i=1}^{\infty} e^{-\lambda_i/\lambda_k} \geq \frac{k}{e}$$

since $\lambda_i/\lambda_k \leq 1$ for $i \leq 1$. The lower estimate follows.

In case $\partial M \neq \emptyset$, the proof is essentially identical and is left to the reader, one would just have to consider the Neumann or Dirichlet heat kernel, and the appropriate Sobolev inequality. □

Acknowledgments The second author is partially supported by the NSF grant DMS-12-06748. The third author gratefully acknowledges the support of the Leibniz Universität Hannover and the Australian National University.

References

1. B. Andrews, J. Clutterbuck, Proof of the fundamental gap conjecture. J. Am. Math. Soc. **24**(3), 899–916 (2011)
2. B. Andrews, L. Ni, Eigenvalue comparison on Bakry-Emery manifolds. Comm. Partial Differential Equations. **37**(11), 2081–2092 (2012)
3. D. Bakry, M. Émery, in *Diffusions Hypercontractives (French)*. Séminaire de probabilités, XIX, 1983/84. Lecture Notes in Math., vol. 1123 (Springer, Berlin, 1985), pp. 177–206
4. D. Bakry, Z. Qian, Some new results on eigenvectors via dimension, diameter, and Ricci curvature. Adv. Math. **155**(1), 98–153 (2000)
5. F. Baudoin, A. Vatamanelu, A note on lower bounds estimates for the Neumann eigenvalues of manifolds with positive Ricci curvature. Potential Anal. **37**(1), 91–101 (2012)
6. P.H. Bérard, *Spectral Geometry: Direct and Inverse Problems*. Monografiías de Matemática, vol. 41 (Instituto de Matemática Pura e Aplicada, Rio de Janeiro, 1986)
7. N. Charalambous, Eigenvalue estimates for the Bochner Laplacian and harmonic forms on complete manifolds. Indiana Univ. Math. J. **59**(1), 183–206 (2010)

8. N. Charalambous, Z. Lu, On the spectrum of the Laplacians. *Math. Ann.* **359**(1–2), 211–238 (2014)
9. N. Charalambous, Z. Lu, Heat kernel estimates and the essential spectrum on weighted manifolds. *J. Geom. Anal.* (to appear)
10. N. Charalambous, Z. Lu, The L^1 Liouville Property on Weighted manifolds. arXiv:1402.6170 (preprint)
11. S.Y. Cheng, Eigenvalue comparison theorems and its geometric applications. Math. Z. **143**(3), 289–297 (1975)
12. S.Y. Cheng, P. Li, Heat kernel estimates and lower bound of eigenvalues. Comment. Math. Helv. **56**(3), 327–338 (1981).
13. Y. Chu, Nonexistence of nontrivial quasi-Einstein metrics. Kodai Math. J. **35**(2), 374–381 (2012)
14. T.H. Colding, New monotonicity formulas for Ricci curvature and applications, I. Acta Math. **209**(2), 229–263 (2012)
15. E.B. Davies, B. Simon, M. Taylor, L^p spectral theory of Kleinian groups. J. Funct. Anal. **78**(1), 116–136 (1988)
16. J. Dodziuk, Eigenvalues of the Laplacian and the heat equation. Amer. Math. Monthly. **88**(9), 686–695 (1981)
17. H. Donnelly, P. Li, Lower bounds for the eigenvalues of Riemannian manifolds. Michigan Math. J. **29**(2), 149–161 (1982)
18. A. El Soufi, S. Ilias, Second eigenvalue of Schrödinger operators and mean curvature. Comm. Math. Phys. **208**(3), 761–770 (2000)
19. A. El Soufi, E.M. Harrell II, S. Ilias, Universal inequalities for the eigenvalues of Laplace and Schrödinger operators on submanifolds. Trans. Am. Math. Soc. **361**(5), 2337–2350 (2009)
20. P. Federbush, Partially alternate derivation of a result of Nelson. J. Math. Phys. **10**, 50–52 (1969)
21. K. Funano, T. Shioya, Concentration, Ricci curvature, and eigenvalues of Laplacian. Geom. Funct. Anal. **23**(3), 888–936 (2013)
22. A. Futaki, Y. Sano, Lower diameter bounds for compact shrinking Ricci solitons. Asian J. Math. **17**(1), 17–31 (2013)
23. A. Grigor'yan, in *Heat Kernel and Analysis on Manifolds*. AMS/IP Studies in Advanced Math. vol. 47 (American Mathematical Society, Providence, (International Press, Boston)), 2009)
24. A. Grigor'yan, in *Heat Kernels on Weighted Manifolds and Applications*. The Ubiquitous Heat Kernel, Contemp. Math. vol. 398 (American Mathematical Society, Providence, 2006), pp. 93–191
25. M. Gromov, *Metric Structures for Riemannian and non-Riemannian Spaces*. Reprint of the 2001 English edition, Modern Birkhäuser Classics (Birkhäuser, Boston, 2007)
26. L. Gross, Logarithmic Sobolev inequalities. Am. J. Math. **97**(4), 1061–1083 (1975)
27. A. Hassannezhad, Eigenvalues of perturbed Laplace operators on compact manifolds. arXiv:1210.7713v2 (preprint)
28. P. Kröger, On the spectral gap for compact manifolds. J. Differential Geom. **36**(2), 315–330 (1992)
29. J. Lott, Some geometric properties of the Bakry-Émery-Ricci tensor. Comment. Math. Helv. **78**(4), 865–883 (2003)
30. P. Li, S.T. Yau, *Estimates of eigenvalues of a compact Riemannian manifold*. Geometry of the Laplace operator. (Proceedings of Symposia in Pure Mathematics, University of Hawaii, Honolulu, Hawaii, 1979). Proceedings of Symposia in Pure Mathematics, vol. XXXVI (American Mathematical Society, Providence, 1980), pp. 205–239
31. P. Li, S.T. Yau, On the parabolic kernel of the Schrödinger operator. Acta Math. **156**(3/4), 153–201 (1986)
32. J. Ling, Lower bounds of the eigenvalues of compact manifolds with positive Ricci curvature. Ann. Global Anal. Geom. **31**(4), 385–408 (2007)

33. Z. Lu, J. Rowlett, Eigenvalues of collapsing domains and drift Laplacians. Math. Res. Lett. **10**(3), 627–648 (2012)
34. O. Munteanu, J. Wang, Analysis of weighted Laplacian and applications to Ricci solitons. Comm. Anal. Geom. **20**(1), 55–94 (2012).
35. E. Nelson, in *A quartic interaction in two dimensions*. Mathematical Theory of Elementary Particles. Proc. Conf., Dedham, MA, 1965 (MIT Press, Cambridge, 1966), pp. 69–73
36. Z.M. Qian, Gradient estimates and heat kernel estimate. Proc. R. Soc. Edinburgh Sect. A. **125**(5), 975–990 (1995)
37. L. Saloff-Coste, in *Aspects of Sobolev-type inequalities*. London Math. Soc. Lecture Note Series, vol. 289 (Cambridge University Press, Cambridge, 2002)
38. R. Schoen, S.-T. Yau, *Lectures on differential geometry*. Conference Proceedings and Lecture Notes in Geometry and Topology I (International Press, Cambridge, 1994)
39. A.G. Setti, Eigenvalue estimates for the weighted Laplacian on a Riemannian manifold. Rend. Sem. Mat. Univ. Padova. **100**, 27–55 (1998)
40. I.M. Singer, B. Wong, S.-T. Yau, S. S.-T. Yau, An estimate of the gap of the first two eigenvalues in the Schrödinger operator. Ann. Scuola Norm. Sup. Pisa Cl. Sci. (4). **12**(2), 319–333 (1985)
41. K.-T. Sturm, On the L^p-spectrum of uniformly elliptic operators on Riemannian manifolds. J. Funct. Anal. **118**(2), 442–453 (1993)
42. L.F. Wang, Rigid properties of quasi-Einstein metrics. Proc. Amer. Math. Soc. **139**(10), 3679–3689 (2011)
43. G. Wei, W. Wylie, Comparison geometry for the Bakry-Emery Ricci tensor. J. Differential Geom. **83**(2), 377–405 (2009)
44. H.C. Yang, Estimates of the first eigenvalue for a compact Riemann manifold. Sci. China Ser. A. **33**(1), 39–51 (1990)
45. S.-T. Yau, Nonlinear analysis in geometry. Enseign. Math. (2). **33**(1/2), 109–158 (1987)
46. S.-T. Yau, *An estimate of the gap of the first two eigenvalues in the Schrödinger operator*. Lectures on partial differential equations, New Studies in Advanced Mathematics, vol. 2 (International Press, Somerville, 2003), pp. 223–235.
47. S.-T. Yau, Gap of the first two eigenvalues of the Schrödinger operator with nonconvex potential. Mat. Contemp. **35**, 267–285 (2008)
48. J.Q. Zhong, H.C. Yang, On the estimate of the first eigenvalue of a compact Riemannian manifold. Sci. Sinica Ser. A. **27**(12), 1265–1273 (1984)

A Note on the Local Well-Posedness for the Whitham Equation

Mats Ehrnström, Joachim Escher and Long Pei

Abstract We prove local well-posedness for the Whitham equation in H^s, $s > \frac{3}{2}$, for both solitary and periodic initial data.

1 Introduction

The Whitham equation,

$$u_t + 2uu_x + Lu_x = 0, \tag{1}$$

with the Fourier multiplier L given by

$$\mathscr{F}(Lf)(\xi) = \left(\frac{\tanh(\xi)}{\xi}\right)^{\frac{1}{2}} \hat{f}(\xi), \tag{2}$$

has recently received renewed attention due to its generically nonlocal properties. Originally introduced by Whitham in 1967 as an alternative to the Korteweg-de Vries (KdV) equation for modelling shallow water waves [17], it is known to feature the exact linear dispersion relation for travelling gravity water waves (KdV, in contrast, features a local approximation of this relation, making it a less-suited model for large wave numbers). Some of the intriguing properties of (1) is the nonlocal, fractional, and inhomogeneous character of the symbol $m(\xi) := (\frac{\tanh(\xi)}{\xi})^{\frac{1}{2}}$, as well as its allowance for qualitative wave breaking, i.e., for bounded solutions whose first spatial

M. Ehrnström (✉) · L. Pei
Department of Mathematical Sciences, Norwegian University of Science and Technology, 7491 Trondheim, Norway
e-mail: mats.ehrnstrom@math.ntnu.no

L. Pei
e-mail: long.pei@math.ntnu.no

J. Escher
Institute for Applied Mathematics, Leibniz University of Hannover, 30167 Hannover, Germany
e-mail: escher@ifam.uni-hannover.de

J. Escher et al., (eds.), *Elliptic and Parabolic Equations,* Springer Proceedings in Mathematics & Statistics 119, DOI 10.1007/978-3-319-12547-3_3

derivate blows up in finite time (we refer to [18] for the concept of wave breaking, and [1, 12] for a proof of this in the case of the Whitham equation). The (1) is also an equation whose balance of dispersive and nonlinear effects admits for the existence of solitary-wave solutions [3].

The operator L in (1) is singular—its convolutional kernel blows up at the origin as $|x|^{-1/2}$, cf. [5]—and not thoroughly understood; even the positivity and single-sided monotonicity of the inverse Fourier transform of $m(\xi)$ are unknown. Although it was conjectured already by Whitham himself that (1) admits cusp-like solutions [17], so far only numerical evidence—and partial analytic results supporting this conjecture—exists, see [4]. In addition, the dispersion determined by (2) is unusually weak, making global estimates and related well-posed results rather difficult. In fact, even for the homogenous symbol $|\xi|^{-1/2}$ very little is known (see [14], and [6] for a related investigation).

The main motivation for this contribution is [3], in which the existence of solitary waves and a conditional stability result were proved for a large class of equations connected to (1). It is noted in [3], as well as in [8], that a local well-posedness result for the Whitham equation can be obtained using Kato's method. Since the proof has never been presented, and since the Whitham equation in important respects differs from many other model equations investigated in the literature, both what concerns its generic nonlocal properties and the weak dispersion associated to L, we give here the details in form of a short, but rigorous, proof of the local well-posedness of the Whitham equation, on the line as well as in the periodic case. The proof is valid for initial data in the Sobolev space H^s, $s > 3/2$, with the natural and best possible regularity for the solutions in terms of their time and space dependence. Note that the wave-breaking results [1, 12] for (1) imply that a general global well-posedness result is excluded. A conditional global existence result is still possible, but so far out of reach. For earlier treatments of the Whitham equation, see also [7] and [19].

The structure of the note is as follows. First, in Section 2, some general properties of the Whitham kernel Eq. 2 and related concepts are presented. Section 2 continues with the main proof for the case of initial data on the line, i.e. for data in $H^s(\mathbb{R})$, $s > 3/2$. The method of proof follows that of [2], which in turn is based on Kato's method, see [10]. Note, however, that the weakly dispersive, nonlocal, term in the Whitham (1) differs from those in the equations treated in [2] and [10], so that a thorough analysis is necessary. Once the arguments are in place, functional-analytic arguments can be employed to establish the result for periodic initial data, in the appropriate spaces. That is the content of Section 3.

2 General Preliminaries and the Case of Initial Data on $\mathbb{R}$

In (1) and (2), the Fourier transform $\mathcal{F}(f)$ of a function f is defined by the formula

$$\mathcal{F}(f)(\xi) = \frac{1}{\sqrt{2\pi}} \int_{\mathbb{R}} e^{-i\xi x} f(x)dx,$$

extended by duality from $S(\mathbb{R}) \subset L^2(\mathbb{R})$ to $S'(\mathbb{R})$, the space of tempered distributions. Here, $S(\mathbb{R})$ is the Schwarz space of rapidly decaying smooth functions on $\mathbb{R}$, and $H^s(\mathbb{R})$, $s \in \mathbb{R}$, is the Sobolev space of tempered distributions whose Fourier transform satisfies

$$\int_{\mathbb{R}} (1 + |\xi|^2)^s |\mathcal{F}(f)(\xi)|^2 \, d\xi < \infty,$$

with the standard inner product. For convenience, we will sometimes omit the domain in the notation for function spaces in the following. Throughout the chapter, the notation $\hat{f}$ will also be used interchangeably with $\mathcal{F}(f)$. Note that although the functions and operators of interest in this chapter are all real-valued, and although the operator L defined in (2) maps real data to real data, the Fourier transform is naturally defined in complex-valued function spaces; hence, the function spaces used in this investigation should in general be understood as the complexifications of corresponding real-valued function spaces.

As usual, C^k denotes the space of k times continuously differentiable functions, and BC^k the corresponding space of functions whose derivatives up to order k are also bounded.

Before considering some properties of the Fourier multiplier operator L, let us state our main theorem.

Theorem 1 *Let $s > \frac{3}{2}$. Given $u_0 \in H^s(\mathbb{R})$ there is a maximal $T > 0$ depending only on $\|u_0\|_{H^s}$, and a unique solution $u = u(\cdot, u_0)$ to (1) in $C([0,T); H^s(\mathbb{R})) \cap C^1([0,T); H^{s-1}(\mathbb{R}))$. The solution depends continuously on the initial data, i.e., the map $u_0 \mapsto u(\cdot, u_0)$ from $H^s(\mathbb{R})$ to $C([0,T); H^s(\mathbb{R})) \cap C^1([0,T); H^{s-1}(\mathbb{R}))$ is continuous.*

Remark 1 It is clear from the method behind Theorem 1 as well as Theorem 3 that the results can be generalised to a larger class of Whitham-like equations (cf. [3]). For example, similar results could be obtained if the linear term L is bounded $L^2 \to L^2$ and the nonlinear term is temperate (slowly growing) and belongs to C^k for $k > s+2$, although the Whitham equation gives some more regularity than this general case. These conditions, however, are not optimal, and a larger class of interest requires results on the Nemytskii operator n that are still only partially available (the periodic case is open, although arguments are likely available [15]). Results in this direction, including the possibility of L being a Fourier integral operator, are under preparation.

In order to prove Theorem 1, we first make some basic assumptions and observations. Let

$$X = L^2(\mathbb{R}) \qquad \text{and} \qquad Y = H^s(\mathbb{R}),$$

for some fixed $s > \frac{3}{2}$. Note that for $f \in L^2(\mathbb{R})$,

$$\|Lf\|_{L^2} = \|\mathscr{F}(Lf)\|_{L^2} = \left\| \left(\frac{\tanh(\xi)}{\xi} \right)^{\frac{1}{2}} \hat{f}(\xi) \right\|_{L^2} \le \|\hat{f}\|_{L^2} = \|f\|_{L^2},$$

so that L is a bounded linear operator on $L^2(\mathbb{R})$. In fact, it is shown in [3] that

$$L \in C^\infty(H^s(\mathbb{R}), H^{s+\frac{1}{2}}(\mathbb{R})) \cap C^\infty(S(\mathbb{R}), S(\mathbb{R})), \tag{3}$$

for all $s \geq 0$. We note that the smooth and even function $\xi \mapsto (\frac{\tanh(\xi)}{\xi})^{\frac{1}{2}}$ is increasing in $(-\infty, 0)$ and decreasing in $(0, \infty)$, reaching its global maximum of unit size at $x = 0$. As $|\xi| \to \infty$, it vanishes with the rate $|\xi|^{-\frac{1}{2}}$.

The operator L is furthermore symmetric on $L^2(\mathbb{R})$, since for $f, g \in L^2(\mathbb{R})$ we have that

$$\begin{aligned}
(Lf, g)_{L^2} &= (\mathcal{F}(Lf), \mathcal{F}(g))_{L^2} \\
&= \int_{\mathbb{R}} \mathcal{F}(Lf)\overline{\mathcal{F}(g)}\, d\xi \\
&= \int_{\mathbb{R}} (\frac{\tanh(\xi)}{\xi})^{\frac{1}{2}} \hat{f}(\xi)\overline{\hat{g}(\xi)}\, d\xi \\
&= \int_{\mathbb{R}} \mathcal{F}(f)\overline{\mathcal{F}(Lg)}\, d\xi \\
&= (f, Lg)_{L^2},
\end{aligned}$$

so it follows that L is a symmetric bounded linear operator on the space X.

To proceed, rewrite Theorem 1 as

$$\begin{aligned}
u_t + A(u)u &= 0, \qquad t \geq 0, \\
u(0) &= u_0.
\end{aligned} \tag{4}$$

Here,

$$A(y) = (2y + L)\partial_x, \qquad \text{dom}(A(y)) := \{v \in L^2 \mid 2yv + Lv \in H^1\},$$

for some $y \in H^s$ with $s > \frac{3}{2}$. In view of (3), and since H^s is a Banach algebra for $s > \frac{1}{2}$, $\text{dom}(A(y))$ is dense in $X = L^2$.

In order to state Kato's theorem in a form suitable for our purposes, we also need the concept of *accretiveness* (cf. [11]). Let T be an operator on a Hilbert space H, and let $\mathcal{L}$ denote the space of all bounded linear operators from some space to another, in this case $H \to H$. We then say that:

- T is *accretive* if $\text{Re}(Tv, v)_H \geq 0$ for all $v \in \text{dom}(T)$.
- T is *quasi-accretive* if $T + \alpha$ is accretive for some scalar $\alpha > 0$.
- T is *m-accretive* if $(T + \lambda)^{-1} \in \mathcal{L}(H)$ with $\text{Re}(\lambda)\|(T + \lambda)^{-1}\| \leq 1$ for $\text{Re}(\lambda) > 0$.
- T is *quasi-m-accretive* if $T + \alpha$ is m-accretive for some scalar $\alpha > 0$.

We shall make use of the following version of Kato's theorem to establish the local well-posedness for the problem (4), adapted from [2].

Theorem 2 *(Cf. [2]) Consider the abstract quasi-linear evolution (4). Let X and Y be as above (Hilbert spaces such that Y is continuously and densely injected into X), and let $Q : Y \to X$ be a topological isomorphism. Assume that:*

(i) $A(y) \in \mathcal{L}(Y, X)$ for $y \in Y$ with

$$\|(A(y) - A(z))w\|_X \le \mu_A \|y - z\|_X \|w\|_Y, \qquad y, z, w \in Y,$$

and $A(y)$ is quasi-m-accretive, uniformly on bounded sets in Y.

(ii) $QA(y)Q^{-1} = A(y) + B(y)$, where $B(y) \in \mathcal{L}(X)$ is bounded, uniformly on bounded sets in Y. Moreover,

$$\|(B(y) - B(z))w\|_X \le \mu_B \|y - z\|_Y \|w\|_X, \qquad y, z \in Y, \quad w \in X.$$

Here, the constants μ_A and μ_B depend only on $\max\{\|y\|_Y, \|z\|_Y\}$.
Then, for any given $v_0 \in Y$, there is a maximal $T > 0$ depending only on $\|v_0\|_Y$ and a unique solution v to (4) such that

$$v = v(\cdot, v_0) \in C([0,T); Y) \cap C^1([0,T); X).$$

The map $v_0 \mapsto v(\cdot, v_0)$ is continuous from Y to $C([0,T); Y) \cap C^1([0,T); X)$.

To continue, we need to study the operator $A(y)$ for a fixed $y \in H^s$, with $s > \frac{3}{2}$. Later y will be taken in a bounded subset of $Y = H^s$ for $s > \frac{3}{2}$, but note that all estimates to come are uniform with respect to any such bounded subset.

For $y \in Y$, we define

$$Dv := (2yv + Lv)_x - 2y_x v, \qquad \operatorname{dom}(D) := \{v \in L^2 \mid 2yv + Lv \in H^1\},$$

and

$$D_0 v := -(2yv + Lv)_x, \qquad \operatorname{dom}(D_0) := \operatorname{dom}(D).$$

Note that the choice of these domains makes both D and D_0 closed operators in $X = L^2$. In view of the embedding $H^s \subset BC$ for $s > \frac{1}{2}$, we furthermore have

$$(2yv + Lv)_x = 2y_x v + 2yv_x + (Lv)_x \quad \text{in } H^{-1}, \tag{5}$$

so that, in that space,

$$Dv = 2yv_x + Lv_x,$$
$$D_0 v = -2y_x v - 2yv_x - (Lv)_x,$$

and

$$2yv_x + (Lv)_x = (2yv + Lv)_x - 2y_x v \in L^2,$$

for y and v as considered here.

We will prove that D satisfies the condition (i) in Theorem 2 with the help of a few consecutive lemmata. To state the first of these, let C_0^∞ be the space of compactly supported smooth functions on some open set, in this case $\mathbb{R}$.

Lemma 1 *Given $v \in \operatorname{dom}(D)$, there exists a sequence $\{v_n\}_n \subset C_0^\infty$ such that*

$$v_n \to v \quad \textit{and} \quad (2yv_n + Lv_n)_x \to (2yv + Lv)_x$$

in L^2 as $n \to \infty$.

Proof Pick $\rho \in C_0^\infty$ with $\rho(x) \geq 0$ and $\int_{\mathbb{R}} \rho\, dx = 1$. For $n \geq 1$, let $\rho_n(x) := n\rho(nx)$ be a mollifier on $\mathbb{R}$. Denoting by v_n the convolution $v * \rho_n$, we have $v_n \in C^\infty$ and $v_n \to v$ in L^2 as $n \to \infty$. This proves the first part of the lemma due to the density of C_0^∞ in L^2.

As what concerns the second part, we have

$$\begin{aligned}
&(2yv_n + Lv_n)_x - (2yv + Lv)_x \\
&= ((2y(v_n)_x + L(v_n)_x) - (2yv_x + Lv_x)) + (2y_xv_n - 2y_xv) \\
&= (2y(v_n)_x + L(v_n)_x - (2yv_x + Lv_x) * \rho_n) \\
&\quad + ((2yv_x + Lv_x) * \rho_n - (2yv_x + Lv_x)) + (2y_xv_n - 2y_xv) \\
&=: \mathrm{I}_n(v) + \mathrm{II}_n(v) + \mathrm{III}_n(v).
\end{aligned}$$

By observing that $2yv_x + Lv_x \in L^2$ with $y_x \in H^{s-1} \subset BC$ for $s > \frac{3}{2}$, one naturally gets

$$\begin{aligned}
\mathrm{II}_n(v) &= (2yv_x + Lv_x) * \rho_n - (2yv_x + Lv_x) \to 0, \\
\mathrm{III}_n(v) &= 2y_xv_n - 2y_xv \to 0,
\end{aligned}$$

in L^2, as $n \to \infty$.

It remains to prove that $\mathrm{I}_n(v) \to 0$ in L^2 as $n \to \infty$. Since this clearly holds for $v \in C_0^\infty$, and since C_0^∞ is densely and continuously embedded in L^2, we only need to prove that $\|I_n(v)\|_{L^2} \leq \|v\|_{L^2}$, for $v \in L^2$. Then the result follows from continuity.

To prove this, note that for any $v \in \operatorname{dom}(D)$, we have

$$\begin{aligned}
\mathcal{F}(I_n(v)) &= \mathcal{F}(2y(v_n)_x + (Lv_n)_x - ((2yv + Lv)_x - 2y_xv) * \rho_n) \\
&= \mathcal{F}(2y(v_n)_x) + \mathcal{F}((Lv_n)_x) - \mathcal{F}((2yv + Lv) * (\rho_n)_x) + \mathcal{F}((2y_xv) * \rho_n) \\
&= \mathcal{F}(2y(v_n)_x) + i\xi\Big(\frac{\tanh(\xi)}{\xi}\Big)^{\frac{1}{2}} \mathcal{F}(v)\mathcal{F}(\rho_n) - \mathcal{F}((2yv) * (\rho_n)_x) \\
&\quad - \Big(\frac{\tanh(\xi)}{\xi}\Big)^{\frac{1}{2}} \mathcal{F}(v) i\xi \mathcal{F}(\rho_n) + \mathcal{F}((2y_xv) * \rho_n) \\
&= \mathcal{F}(2y(v_n)_x - (2yv) * (\rho_n)_x + (2y_xv) * \rho_n),
\end{aligned}$$

where we have used $(2yv + Lv)_x * (\rho_n) = (2yv + Lv) * (\rho_n)_x$ for $v \in \operatorname{dom}(D)$.

Then

$$\begin{aligned}
\mathrm{I}_n(v) &= 2y(v * (\rho_n)_x) - (2yv) * (\rho_n)_x + (2y_x v) * \rho_n \\
&= 2\int_{\mathbb{R}} (y(x) - y(x-s))v(x-s)(\rho_n)_x(s)\,ds + (2y_x v) * \rho_n \\
&= 2n^2 \int_{\mathbb{R}} (y(x) - y(x-s))v(x-s)\rho_x(ns)\,ds + (2y_x v) * \rho_n \\
&=: \hat{\mathrm{I}}_n(v) + (2y_x v) * \rho_n.
\end{aligned}$$

Note that $2y_x v \in L^2$ implies that $(2y_x v) * \rho_n \to 2y_x v$ in L^2, whence we only need to prove $\|\hat{\mathrm{I}}_n(v)\|_{L^2} \le \|v\|_{L^2}$. For this purpose, suppose $\mathrm{supp}(\rho) \subset [-\lambda, \lambda]$ for some $\lambda > 0$. Then, by Hölder's inequality, we have

$$\begin{aligned}
\|\hat{\mathrm{I}}_n(v)\|_{L^2} &= \|2n \int_{-\lambda}^{\lambda} (y(x) - y(x-s/n))v(x-s/n)\rho_x(s)\,ds\|_{L^2} \\
&\le \|2 \sup_{s\in\mathbb{R}} |y_x(s)| \int_{-\lambda}^{\lambda} |sv(x-s/n)\rho_x(x-s)|\,ds\|_{L^2} \\
&\le 2 \sup_{s\in\mathbb{R}} |y_x(s)| \|(\int_{-\lambda}^{\lambda} |s\rho_x(s)|^2\,ds)^{\frac{1}{2}} (\int_{-\lambda}^{\lambda} |v(x-s/n)|^2\,ds)^{\frac{1}{2}}\|_{L^2}
\end{aligned}$$

Let $M = 2\sup_{s\in\mathbb{R}} |y_x(s)|(\int_{-\lambda}^{\lambda} |s\rho_x(s)|^2\,ds)^{\frac{1}{2}} < \infty$. Then, by Fubini's theorem, we have

$$\begin{aligned}
\|\hat{\mathrm{I}}_n(v)\|_{L^2} &\le M\Big(\int_{\mathbb{R}} \int_{-\lambda}^{\lambda} |v(x-s/n)|^2\,ds\,dx\Big)^{\frac{1}{2}} \\
&= M\Big(\int_{-\lambda}^{\lambda} \int_{\mathbb{R}} |v(x-s/n)|^2\,dx\,ds\Big)^{\frac{1}{2}} \\
&= (2\lambda)^{\frac{1}{2}} M \|v\|_{L^2},
\end{aligned}$$

which completes the proof. □

Lemma 2 *The operators D and D_0 are both quasi-accretive in L^2.*

Proof By definition, D is quasi-accretive in L^2 if and only if

$$\mathrm{Re}((D + \alpha I)v, v)_{L^2} \ge 0,$$

for all $v \in \mathrm{dom}(D)$ and some scalar $\alpha > 0$. In view of Lemma 1 and (5), we can find a sequence $v_n \in C_0^\infty$ such that

$$(Dv, v)_{L^2} = \lim_{n\to\infty} (Dv_n, v_n)_{L^2} = \lim_{n\to\infty} \int_{\mathbb{R}} (2y(v_n)_x + L(v_n)_x)v_n\,dx.$$

As L is symmetric the operator $L\partial_x$ is skew-symmetric, and we have $(L\partial_x v_n, v_n)_{L^2} = -(v_n, L\partial_x v_n)_{L^2}$. Since both L and v_n are real-valued, the term $\int_{\mathbb{R}} L(v_n)_x v_n\,dx$ thus vanishes completely for all $n \ge 1$.

Since, in addition,

$$\int_{\mathbb{R}} 2y(v_n)_x v_n dx = \int_{\mathbb{R}} y(v_n^2)_x \, dx = -\int_{\mathbb{R}} y_x v_n^2 \, dx,$$

we have

$$(Dv_n, v_n)_{L^2} = -\int_{\mathbb{R}} y_x v_n^2 \, dx. \tag{6}$$

Now, for $y_x \in H^{s-1} \subset BC$, $s > \frac{3}{2}$, we can select $\alpha > 0$ such that $\alpha \geq \|y_x\|_{L^\infty}$. Then

$$\mathrm{Re}((D + \alpha I)v_n, v_n)_{L^2} = \int_{\mathbb{R}} (\alpha - y_x)v_n^2 \, dx \geq 0,$$

implying that

$$\mathrm{Re}((D + \alpha I)v, v)_{L^2} = \lim_{n \to \infty} \mathrm{Re}((D + \alpha I)v_n, v_n)_{L^2} \geq 0.$$

Hence, D is quasi-accretive.

As what concerns the operator D_0, note that $D_0 v = Dv + 2y_x v$ and $(2y_x v, v)_{L^2}$ can be bounded with same technique as used in connection to (6). Thus, the quasi-accretiveness of D_0 follows from that of D. □

For use in the following, denote by

$$[T_1, T_2] = T_1 T_2 - T_2 T_1$$

the commutator of two operators T_1 and T_2. Note that both ∂_x and L are Fourier multiplier operators, so that $[\partial_x, L] = 0$ on H^s for all $s \in \mathbb{R}$.

Lemma 3 *The adjoint of D in L^2 is D_0.*

Proof For $v \in C_0^\infty \subset \mathrm{dom}(D)$ and any $\omega \in \mathrm{dom}(D^*)$, we have

$$\begin{aligned}
(v, D^*\omega)_{L^2} &= (Dv, \omega)_{L^2} \\
&= \int_{\mathbb{R}} [(2yv + Lv)_x - 2y_x v]\omega \, dx \\
&= \int_{\mathbb{R}} (2yv_x + Lv_x)\omega \, dx \\
&= \int_{\mathbb{R}} (2y\omega + L\omega)v_x \, dx,
\end{aligned}$$

so that $D^*\omega \in L^2$ is the weak derivative of $2y\omega + L\omega \in L^2$. Consequently, $2y\omega + L\omega \in H^1$ and $\omega \in \mathrm{dom}(D_0)$. Then

$$\begin{aligned}
(v, D^*\omega)_{L^2} &= \int_{\mathbb{R}} -(2y\omega + L\omega)_x v \, dx \\
&= (v, D_0\omega)_{L^2},
\end{aligned}$$

so that $D^* \subset D_0$.

Assume now that $v \in \mathrm{dom}(D_0) \subset L^2$ and note that for any $\tilde{v} \in \mathrm{dom}(D^*)$, which by the above calculation belongs to $\mathrm{dom}(D_0) = \mathrm{dom}(D)$, we can always find a sequence $\{\tilde{v}_n\} \subset C_0^\infty$ such that Lemma 1 holds. Therefore, we have that

$$\begin{aligned}(D\tilde{v}, v)_{L^2} &= \lim_{n\to\infty} (D\tilde{v}_n, v)_{L^2}\\ &= \lim_{n\to\infty} \int_{\mathbb{R}} ((2y(\tilde{v}_n)_x + L(\tilde{v}_n)_x)v\,dx\\ &= \lim_{n\to\infty} \int_{\mathbb{R}} (2yv + Lv)(\tilde{v}_n)_x\,dx.\end{aligned}$$

Since $v \in \mathrm{dom}(D_0)$ implies that $2yv + Lv \in H^1$ by our assumptions, we also have that

$$\begin{aligned}(D\tilde{v}, v)_{L^2} &= \lim_{n\to\infty} \int_{\mathbb{R}} (2yv + Lv)(\tilde{v}_n)_x\,dx\\ &= -\lim_{n\to\infty} \int_{\mathbb{R}} (2yv + Lv)_x \tilde{v}_n\,dx\\ &= (\tilde{v}, -(2yv + Lv)_x)_{L^2}\,dx\\ &= (\tilde{v}, D_0 v)_{L^2}.\end{aligned}$$

Thus, it follows that $v \in \mathrm{dom}(D^*)$ and $D_0 \subset D^*$. In view of the above, it is then clear that $D_0 = D^*$.

By Lemmata 2 and 3, both D and D^* are quasi-accretive. A classical argument (cf. [13], Corollary 4.4) then gives the following.

Lemma 4 *For the closed linear operator D, densely defined on the Banach space X, with both D and its adjoint D^* quasi-accretive, there exists a scalar $\alpha \in \mathbb{R}$ such that the operator $-(D+\alpha)$ is the infinitesimal generator of a C_0-semigroup of contractions on X, i.e., D is a quasi-m-accretive operator.*

This proposition shows that the operator A in our Cauchy problem satisfies condition (i) in Theorem 2.

We next turn to condition (ii). Let

$$B(y) := Q(A(y))Q^{-1} - A(y) = [Q, A(y)]Q^{-1},$$

where $A(y) = (2y + L)\partial_x$ and $Q := \Lambda^s = (1 - \partial_x^2)^{\frac{s}{2}}$ is an isomorphism from H^s to L^2. We then have the following lemma.

Lemma 5 *For $y \in Y$, the operator $B(y)$ satisfies* condition (ii) *in* Theorem 2.

Proof Note that

$$\begin{aligned}[Q, A(y)] &= [\Lambda^s, A(y)]\\ &= [\Lambda^s, (2y + L)\partial_x]\\ &= 2[\Lambda^s, y]\partial_x + [\Lambda^s, L]\partial_x\\ &= 2[\Lambda^s, y]\partial_x,\end{aligned} \tag{7}$$

where we have used the commutation properties $[\Lambda^s, \partial_x] = 0$ and $[\Lambda^s, L] = 0$.

In order to prove uniform boundedness of $B(y)$ for y in a bounded subset of H^s, we assume without loss of generality that $y \in \mathscr{W} \subset H^s$, where $\mathscr{W}$ is an open ball in H^s with radius $R > 0$. Using classical estimates for (7) (cf. [10]), we get

$$\|[\Lambda^s, y]\Lambda^{1-s}\| \leq C_0 \|\partial_x y\|_{H^{s-1}} \leq C_0 \|y\|_{H^s} \leq \alpha_0(R),$$

where C_0 is a constant relying only on s, and $\alpha_0(R)$ is a constant relying only on $\mathscr{W}$. Then, for any $z \in L^2$, we have

$$\begin{aligned} \|B(y)z\|_{L^2} &= \|[\Lambda^s, y]\Lambda^{1-s}\Lambda^{s-1}\partial_x \Lambda^{-s} z\|_{L^2} \\ &\leq \|[\Lambda^s, y]\Lambda^{1-s}\| \, \|\Lambda^{s-1}\partial_x \Lambda^{-s} z\|_{L^2} \\ &\leq \alpha_0(R)\|\partial_x \Lambda^{-1} z\|_{L^2} \\ &\leq \alpha_0(R)\|z\|_{L^2}, \end{aligned}$$

where the last step is due to the fact that

$$\|\partial_x \Lambda^{-1} z\|_{L^2} \leq \|\Lambda^{-1} z\|_{H^1} = \|z\|_{L^2}.$$

Hence, $B(y)$ is a bounded linear operator on L^2 for $y \in Y$. In addition, for any $y, z \in \mathscr{W}$ and $w \in X$,

$$\begin{aligned} \|B(y)w - B(z)w\|_{L^2} &= \|[\Lambda^s, y - z]\partial_x \Lambda^{-s} w\|_{L^2} \\ &\leq \alpha_1(R)\|[\Lambda^s, y - z]\Lambda^{1-s}\| \, \|\Lambda^{s-1}\partial_x \Lambda^{-s} w\|_{L^2} \\ &\leq \alpha_2(R)\|y - z\|_{H^s}\|w\|_{L^2}, \end{aligned}$$

where $\alpha_1(R)$ and $\alpha_2(R)$ are constants depending only on $\mathscr{W}$. Thus, $B(y)$ satisfies condition (ii) for all $y \in \mathscr{W} \subset H^s$. □

We are now ready to prove the main theorem for initial data $u_0 \in H^s(\mathbb{R})$.

Proof of Theorem 1 According to Lemmata 1–5 we can apply Theorem 2 to find a solution u as described in Theorem 1, although in the solution class $C([0,T); H^s(\mathbb{R})) \cap C^1([0,T); L^2(\mathbb{R}))$. In view of that $H^{s-1}(\mathbb{R})$ is an algebra with respect to pointwise multiplication, and that L maps $H^s(\mathbb{R})$ continuously into $H^{s+1/2}(\mathbb{R})$, one, however, sees that for the Whitham equation,

$$u_t = -2uu_x - Lu_x \in H^{s-1}(\mathbb{R}).$$

Hence, $u \in C^1([0,T); H^{s-1}(\mathbb{R}))$.

Also, since $[u_0 \mapsto u] \in C(H^s(\mathbb{R}), C([0,T), H^s(\mathbb{R}))$, and ∂_x maps $H^s(\mathbb{R})$ continuously into $H^{s-1}(\mathbb{R})$, the same argument can be used to conclude that $[u_0 \mapsto u] \in C(H^s(\mathbb{R}), C^1([0,T), H^{s-1}(\mathbb{R})))$. This yields the desired solution class and concludes the proof of Theorem 1.

3 The Periodic Case

Consider now the Cauchy problem (1) in the periodic setting. We first define the periodic Sobolev space $H^s_{2\pi\text{-per}}$ for $s \in \mathbb{R}$ as the set of all 2π-periodic distributions $f \in S'$ such that

$$\|f\|_{H^s_{2\pi\text{-per}}} = \Big(\sum_{k\in\mathbb{Z}} (1+|k|^2)^s |\hat{f}(k)|^2\Big)^{\frac{1}{2}} < \infty,$$

with inner product

$$(f|g)_{H^s_{2\pi\text{-per}}} = \sum_{k\in\mathbb{Z}} (1+|k|^2)^s \hat{f}(k)\overline{\hat{g}(k)}.$$

Here $\hat{f}(k)$ are generalized Fourier coefficients, see [16]. In what follows, we will be working with H^s for $s \geq 0$ and $H^0_{2\pi\text{-per}}$ will be denoted by $L^2_{2\pi\text{-per}}$. If we define the Fourier coefficients of $f \in H^s_{2\pi\text{-per}}$ to be

$$c_k = \hat{f}(k) = \frac{1}{\sqrt{2\pi}}(f|e^{ikx})_{L^2((-\pi,\pi))} = \frac{1}{\sqrt{2\pi}}\int_{-\pi}^{\pi} f(x)e^{-ikx}\,dx,$$

then each $f \in H^s_{2\pi\text{-per}}$ is uniquely determined by its Fourier coefficients, and we have

$$f(x) = \sum_{k\in\mathbb{Z}} \hat{f}(k)e^{ikx} \qquad \text{in} \quad L^2_{2\pi\text{-per}}. \tag{8}$$

Note that since $H^s_{2\pi\text{-per}} \subset S'$, the Fourier transform $\mathcal{F}f \in S'$ is well-defined for all $f \in H^s_{2\pi\text{-per}}$, but in view of (8) it can also be identified with an element in l^s for all $s \geq 0$. From Parseval's identity,

$$\|\hat{f}\|^2_{l^2} = \|f\|^2_{L^2((-\pi,\pi))}, \quad f \in H^s_{2\pi\text{-per}},$$

and

$$(\hat{f}|\hat{g})_{l^2} = \sum_{k\in\mathbb{Z}} \hat{f}(k)\overline{\hat{g}(k)} = \int_{-\pi}^{\pi} f(x)\overline{g(x)}\,dx = (f|g)_{L^2((-\pi,\pi))},$$

for all $f, g \in H^s_{2\pi\text{-per}}$. It follows that all Fourier integral formulas in the case on the line immediately translate into Fourier series in the periodic case. In particular (cf. [3]),

$$\mathscr{F}(Lf)(k) = \Big(\frac{\tanh(k)}{k}\Big)^{\frac{1}{2}} \hat{f}(k), \quad k \in \mathbb{Z},$$

$$L \in C^\infty(H^s_{2\pi\text{-per}}, H^{s+\frac{1}{2}}_{2\pi\text{-per}}) \cap C^\infty(\mathcal{S}'(\mathbb{R}), \mathcal{S}'(\mathbb{R})), \quad s \geq 0,$$

and L is a bounded linear symmetric operator on $L^2_{2\pi\text{-per}}$. For more information on the periodic Sobolev spaces, see also the monograph [9].

As a consequence of this relationship between the periodic Sobolev spaces $H^s_{2\pi\text{-per}}$ and their non-periodic counterparts $H^s(\mathbb{R})$, the proof of Theorem 1 can be followed in detail with $H^s(\mathbb{R})$ replaced by $H^s_{2\pi\text{-per}}$, and the Fourier transform replaced accordingly, as described above[1]. Note here that the crucial embedding $H^s_{2\pi\text{-per}} \hookrightarrow BC^1$ for $s > 3/2$ is equally valid in the periodic case. Analogous to the case on $\mathbb{R}$, we thus obtain the following result.

Theorem 3 *Let $s > \frac{3}{2}$. Given $u_0 \in H^s_{2\pi\text{-per}}$, there is a maximal $T > 0$ and a unique solution u to 1 such that $u = u(\cdot, u_0) \in C([0,T); H^s_{2\pi\text{-per}}) \cap C^1([0,T); H^{s-1}_{2\pi\text{-per}})$. The solution depends continuously on the initial data, i.e., the map $u_0 \mapsto u(\cdot, u_0)$ is continuous from $H^s_{2\pi\text{-per}}$ to $C([0,T); H^s_{2\pi\text{-per}}) \cap C^1([0,T); H^{s-1}_{2\pi\text{-per}})$.*

Acknowledgments The authors would like to thank the referee for valuable comments and suggestions that helped to improve the paper's final form.

References

1. A. Constantin, J. Escher, Wave breaking for nonlinear nonlocal shallow water equations, Acta Math. **181**, 229–243 (1998)
2. A. Constantin, J. Escher, Well-posedness, global existence, and blowup phenomena for a periodic quasi-linear hyperbolic equation, Commun. Pure Appl. Math. **51**, 475–504 (1998)
3. M. Ehrnström, M.D. Groves, E. Wahlén, On the existence and stability of solitary-wave solutions to a class of evolution equations of Whitham type, Nonlinearity, **25**, 1–34 (2012)
4. M. Ehrnström, H. Kalisch, Global bifurcation for the Whitham equation. Math. Model. Nat. Phenom. **7**, 13–30 (2013)
5. M. Ehrnström, H. Kalisch, Traveling waves for the Whitham equation. Differential Integral Equations. **22**, 1193–1210 (2009)
6. J. Escher, B. Kolev, Right-invariant Sobolev metrics of fractional order on the diffeomorphisms group of the circle. J. Geom. Mech., **6**, 335–372 (2014)
7. S. Gabov, On Whitham's equation. Sov. Math., Dokl. [Translation from Dokl. Akad. Nauk SSSR 242, 993–996 (1978)]. **19**, 1225–1229 (1978)
8. V.M. Hur, M.A. Johnson, Stability of periodic traveling waves for nonlinear dispersive equations. arXiv:1303.4765 (2013)
9. R.J.J. Iorio, V.D.M. Iorio, *Fourier Analysis and Partial Differential Equations* (Cambridge University Press, Cambridge, 2001)
10. T. Kato, *Quasi-linear equations of evolution, with applications to partial differential equations*, in Spectral theory and differential equations. Proc. Sympos., Dundee, 1974 (dedicated to Konrad Jörgens). Lecture Notes in Math., vol. 448 (Springer, Berlin, 1975), pp. 25–70
11. T. Kato, *Perturbation Theory for Linear Operators* (Springer, Berlin-New York, 1995)
12. P.I. Naumkin, I.A. Shishmarev, *Nonlinear Nonlocal Equations in the Theory of Waves*, Translations of Mathematical Monographs, vol. 133 (American Mathematical Society, Providence, 1994)
13. A. Pazy, *Semigroups of Linear Operators and Applications to Partial Differential Equations* (Springer, Berlin-New York, 1983)

[1] For questions related to global existence questions, the difference between the periodic and non-periodic case might play a bigger role.

14. F. Linares, D. Pilod, J.-C. Saut, Dispersive perturbations of burgers and hyperbolic equations I: Local theory. SIAM J. Math. Anal. **46**(2), 1505–1537 (2014)
15. W. Sickel (personal communication) (2013)
16. H. Triebel, *Theory of Function Space*. Modern Birkhäuser classics, vol. IV. (Springer, Birkhäuser Verlag, 1983)
17. G.B. Whitham, Variational methods and applications to water waves. Proc. R. Soc. Lond., Ser. A. **299**, 6–25 (1967)
18. G.B. Whitham, in *Linear and Nonlinear Waves*. Pure and Applied Mathematics (Wiley, New York, 1974)
19. A.A. Zaitsev, *Stationary Whitham waves and their dispersion relation*. Dokl. Akad. Nauk SSSR. **286**, 1364–1369 (1986)

On the Lifetime of a Conditioned Brownian Motion in Domains Connected Through Small Gaps

Matthias Erven and Guido Sweers

Abstract For Ω a domain in $\mathbb{R}^n$ the expression $\mathbb{E}_x^y(\tau_\Omega)$ denotes the expected lifetime of Brownian motion starting at $x \in \Omega$, conditioned to converge to $y \in \Omega$ and with paths that are killed when touching $\partial\Omega$ before reaching (an ε-neighborhood of) y. This expectation coincides with the integral expression in the so-called $3G$-theorem of Cranston et al. [7]. In general, it is hard to find sharp estimates for the bound in the $3G$-theorem. In this chapter, we study the special case that the domain Ω consists of two parts which are connected through small gaps. We will derive a formula for $\mathbb{E}_x^y(\tau_\Omega)$, that depends on the expected times to reach each individual gap and which involves weights that depend asymptotically on the Poisson kernels at these gaps for the two subdomains. Since conformal mappings between Jordan domains are used, we restrict ourselves to domains in $\mathbb{R}^2$.

1 Introduction

Let Ω be a domain in $\mathbb{R}^n$ and consider Brownian motion that starts at $x \in \Omega$ and is conditioned to be killed when reaching the boundary $\partial\Omega$. The expectation for the lifetime of this Brownian motion that converges to y can be expressed by a so-called $3G$-integral (see [3]):

$$\mathbb{E}_x^y(\tau_\Omega) = \int_\Omega G_\Omega(x,z)\frac{G_\Omega(y,z)}{G_\Omega(y,x)}dz.$$

Here G_Ω is the Green function for

$$\begin{cases} -\Delta u = f & \text{in } \Omega, \\ u = 0 & \text{on } \partial\Omega, \end{cases} \tag{4.1}$$

G. Sweers (✉) · M. Erven
Mathematical Institute, University of Cologne, 50931 Cologne, Germany
e-mail: gsweers@math.uni-koeln.de

M. Erven
e-mail: merven@math.uni-koeln.de

J. Escher et al., (eds.), *Elliptic and Parabolic Equations*, Springer Proceedings in Mathematics & Statistics 119, DOI 10.1007/978-3-319-12547-3_4

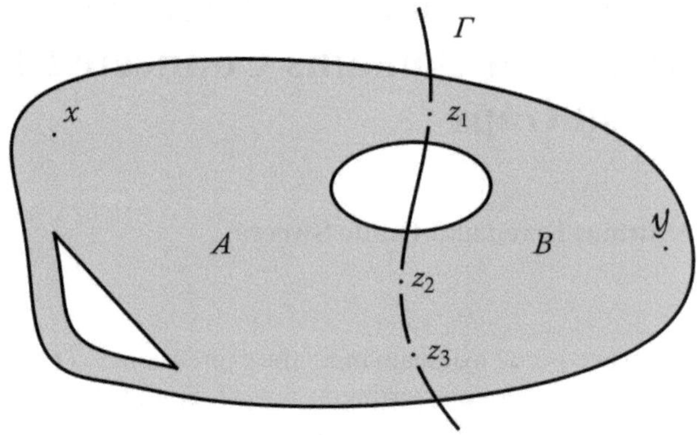

Fig. 4.1 We consider domains Ω_ℓ connected through small gaps, i.e., $\Omega_\ell = A \cup B \cup$ "gates," where A and B are disjoint domains and where the curve $\Gamma \supset (\partial A \cap \partial B)$ contains small gates, here around z_1, z_2, and z_3, that make Ω_ℓ connected

that is, $u(x) = \int_\Omega G_\Omega(x, y) f(y)\, dy$ solves (4.1). The $3G$-theorem of Cranston, Fabes and Zhao [7] states that for any bounded Lipschitz domain Ω in $\mathbb{R}^n$ there exists a constant C_Ω such that

$$\int_\Omega G_\Omega(x, z) \frac{G_\Omega(y, z)}{G_\Omega(y, x)} dz \leq C_\Omega.$$

Cranston and McConnell in [6] showed that in two dimensions there even exists a uniform constant c such that

$$\mathbb{E}_x^y(\tau_\Omega) \leq c\ |\Omega|.$$

See also [5]. Further estimates of $3G$-type can be found for example in [15, 19] and in the book by Chung and Zhao [3]. Zhao in [20, 21] was one of the first to derive sharp two-side estimates for the Green function.

Our interest in this bound came up since the inverse of this bound from above is also a bound for the parameter in noncooperative elliptic system below which the system remains positivity preserving. See [17] and [18].

It is expected, see [1], that the maximal value of $\mathbb{E}_x^y(\tau_\Omega)$, as a function of x and y, is attained at opposite boundary points. In the case of a disk, respectively a ball, indeed the result is true. See [11] and [9]. For other domains, the question still remains open. A partial result was proven in [15]. There a crucial condition is the assumption that the domain is simply connected. In this chapter, we consider another extreme, namely the case where the domain is multiply connected but only through small passages.

We consider the lifetime on such multiply connected domains as in Fig. 4.1 with connections through small passages that become arbitrary small. To be precise, let us fix the following class of domains. Recall that a domain in $\mathbb{R}^n$ is an open and connected subset of $\mathbb{R}^n$.

Definition 4.1 We call $\Omega \subset \mathbb{R}^2$ a finite Jordan-type domain, if Ω is bounded and if its boundary consists of finitely many nonintersecting Jordan curves.

The assumption for our initial domain Ω_1 and separating curve Γ are as follows:

Assumption 4.1 *Suppose the following holds:*

1. $\Omega_1 \subset \mathbb{R}^2$ *is a finite Jordan-type domain.*
2. $\Gamma = \gamma[0,T]$, *where* $\gamma : [0,T] \to \mathbb{R}^2$ *is an injective analytical curve parametrised by arclength with* $\gamma(0), \gamma(T) \notin \Omega_1$ *and such that* $\Omega_1 \backslash \Gamma$ *consists of exactly two disjoint domains, which we call A and B. The set* Γ *may be called the separating wall.*

The family of domains Ω_ℓ is as follows:

Definition 4.2 Let Ω_1 and Γ be as in Assumption 4.1 and let $k \in \mathbb{N}^+$. Suppose that $z_i = \gamma(t_i) \in \Gamma \cap \Omega_1$ with $i \in \{1, \ldots, k\}$. There exist k "gates in the wall" Γ defined by

$$\Gamma_{i,\ell} := \left\{\gamma(t); t \in [0,T] \text{ with } |t - t_i| < \tfrac{1}{2}\ell\right\} \text{ for } i \in \{1, \ldots, k\}$$

and we set

$$\Omega_\ell = \Omega_1 \backslash \left\{\gamma(t); t \in [0,T] \text{ with } |t - t_i| \geq \tfrac{1}{2}\ell \text{ for all } i \in \{1, \ldots, k\}\right\}.$$

We assume that $\ell_0 > 0$ is such that Γ_{i,ℓ_0}, $i \in \{1, \ldots, k\}$, are mutually disjoint and lie in Ω_1. Note that the gaps have arclength ℓ.

The pointwise version of our main result is as follows:

Theorem 4.1 *Let* Ω_ℓ *be as in Definition 4.2. Then for all* $x \in A$ *and* $y \in B$*:*

$$\lim_{\ell \downarrow 0} \mathbb{E}_x^y\left(\tau_{\Omega_\ell}\right) = \sum_{j=1}^{k} \frac{K_A\left(x, z_j\right) K_B\left(y, z_j\right)}{\sum_{i=1}^{k} K_A\left(x, z_i\right) K_B\left(y, z_i\right)} \left(\mathbb{E}_x^{z_j}\left(\tau_A\right) + \mathbb{E}_{z_j}^y\left(\tau_B\right)\right), \tag{4.2}$$

where K_A, *respectively* K_B *are the Poisson kernels for A respectively B.*

This theorem is a direct consequence of Theorem 4.2 part 2.

Remark 4.1 The Poisson kernel $K_A : A \times \partial A \to [0, \infty]$ is such that

$$u(x) = \int_{\partial A} K_A(x, w)\, \varphi(w)\, d\sigma_w \tag{4.3}$$

solves

$$\begin{cases} -\Delta u = 0 & \text{in } A, \\ u = \varphi & \text{on } \partial A. \end{cases} \tag{4.4}$$

The formula in (4.2) will allow us to show in a consecutive chapter, that the maximal expectation, namely $\sup_{x,y \in \Omega} \mathbb{E}_x^y(\tau_\Omega)$ is not necessarily reached at boundary points. Indeed, one would expect that for such a maximum the points x and y should be far away and this might hint at opposite boundary points. A special domain connected through small passages will contradict this expectation. The subject of this chapter has been addressed in the PhD thesis [14].

In Sect. 2, we will recall the connection between conditioned Brownian motion, heat kernel, and Green function. This part may be skipped by the specialist. In Sect. 3, we will formulate the precise statement of our main result (4.2). From Sect. 4 onwards, we will prove that result for smooth domains and show how we may extend the result to less smooth domains. In the appendix, we will collect some results based on conformal mappings.

2 The Link Between Brownian Motion and Kernel Functions

For the nonspecialist, we recall some relations mentioned in the title of this section that originate from [12]. See also [13] or [3]. A explanation of these relations can also be found in [18].

Standard Brownian Motion

For a domain in $\mathbb{R}^n$, one may consider the lifetime τ_Ω of Brownian motion that starts at $x \in \Omega$ and that is killed when reaching the boundary. The probability that this Brownian motion $\{X(t)\}_{t\in\mathbb{R}^+}$ is still alive at time t and inside $D \subset \Omega$ is given by the following formula

$$\mathbb{P}_x [X(t) \in D] = \int_{z\in D} p_\Omega (t,z,x)\, dz. \tag{4.5}$$

The expression $p_\Omega(t,z,x)$ denotes the heat kernel under zero Dirichlet boundary conditions, that is, the fundamental solution that satisfies for $x \in \Omega$ the following initial value problem:

$$\begin{cases} (\partial_t - \Delta_z)\, p_\Omega (t,z,x) = 0 & \text{for } (z,t) \in \Omega \times \mathbb{R}^+, \\ \lim_{t\downarrow 0} p_\Omega (t,\cdot,x) = \delta_x & \text{as distribution on } \Omega, \\ p_\Omega (t,z,x) = 0 & \text{for } (z,t) \in \partial\Omega \times \mathbb{R}^+, \end{cases}$$

where δ_x is the Dirac-δ-distribution in x.

Integrating the heat kernel over time t in $\mathbb{R}^+$ yields the Green function G_Ω: $\overline{\Omega} \times \overline{\Omega} \to [0,\infty]$ for

$$\begin{cases} -\Delta u = f & \text{in } \Omega, \\ u = 0 & \text{on } \partial\Omega. \end{cases} \tag{4.6}$$

Indeed, we find formally for

$$G_\Omega (x,z) = \int_0^\infty p_\Omega (t,z,x)\, dt \tag{4.7}$$

that

$$-\Delta_x G_\Omega(x,z) = \int_0^\infty -\Delta_x p_\Omega(t,z,x)\,dt$$

$$= \int_0^\infty -\partial_t p_\Omega(t,z,x)\,dt = \lim_{t\downarrow 0} p_\Omega(t,z,x) - \lim_{t\to\infty} p_\Omega(t,z,x) = \delta_x(z) - 0,$$

i.e. $u(x) = \int_\Omega G_\Omega(x,z) f(z)\,dz$ solves (4.6). Combining (4.5) and (4.7) gives us the expectation for the lifetime of Brownian motion that starts at $x \in \Omega$ and is killed when hitting the boundary:

$$\mathbb{E}_x(\tau_\Omega) = \int_0^\infty t\,\partial_t \mathbb{P}_x[X(t) \notin \Omega] = \int_0^\infty t\left(-\partial_t \int_{z\in\Omega} p_\Omega(t,z,x)\,dz\right)dt$$
$$= \int_0^\infty \int_{z\in\Omega} p_\Omega(t,z,x)\,dzdt = \int_{z\in\Omega} G_\Omega(x,z)\,dz.$$

Conditioned Brownian Motion

In a next step, which uses Doob's conditioned Brownian motion [12], one may consider only those paths that are conditioned to be killed on a certain part of the boundary. Indeed, taking $\Gamma \subset \partial\Omega$, the probability that the Brownian motion $X(t) \in D \subset \Omega$ is killed on Γ, is given by

$$\mathbb{P}_x^\Gamma[X(t) \in D] = \int_D p_\Omega(t,z,x)\,\frac{v_{\Omega,\Gamma}(z)}{v_{\Omega,\Gamma}(x)}dz.$$

Here, $v_{\Omega,\Gamma}$ is the harmonic function satisfying $v_{\Omega,\Gamma} = 1$ on Γ and $v_{\Omega,\Gamma} = 0$ on $\partial\Omega\backslash\Gamma$ (allow us to be sloppy where both boundary parts meet). With a similar reasoning as before one finds

$$\mathbb{E}_x^\Gamma(\tau_\Omega) = \int_\Omega G_\Omega(x,z)\,\frac{v_{\Omega,\Gamma}(z)}{v_{\Omega,\Gamma}(x)}dz. \tag{4.8}$$

Next one may look at conditioned Brownian motion, which lives on a domain that we get by removing a small ball $B_\varepsilon(y)$ around $y \in \Omega$, and consider the expectation as in (4.8):

$$\mathbb{E}_x^{\partial B_\varepsilon(y)}\left(\tau_{\Omega\backslash B_\varepsilon(y)}\right) = \int_{\Omega\backslash B_\varepsilon(y)} G_{\Omega\backslash B_\varepsilon(y)}(x,z)\,\frac{v_{\Omega\backslash B_\varepsilon(y),\partial B_\varepsilon(y)}(z)}{v_{\Omega\backslash B_\varepsilon(y),\partial B_\varepsilon(y)}(x)}dz.$$

One may show that for $n \geq 2$

$$\lim_{\varepsilon\downarrow 0} \frac{v_{\Omega\backslash B_\varepsilon(y),\partial B_\varepsilon(y)}(z)}{v_{\Omega\backslash B_\varepsilon(y),\partial B_\varepsilon(y)}(x)} = \frac{G_\Omega(y,z)}{G_\Omega(y,x)}, \tag{4.9}$$

$$\lim_{\varepsilon\downarrow 0} G_{\Omega\backslash B_\varepsilon(y)}(x,z) = G_\Omega(x,z) \tag{4.10}$$

holds pointwise and even uniformly for x and z away from y. Using (4.9) and (4.10) and the uniform integrability, one finds

$$\lim_{\varepsilon\downarrow 0} \mathbb{E}_x^{\partial B_\varepsilon(y)}\left(\tau_{\Omega\backslash B_\varepsilon(y)}\right) = \int_\Omega G_\Omega(x,z)\,\frac{G_\Omega(y,z)}{G_\Omega(y,x)}dz. \tag{4.11}$$

Let us recall the following estimate from ([3], Theorem 6.15) for any finite Jordan-type domain $\Omega \subset \mathbb{R}^2$: there exists a constant C_Ω such that

$$\frac{G_\Omega(x,z)\,G_\Omega(z,y)}{G_\Omega(x,y)} \le C_\Omega \max\left(1, \ln\left(\frac{1}{|x-z|}\right), \ln\left(\frac{1}{|y-z|}\right)\right).$$

Since the limit in (4.11) exists for quite general domains, we come to the following definition.

Definition 4.3 Let Ω be a finite Jordan-type domain in $\mathbb{R}^n$ with $n \ge 2$. Then we define the expectation of Brownian motion starting at $x \in \Omega$, that is to be killed on the boundary $\partial\Omega$ and is conditioned to converge to $y \in \Omega$, as follows:

$$\mathbb{E}_x^y(\tau_\Omega) := \int_\Omega G_\Omega(x,z)\,\frac{G_\Omega(y,z)}{G_\Omega(y,x)}\,dz. \tag{4.12}$$

See also [7] or [3].

Extending to the Boundary

For a point y at a smooth part of the boundary, with ν_y as outside normal direction, one finds for the Poisson kernel:

$$K_\Omega(z,y) = -\nu_y \cdot \nabla_y G_\Omega(y,z) = \lim_{t\downarrow 0} \frac{G_\Omega\left(y - t\,\nu_y, z\right)}{t}.$$

So one may extend the definition of $\mathbb{E}_x^y(\tau_\Omega)$ to y on smooth parts of the boundary. Indeed, at such a point it holds that

$$\lim_{t\downarrow 0} \frac{G_\Omega\left(y - t\,\nu_y, z\right)}{G_\Omega\left(y - t\,\nu_y, x\right)} = \frac{K_\Omega(z,y)}{K_\Omega(x,y)}$$

pointwise and even uniformly for x, z away from y. Since $G_\Omega(x,y) = G_\Omega(y,x)$, one finds that $\mathbb{E}_x^y(\tau_\Omega) = \mathbb{E}_y^x(\tau_\Omega)$ holds. So we consider

$$\mathbb{E}_x^y(\tau_\Omega) = \begin{cases} \displaystyle\int_\Omega \frac{G_\Omega(x,z)\,G_\Omega(z,y)}{G_\Omega(x,y)}\,dz & \text{for } x, y \in \Omega, \\ \displaystyle\int_\Omega \frac{G_\Omega(x,z)\,K_\Omega(z,y)}{K_\Omega(x,y)}\,dz & \text{for } x \in \Omega, y \in \partial\Omega. \end{cases} \tag{4.13}$$

By symmetry, one finds a formula for $x \in \partial\Omega$, $y \in \Omega$. If both x and y lie on smooth parts of the boundary, then also the limit

$$\mathbb{E}_x^y(\tau_\Omega) = \lim_{\Omega \ni w \to y} \mathbb{E}_x^w(\tau_\Omega)$$

is well defined. This even holds when x equals y. Moreover, for all smooth (bounded) domains Ω, it holds that $\lim_{y\to x} \mathbb{E}_x^y(\tau_\Omega) = 0$ for all $x \in \bar{\Omega}$. Optimal rates for this convergence can be found in [2].

Conformal Mappings
In two dimensions the Riemann mapping is a helpful tool. A related theorem by Caratheodory (see [16]) states that for all domains Ω_1 and Ω_2, for which the boundary $\partial\Omega_1$ respectively $\partial\Omega_2$ is a Jordan curve, there exists an extended biconformal mapping $h : \overline{\Omega_1} \to \overline{\Omega_2}$. A mapping h is called biconformal, if it is conformal on the interior and a homeomorphism from $\overline{\Omega_1} \to \overline{\Omega_2}$.

A conformal mapping h in two dimensions has the following property:

$$\Delta (u \circ h) = \left|h'\right|^2 (\Delta u) \circ h,$$

where h' is the complex derivative when we identify $\mathbb{R}^2$ with $\mathbb{C}$. Since in two dimensions the Jacobian also equals $\left|h'\right|^2$, one finds the following relation between G_{Ω_1} and G_{Ω_2}, i.e., the Green functions on Ω_1, respectively Ω_2:

$$G_{\Omega_1}(x, y) = G_{\Omega_2}(h(x), h(y)) \text{ for all } x, y \in \Omega_1. \tag{4.14}$$

If Ω_1 is simply connected, one may choose $\Omega_2 = B_1(0)$, the unit disk, in order to use the explicitly known Green function on $B_1(0)$. Or one may use a mapping to a smooth domain and use estimates for Green functions on smooth domains. In [3], Sect. 6.3 or [8], one finds a generalization of the Caratheodory theorem for finite Jordan-type domains Ω_1 and $\Omega_2 \in C^\infty$. By the way, assuming some smoothness one finds for the Poisson kernel (see Theorem 4.3) that

$$K_{\Omega_1}(x, w) = \left|h'(w)\right| K_{\Omega_2}(h(x), h(w)) \text{ for all } x \in \Omega_1, w \in \partial\Omega_1. \tag{4.15}$$

3 Main Results

We assume that the domains Ω_ℓ are as in Definition 4.2. However, in order to avoid some additional technicalities, we will assume that x and y stay away both from non-Dini-smooth boundary points and from the gaps. Let us set

$$\begin{aligned} \partial A^* &= \text{the non-Dini-smooth part of } \partial A, \\ \partial B^* &= \text{the non-Dini-smooth part of } \partial B. \end{aligned}$$

We define the following neighborhoods of size $\rho > 0$:

$$\partial A^*_\rho = \partial A^* + B_\rho(0), \partial B^*_\rho = \partial B^* + B_\rho(0) \tag{4.16}$$

and

$$U_\rho = \{z_1, \ldots, z_k\} + B_\rho(0) = \bigcup_{i=1}^{k} B_\rho(z_i) \tag{4.17}$$

Our main result is in the next theorem.

Theorem 4.2 *Suppose Assumption* 4.1 *holds true an let* Ω_ℓ *be as in Definition* 4.2. *Let* $\rho > 0$ *and assume that* $x, y \notin U_\rho \cup \partial A^*_\rho \cup \partial B^*_\rho$. *Then there exist* $C_{\mathrm{I}} = C(\Omega_1, \Gamma, \rho) > 0$ *such that the following holds:*

1. *If* $x, y \in A$, *then for all* $\ell \in (0, \ell_0)$*:*

$$\left|\mathbb{E}^y_x\left(\tau_{\Omega_\ell}\right) - \mathbb{E}^y_x\left(\tau_A\right)\right| \leq C_{\mathrm{I}}\, \ell\, |\log \ell|\,. \tag{4.18}$$

2. *If* $x \in A$ *and* $y \in B$, *then for all* $\ell \in (0, \ell_0)$*:*

$$\left|\mathbb{E}^y_x\left(\tau_{\Omega_\ell}\right) - \sum_{j=1}^{k} \theta_j(x, y)\left(\mathbb{E}^{z_j}_x(\tau_A) + \mathbb{E}^y_{z_j}(\tau_B)\right)\right| \leq C_{\mathrm{I}}\, \ell^{\frac{1}{3}}\, |\log \ell|\,, \tag{4.19}$$

with

$$\theta_j(x, y) = \frac{K_A\left(x, z_j\right) K_B\left(y, z_j\right)}{\sum_{m=1}^{k} K_A(x, z_m)\, K_B(y, z_m)}.$$

Here, K_A *is the Poisson kernel defined in (4.3).*

Remark 4.2 We expect that the rate of convergence in (4.19) can be improved to $\ell\, |\log \ell|$. To prove such an estimate, one would have to do a bootstrapping argument for the asymptotic estimates, which would involve additional technical steps.

Remark 4.3 The condition that x and y are away from non-Dini-smooth boundary parts is almost certainly superfluous. However, the estimates for the Green functions near non-Dini-smooth boundary parts are much more involved, and the consequent estimates would become rather cumbersome. See also Remarks 4.7 and 4.8. The alternative, transforming conformally to a smooth domain, would turn all integrals in weighted integrals, which is not much better.

Remark 4.4 Generically, the intersection of Γ and $\partial\Omega$, even when both are smooth, would introduce some corners. For isolated corners there exists a simple conformal smoothening by an ε-perturbation, namely by using mappings like $z \mapsto \left(z^{\pi/\alpha} + \varepsilon\right)^{\alpha/\pi}$, where α is the angle of the corner measured from inside.

Proof The first item will be proved in Sect. 5. The second estimate follows from Theorem 4.6 in Sect. 6. □

4 Estimates on Green Functions and Poisson Kernels

In the proof of the convergence results, we need estimates for the Green function, the Poisson kernel and their derivatives on A and B. To simplify arguments, we will assume that the subdomains A and B have a Dini-smooth boundary which would mean that instead of Fig. 4.1 the domain will look as in Fig. 4.2. The modifications, necessary when $\partial A \backslash \Gamma$ is Jordan but not Dini-smooth, are explained in Remark 4.7.

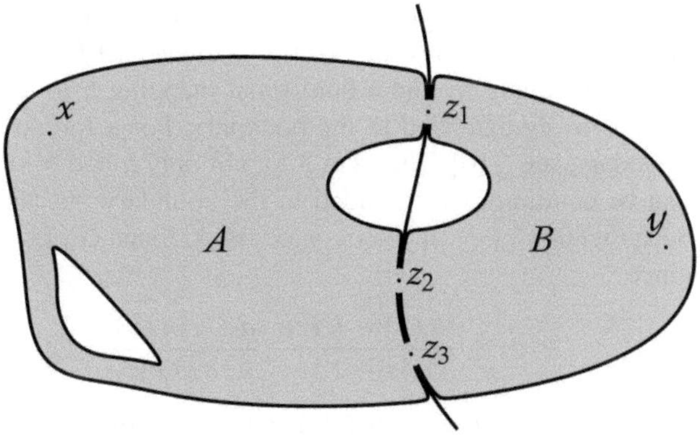

Fig. 4.2 Assuming that the subdomains A and B are Dini-smooth

4.1 Recalling and Adjusting Known Estimates

Estimates for Green function and Poisson kernel for a domain A are determined by the distance to the singularity and the distance from the boundary d_A. This last distance is defined by

$$d_A(x) := \inf\left\{\left|x - x^*\right|; x^* \in \partial A\right\}.$$

Theorem 4.3 *Suppose that A is a Dini-smooth finite Jordan-type domain. Then there exist $C_{\mathrm{II}.i} = C_i(A) > 0$, such that the following holds.*

- *For all $x, z \in A$:*

$$\frac{1}{4\pi}\log\left(1 + C_{\mathrm{II}.1}\frac{d_A(x)\, d_A(z)}{|x - z|^2}\right) \le G_A(x, z) \le \frac{1}{4\pi}\log\left(1 + C_{\mathrm{II}.2}\frac{d_A(x)\, d_A(z)}{|x - z|^2}\right). \tag{4.20}$$

- *For all $x \in A$, $z \in \partial A$*

$$C_{\mathrm{II}.1}\frac{d_A(x)}{|x - z|^2} \le K_A(x, z) \le C_{\mathrm{II}.2}\frac{d_A(x)}{|x - z|^2}. \tag{4.21}$$

Remark 4.5 Since $d_A(x) \le |x - z|$ one finds for all $x \in A$, $z \in \partial A$ that

$$K_A(x, z) \le C_{\mathrm{II}.2}\, |x - z|^{-1}. \tag{4.22}$$

Remark 4.6 For the Green function, respectively Poisson kernel, both on the unit disk $B_1(0)$, one may use the explicit formulas:

$$G_{B_1(0)}(x, z) = \frac{1}{4\pi}\log\left(1 + \frac{(1-|x|^2)(1-|z|^2)}{|x-z|^2}\right) \text{ for } x, z \in B_1(0), \tag{4.23}$$

$$K_{B_1(0)}(x, z) = \frac{1}{2\pi}\frac{1 - |x|^2}{|x - z|^2} \text{ for } x \in B_1(0) \text{ and } z \in \partial B_1(0). \tag{4.24}$$

Proof The estimate (4.20) on G_A can be found in ([17], Lemma 4.2) under extra smoothness conditions. For (simply connected) Jordan domains, one may use Riemann's mapping to find a conformal mapping h to the disk $B_1(0)$ and which can be injectively extended to the boundary. For a Jordan domain with a Dini-smooth boundary, see ([16], Theorem 3.5), not only h and h' (the complex derivative on $\mathbb{C}$) can be continuously extended to the boundary but also there exist even constants such that $0 < c_1 \leq |h'| \leq c_2 < \infty$ on $\bar{A}$. Since $G_A(x,z) = G_{h(A)}(h(x), h(z))$ and since

$$c_1 \leq \frac{|h(x) - h(z)|}{|x - z|}, \frac{d_{h(A)}(h(x))}{d_A(x)} \leq c_2 \text{ for all } x, z \in A,$$

the estimate for $G_A(x,z)$ follows from the one on the disk. Almost similarly one finds an estimate for the Poisson kernel. Indeed, since $K_A(x,z) = -\nu_z \cdot \nabla_z G_\Omega(x,z)$ and

$$\nabla_z G_A(x,z) = J_h(z)\ \nabla_y G_{h(A)}(h(x), y)_{|y=h(z)},$$
$$\nu_{h(z)} = |J_h(z)\ \nu_z|^{-1}\ J_h(z)\ \nu_z$$

with

$$J_h(\cdot) = \begin{pmatrix} \partial_1 h_1 & \partial_1 h_2 \\ \partial_2 h_1 & \partial_2 h_2 \end{pmatrix} \text{ and } \det(J_h(\cdot)) = |h'|^2,$$

one finds that

$$K_A(x,z) = -\nu_z \cdot \nabla_z G_{h(A)}(h(x), h(z)) =$$
$$= -|h'(z)|\ \nu_{h(z)} \cdot \nabla_y G_{h(A)}(h(x), y)_{|y=h(z)} = |h'(z)|\ K_{h(A)}(h(x), h(z)).$$

For multiply (but finitely) connected domains, one may compare with simply connected domains; as is done in ([17], Lemma 4.2) to find these estimates.

Let us recall two estimates from [17] for the Green function.

Corollary 4.1 *Suppose that the domain A has a Dini-smooth boundary. Then there are* $C_{4.1.i} > 0$ *such that*

$$G_A(x,z) \geq C_{\text{III}.1}\, d_A(x)\, d_A(z) \tag{4.25}$$

and

$$G_A(x,z) \leq C_{\text{III}.2} \frac{d_A(x)}{|x - z|}. \tag{4.26}$$

Remark 4.7 Assuming that A is a finite Jordan-type domain, but without additional regularity of the boundary, one may use the conformal mapping h to a ∞-smooth domain $h(A)$ and (4.14) to find estimates like in (4.20), (4.21), (4.25), and (4.26) using $\widetilde{d_A}(\cdot)$ and $\widetilde{|\cdot - \cdot|}$ instead of $d_A(\cdot)$ and $|\cdot - \cdot|$. These functions are defined by

$$\widetilde{d_A}(x) = d_{h(A)}(h(x)),$$
$$\widetilde{|x-y|} = |h(x) - h(y)|.$$

Although $\widetilde{d_A}(\cdot)$ and $\widetilde{|\cdot - \cdot|}$ depend on h, two different h give equivalent metrics.

Remark 4.8 If part of the boundary is not Dini-smooth, say $\partial A \backslash \Gamma_A$, then away from that boundary part the metrics without and with wiggle are equivalent. That is, for every compact set $A^\diamond$ of $A \cup \Gamma_A$, one finds that there exist constants $C_{A^\diamond,i} > 0$ such that:

1. For $x \in A^\diamond$, it holds that $C_{A^\diamond,1}\, d_A(x) \le \widetilde{d_A}(x) \le C_{A^\diamond,2}\, d_A(x)$.
2. For $x \in A^\diamond$ or $y \in A^\diamond$, it holds that $C_{A^\diamond,1}\, |x-y| \le \widetilde{|x-y|} \le C_{A^\diamond,2}\, |x-y|$.

Let us sketch the proof for a (simply connected) Jordan domain. One takes an ε-neighborhood of the non-smooth boundary part

$$\mathcal{N}_\varepsilon = (\partial A \backslash \Gamma_A) + B_\varepsilon(0),$$

such that $A^\diamond \cap \mathcal{N}_\varepsilon = \emptyset$, and one constructs a ∞-smooth Jordan domain D with $h(A \backslash \mathcal{N}_\varepsilon) \subset D \subset h(A)$ and $h(\overline{\partial A \backslash \Gamma_A}) \cap \overline{D} = \emptyset$. Letting $g : B_1(0) \to D$ be a smooth biconformal mapping, a theorem from Kellogg and Warschawski (see [16, Theorem 3.5]) allows one to extend the mapping $h^{-1} \circ g : B_1(0) \to h^{-1}(D)$ to $\overline{B_1(0)}$ with $(h^{-1} \circ g)'$ continuous on $\overline{B_1(0)}$ and nonzero. Since g' is continuous and nonzero on $\overline{B_1(0)}$ the function $(h^{-1})'$ is continuous and nonzero on $\overline{D}$. Then also $h_{|h^{-1}(D) \cap A^\diamond}$ is a $C^1(A^\diamond)$ function with nonzero derivative.

4.2 Estimates for Higher-Order Green Function Derivatives

In [10], Theorem 12(1)(a), estimates were presented for higher-order derivatives of general Green functions in arbitrary dimensions. For technical reasons, one assumed that the boundary satisfied a strong regularity. In the present case, due to the regularity for the conformal mapping h, one finds that the following estimates are valid for $x \in A$ and z on any compact subset of $A \cup \Gamma_A$, where $\Gamma_A \subset \partial A$ is sufficiently smooth.

Theorem 4.4 *Let A be a finite Jordan-type domain and let $m = (m_1, m_2) \in \mathbb{N}^2 \backslash (0,0)$. Suppose that $\Gamma_A \subset \partial A$ is relatively open in ∂A and satisfies $\Gamma_A \in C^{m_1+m_2+1}$. Then for any compact subset $A^\diamond$ of $A \cup \Gamma_A$ there is a $C_{IV} := C(A^\diamond) > 0$ such that*

$$\left|D_z^m G_A(x,z)\right| \le C_{IV}\, |x-z|^{-|m|} \min\left\{1, \frac{d_A(x)}{|x-z|}\right\} \tag{4.27}$$

for all $x \in A$ and $z \in A^\diamond$.

Proof As in the proof of Theorem 4.3, it is sufficient to consider simply connected A and the conformal mapping h from A to $B_1(0)$. By [10], Theorem 12(1)(a), the estimate in (4.27) holds true for $G_{B_1(0)}(\cdot,\cdot)$ and with the bounds for the derivatives of h on $A^\diamond$ these estimates transfer to $G_A(\cdot,\cdot)$.

4.3 Asymptotic Estimates for G_{Ω_ℓ}

A direct consequence of Theorem 4.4 is a Lipschitz-type estimate for the Poisson kernel near the gaps. Remember that the separating curve Γ is supposed to be analytic.

Corollary 4.2 *Suppose Assumption* 4.1 *holds true and let Ω_ℓ be as in Definition* 4.2. *Let $\rho > 0$ and fix z_j. Then there is a $C_V := C(A,\rho) > 0$ such that for all $\ell \in (0,\rho)$, $x \in A$, with $|x - z_j| > \rho$, and $z \in \Gamma_{j,\ell}$ one finds:*

$$\left|K_A(x,z) - K_A(x,z_j)\right| \le C_V \min\left\{(\rho-\ell)^{-2}, d_A(x)(\rho-\ell)^{-3}\right\}\left|z - z_j\right|.$$

Proof By the mean value theorem, there is a θ_t between t_j and t such that

$$\begin{aligned} K_A(x,z) &= K_A(x,z_j) + (t - t_j)\left(\tfrac{d}{ds}K_A(x,\gamma(s))\right)_{|s=\theta_t} \\ &= K_A(x,z_j) - (t-t_j)\,\partial_{\tau_y}\partial_{\nu_y}G_A(x,y)_{|y=\gamma(\theta_t)}. \end{aligned}$$

By (4.27) one finds

$$\left|(t-t_j)\,\partial_{\tau_y}\partial_{\nu_y}G_A(x,y)_{|y=\gamma(\theta_t)}\right| \le c\,\left|z - z_j\right|\,|x-\gamma(\theta_t)|^{-2}\min\left\{1, \frac{d_A(x)}{|x-\gamma(\theta_t)|}\right\}$$

and the result follows, since $|x-\gamma(\theta_t)| \ge |x - z_j| - |z_j - \gamma(\theta_t)| \ge \rho - \ell$. □

Before being able to show the limit behavior of $\mathbb{E}_x^y(\tau_{\Omega_\ell})$ for $l \to 0$, we need some uniform asymptotic results for G_{Ω_ℓ}. In doing so, we need to assume additionally smoothness.

Theorem 4.5 *Suppose Assumption 4.1 is satisfied and that ∂A and ∂B are Dini-smooth. Let Ω_ℓ be as in Definition 4.2. Let $\rho, \rho_1 > 0$ and $0 \le \alpha < 1$. Then there is $C_{VI} := C(\Omega, A, \rho, \rho_1) > 0$ and $\ell_1 > 0$, such that the following holds for all $\ell \in (0,\ell_1)$ and $x \in A\backslash U_\rho$ (see* (4.17)*):*

1. *If $z \in A\backslash\{x\}$, then*

$$0 \le G_{\Omega_\ell}(x,z) - G_A(x,z) \le C_{VI}\, d_A(x)\, \ell\, |\log \ell|.$$

2. *If $z \in A\backslash U_\rho$ and $z \ne x$, then*

$$0 \le G_{\Omega_\ell}(x,z) - G_A(x,z) \le C_{VI}\, d_A(x) \min\{d_A(z), \ell^\alpha\}\, \ell^{2-2\alpha}\, |\log \ell|.$$

3. *If* $z \in B$*, then*

$$0 \leq G_{\Omega_\ell}(x,z) \leq C_{\mathrm{VI}}\, d_A(x)\ \ell\, |\log \ell|\,.$$

4. *If* $z \in B$ *with* $\left|z - z_j\right| \geq \rho_1\, \ell^\alpha$ *for* $j = 1, \ldots, k$*, then*

$$0 \leq G_{\Omega_\ell}(x,z) \leq C_{\mathrm{VI}}\, d_A(x) \min\{d_B(z), \ell^\alpha\}\, \ell^{2-2\alpha}\, |\log \ell|\,.$$

Remark 4.9 $G_A(x,\cdot)$ is zero everywhere on ∂A including the $\Gamma_{j,\ell}$, while $G_{\Omega_\ell}(x,\cdot) > 0$ on the gaps $\Gamma_{j,\ell}$. The smaller ℓ is, the less nonzero boundary there is. Hence, G_{Ω_ℓ} tends to G_A, but we only get uniform convergence, if at least one point, say x, stays away from the gaps. The estimates in 2 and 4 sharpen the estimates of 1 and 3.

Proof Green functions for $-\Delta$ are nonnegative. Moreover, as $A \subset \Omega_\ell$, the maximum principle implies that $0 \leq G_A(x,z) \leq G_{\Omega_\ell}(x,z)$ for all $x, z \in A$. Hence, we only have to show the upper bounds of these inequalities.

Item 1 Let $z \in A$ be fixed and define the function $u_z : A \to \mathbb{R}$ by

$$u_z(x) := G_{\Omega_\ell}(x,z) - G_A(x,z)\,.$$

As the singularity at $x = z$ cancels out, u_z is harmonic in A and satisfies the boundary condition

$$u_z(x) = \begin{cases} G_{\Omega_\ell}(x,z) & \text{if } x \in \Gamma_{j,\ell} \text{ for some } j \in \{1,\ldots,k\}, \\ 0 & \text{elsewhere on } \partial A. \end{cases}$$

Hence, u_z can be represented by means of the Poisson kernel as

$$u_z(x) = \int_{\bigcup_{j=1}^k \Gamma_{j,\ell}} K_A(x,w)\, G_{\Omega_\ell}(w,z)\, d\sigma_w. \tag{4.28}$$

As Ω_ℓ is bounded, there is some $M > 0$ such that $\Omega_\ell \subset B_M(w)$ for all $w \in \bigcup_{j=1}^k \Gamma_{j,\ell}$ (and all $l \in \left(0, \frac{1}{2}\rho\right)$). We use the estimate on the Poisson kernel of (4.21) and the fact that $G_{\Omega_\ell}(w,z) \leq G_{B_M(w)}(w,z) = \frac{1}{2\pi}\log\left(\frac{M}{|w-z|}\right)$ to obtain that there is a $c_1 > 0$ such that

$$u_z(x) \leq c_1 \sum_{j=1}^k \int_{\Gamma_{j,\ell}} \frac{d_A(x)}{|x-w|^2} \log\left(\frac{M}{|w-z|}\right) d\sigma_w.$$

We make two observations. First, for each $z \in A$ and $j \in \{1,\ldots,k\}$, there is a $\tilde{z} \in \Gamma_{j,\ell}$ such that $|z - \tilde{z}| = \min\left\{|z - v|\,; v \in \Gamma_{j,\ell}\right\}$. Hence, $|w - \tilde{z}| \leq |w - z| + |z - \tilde{z}| \leq 2\,|w - z|$ for all $w \in \Gamma_{j,\ell}$. Secondly, for $w \in \Gamma_{j,\ell}$, $|x - w| \geq \left|x - z_j\right| - \left|z_j - w\right| \geq \rho - \frac{1}{4}\rho = \frac{3}{4}\rho$. Both put together yield

$$u_z(x) \le c_1 d_A(x) \left(\frac{4}{3\rho}\right)^2 \sum_{j=1}^{k} \int_{\Gamma_{j,\ell}} \log\left(\frac{2M}{|w-\tilde{z}|}\right) d\sigma_w$$

$$\le c_2 d_A(x) \int_{-\frac{\ell}{2}}^{\frac{\ell}{2}} (1-\log s)\, ds \le c_3 d_A(x)\, \ell\, |\log \ell|$$

for appropriate constants $c_2, c_3 > 0$ and small $\ell > 0$. In the last line, we used that $\ell\,|\log \ell|$ dominates ℓ for small $\ell > 0$.

Item 2 We look again at the difference of the Green functions, but this time we fix $x \in A$ and define

$$u_x(z) := G_{\Omega_\ell}(x,z) - G_A(x,z).$$

As also $z \mapsto u_x(z)$ is harmonic in A, we can represent it with help of the Poisson kernel. Then we use the Poisson kernel estimate of (4.21) and the result of case 1, which also holds, by continuity, for z being a boundary point, and find for some $c_1 > 0$, that

$$u_x(z) = \sum_{j=1}^{k} \int_{\Gamma_{j,\ell}} K_A(z,w)\, G_{\Omega_\ell}(x,w)\, d\sigma_w \le c_1 \sum_{j=1}^{k} \int_{\Gamma_{j,\ell}} \frac{d_A(z)}{|z-w|^2} d_A(x)\, \ell\, |\log \ell|\, d\sigma_w.$$

We give two upper bounds for the integrand. First, $|z-w| \ge |z-z_j| - |z_j - w| \ge \rho_1 \ell^\alpha - \frac{1}{2}\ell = \rho_1 \ell^\alpha \left(1 - \frac{1}{2\rho_1}\ell^{1-\alpha}\right)$. For small $\ell > 0$, this is greater than $\frac{1}{2}\rho_1 \ell^\alpha$. Secondly, $d_A(z) \le |z-w|$ by definition and hence $d_A(z)/|z-w| \le 1$. Summing up, we get that

$$\frac{d_A(z)}{|z-w|^2} \le \min\left\{4\rho_1^{-2} \frac{d_A(z)}{\ell^{2\alpha}}, 2\rho_1^{-1} \frac{1}{\ell^\alpha}\right\},$$

which yields for some $c_2 > 0$, that

$$u_x(z) \le c_2 d_A(x) \min\{d_A(z), \ell^\alpha\}\, \ell^{2-2\alpha}\, |\log \ell|.$$

Item 3 Fix $x \in A$. The function $z \mapsto G_{\Omega_\ell}(x,z)$ is harmonic in B and therefore attains its maximum at the boundary of B. In fact, it is attained at the gaps $\Gamma_{j,\ell}$, as $G_{\Omega_\ell}(x,.) = 0$ on $\partial B \cap \partial \Omega_\ell$. By continuity, the estimate of 1 also holds for $z \in \bigcup_{j=1}^{k} \Gamma_{j,\ell}$, which gives the stated upper bound.

Item 4 In analogy to the proof of case 2, we use the Poisson kernel representation, this time on B. We write

$$G_{\Omega_\ell}(x,z) = \sum_{j=1}^{k} \int_{\Gamma_{j,\ell}} K_B(z,w)\, G_{\Omega_\ell}(x,w)\, d\sigma_w$$

and apply the estimate on the Poisson kernel of (4.21) and the upper bound on G_{Ω_ℓ} of case 1 as before. □

5 The Case that Both x and y Lie in A

Proof of Part 1 of Theorem 4.2. We express both lifetimes with help of the 3G formula and get

$$\left|\mathbb{E}_x^y\left(\tau_{\Omega_\ell}\right) - \mathbb{E}_x^y\left(\tau_A\right)\right| = \left|\int_{\Omega_\ell} \frac{G_{\Omega_\ell}(x,z)\,G_{\Omega_\ell}(z,y)}{G_{\Omega_\ell}(x,y)}dz - \int_A \frac{G_A(x,z)\,G_A(z,y)}{G_A(x,y)}dz\right|$$

$$\leq \underbrace{\left|\int_A \left(\frac{G_{\Omega_\ell}(x,z)\,G_{\Omega_\ell}(z,y)}{G_{\Omega_\ell}(x,y)} - \frac{G_A(x,z)\,G_A(z,y)}{G_A(x,y)}\right)dz\right|}_{=:I_A} + \underbrace{\left|\int_B \frac{G_{\Omega_\ell}(x,z)\,G_{\Omega_\ell}(z,y)}{G_{\Omega_\ell}(x,y)}dz\right|}_{=:I_B}.$$

We show that both integrals get small and start with I_B. For the enumerators in the integrand, we use the estimate on G_{Ω_ℓ} of Theorem 4.5–3. $G_{\Omega_\ell}(x,y)$ in the denominator is greater than $G_A(x,y)$, which is bounded from below by $C_{\mathrm{III}.1}\, d_A(y)$ by (4.25). Hence, there is some $c_1 > 0$ such that

$$I_B \leq c_1 \int_B \frac{d_A(x)\,\ell\,|\log \ell|\, d_A(y)\,\ell\,|\log \ell|}{d_A(x)\, d_A(y)}dz \leq c_1 \operatorname{Vol}(B)\,\ell^2\,|\log \ell|^2\,.$$

Next, we rewrite the integrand in I_A:

$$\frac{G_{\Omega_\ell}(x,z)\,G_{\Omega_\ell}(z,y)}{G_{\Omega_\ell}(x,y)} - \frac{G_A(x,z)\,G_A(z,y)}{G_A(x,y)}$$

$$= \underbrace{\tfrac{\left(G_{\Omega_\ell}(x,z)-G_A(x,z)\right)G_{\Omega_\ell}(z,y)}{G_{\Omega_\ell}(x,y)}}_{=:J_1} + \underbrace{\tfrac{G_A(x,z)\left(G_{\Omega_\ell}(z,y)-G_A(z,y)\right)}{G_{\Omega_\ell}(x,y)}}_{=:J_2} + \underbrace{\tfrac{G_A(x,z)G_A(z,y)\left(G_A(x,y)-G_{\Omega_\ell}(x,y)\right)}{G_{\Omega_\ell}(x,y)G_A(x,y)}}_{=:J_3}.$$

The Green function G_{Ω_ℓ} can be approximated by G_A with the help of Theorem 4.5–1, and for the latter, we find lower and upper bounds, respectively, in (4.25) and (4.26). Hence, there are constants $c_2, \dots, c_6 > 0$ such that

$$|J_1| \leq c_2\, \frac{d_A(x)\,\ell\,|\log \ell|\,(G_A(z,y) + d_A(y)\,\ell\,|\log \ell|)}{G_A(x,y)}$$

$$\leq c_2\, \frac{d_A(x)\left(c_3 \frac{d_A(y)}{|y-z|} + d_A(y)\,\ell\,|\log \ell|\right)}{c_4 d_A(x)\, d_A(y)}\ell\,|\log \ell|$$

$$\leq c_5\, (1 + |y-z|\,\ell\,|\log \ell|)\, \frac{1}{|y-z|}\ell\,|\log \ell| \leq c_6\, \frac{1}{|y-z|}\ell\,|\log \ell|$$

for small $\ell > 0$, where we have neglected the term $|y-z|\,\ell\,|\log \ell|$ in the last line, as it is bounded from above by a constant times something that gets small for $\ell \to 0$.

J_2 can be treated by similar estimates to obtain

$$|J_2| \le c_7 \frac{1}{|x-z|} \ell \, |\log \ell|$$

for small $\ell > 0$ and some $c_7 > 0$. For J_3, we use Theorem 4.5–2 instead of Theorem 4.5–1 and get

$$|J_3| \le c_8 \frac{G_A(x,z)\, G_A(z,y)}{G_A(x,y)} \ell^2 \, |\log \ell| \, .$$

Finally, we integrate over z. The remaining upper bounds for J_1, J_2, and J_3 are integrable over z, and these integrals in return have upper bounds independent of x and y, as A is bounded and $\mathbb{E}_x^y(\tau_A) \le c \operatorname{Vol}(A)$ (see [6]). We thus get that there is some constant $c_9 > 0$ such that, for all small $\ell > 0$,

$$I_A \le \int_A (|J_1| + |J_2| + |J_3|) \, dz \le c_9 \, l \, |\log l| \, .$$

6 The Main Proof: x Lies in A and y in B

This section establishes the proof of the second part of Theorem 4.2.

6.1 *The Halfway Statement*

Now we turn to the second case, where start and endpoint lie in different subdomains. The next theorem estimates the expected time spent in A. Still U_ρ is as in (4.17).

Theorem 4.6 *Suppose that Assumption 4.1 is satisfied and that ∂A and ∂B are Dini-smooth. Let Ω_ℓ be as in Definition 4.2 and let $\rho > 0$. Then there is $C_{\mathrm{VII}} := C(\Omega, \Gamma, \rho) > 0$ such that the following holds. If $x \in A \backslash U_\rho$ and $y \in B \backslash U_\rho$ with $|y - z_j| > \rho$ for $j = 1, \dots, k$, then*

$$\left| \int_A \frac{G_{\Omega_\ell}(x,z)\, G_{\Omega_\ell}(z,y)}{G_{\Omega_\ell}(x,y)} \mathrm{d}z - \sum_{j=1}^{k} \frac{K_A(x,z_j)\, K_B(y,z_j)}{\sum_{m=1}^{k} K_A(x,z_m)\, K_B(y,z_m)} \mathbb{E}_x^{z_j}(\tau_A) \right|$$
$$\le C_{\mathrm{VII}} \, \ell^{\frac{1}{3}} \, |\log \ell| \qquad (4.29)$$

for all $\ell \in (0, \ell_0)$.

The $3G$-term on the left of (4.29) denotes the expected time the Brownian motion spends in A; the term with the 4 K and $\mathbb{E}$ is the weighted lifetime for paths to reach a gap center. The result in (4.19) of Theorem 4.2, follows by adding the similar result for B.

Before we are able to go to the details of the proof of Theorem 4.6, we need some technical steps that use the Riemann mapping theorem. We map A and B separately on domains such that the curve that separated these two domains becomes part of a straight line. We need precise scaling results near the gaps, and the simplest way seems to be the case where the separating curve is such a straight line. Note that by the Riemann mapping theorem there exists a map h_1 from $A \cup \{\text{holes in} A\}$ to $B_1(0)$ and that by a consecutive map we can take care that there is a conformal map h_A from A to a bounded domain $h_A(A)$ such that $h_A(\Gamma) \subset \{(x_1, x_2); x_1 = 0\}$. Similarly, there is a conformal map h_B from B to a bounded domain $h_B(B)$ such that $h_B(\Gamma) \in \{(x_1, x_2); x_1 = 0\}$. In general, one cannot extend this conformal map outside of A (or B). However, since we assumed that Γ is an analytical curve, the Cauchy–Kovalevskaya theorem implies that h_A can be extended outside of A in a neighborhood of the points $\{z_1, \ldots, z_k\}$. So far the good news. The bad news is that although we can fix the position of at most three boundary points, there is no way to fix more points or to fix the width of the gaps in the image. We formulate these transformation results in the next lemma without further proof.

Lemma 4.1 *Let Ω_1, A, Γ be as in Assumption 4.1. Then there is conformal mapping h_A from A to $h_A(A) \subset \mathbb{R}^- \times \mathbb{R}$ such that:*

1. $h_A(\Gamma \cap \partial A) \subset \{0\} \times \mathbb{R}$.
2. *h_A can be extended to a homeomorphism $h_A : \overline{A} \to \overline{h(A)}$.*

Moreover, since Γ is analytic there is $\varepsilon > 0$ such that:

3. *h_A can be extendend to $\overline{A} \cup \bigcup_{i=1}^k \overline{B_\varepsilon(z_i)}$*
4. *For any compact $A^\diamond \subset A \cup (\Gamma \cap \Omega_1)$ the mapping h_A restricted to $A^\diamond \cup \bigcup_{i=1}^k \overline{B_\varepsilon(z_i)}$ is a diffeomorphism.*

A similar h_B exists for B. There we will assume that $h_B(B) \subset \mathbb{R}^+ \times \mathbb{R}$. The results are sketched in Fig. 4.3. Although the transformed subdomains A and B have no physical connection anymore, the additional disks around the z_i allow an asymptotic identification of the Green and Poisson kernels near those z_i.

Next, we will fix a weight factor for the connection through the gaps.

Notation 4.1 For each $\ell \in (0, \ell_0)$ the gaps are transformed to the intervals

$$\tilde{\Gamma}_{i,\ell} := \left(z_{i,\ell}^-, z_{i,\ell}^+\right) \text{ with } z_{i,\ell}^\pm = \left(0, \left(z_{i,\ell}^\pm\right)_2\right) := h_A\left(\gamma\left(t_i \pm \tfrac{1}{2}\ell\right)\right).$$

Moreover, we set $\ell_i := \left|z_{i,\ell}^- - z_{i,\ell}^+\right|$ and $z_{i,\ell} := \frac{1}{2}\left(z_{i,\ell}^- + z_{i,\ell}^+\right)$. For $\alpha, \beta \in (0,1)$ we fix

$$W_{i,\ell} := z_{i,\ell} + \tfrac{1}{2} l_i \mathcal{W}_{R_1,R_2} \text{ with } R_1 = \left(\tfrac{1}{2}\ell_i\right)^{\alpha-1} \text{ and } R_2 = \left(\tfrac{1}{2}\ell_i\right)^{\beta-1} \tag{4.30}$$

and with

$$\mathcal{W}_{R_1,R_2} = \left\{x \in B_{R_1}(0); x_1 < 0\right\} \cup \left\{x \in B_{R_2}(0); x_1 > 0\right\} \cup B_1(0).$$

See Figs. 4.4 and 4.5 in the appendix. Also, we will use

$$W_{i,\ell}^+ = \left\{z \in W_{i,\ell}; z_1 > 0\right\} \quad \text{and} \quad \left(\partial W_{i,\ell}\right)^+ = \left\{z \in \partial W_{i,\ell}; z_1 > 0\right\},$$
$$W_{i,\ell}^- = \left\{z \in W_{i,\ell}; z_1 < 0\right\} \quad \text{and} \quad \left(\partial W_{i,\ell}\right)^- = \left\{z \in \partial W_{i,\ell}; z_1 < 0\right\}.$$

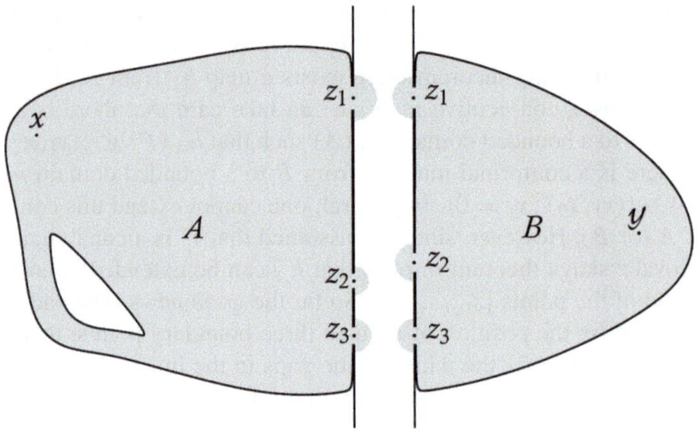

Fig. 4.3 Both subdomains are conformally mapped such that Γ becomes part of a straight line. Since the original Γ is analytic, one may extend the conformal mapping in an ε-neighborhood outside of A, respectively B. The corresponding transformed neighborhoods by these two conformal mappings cannot be superimposed, i.e., the holes do not fit

Fig. 4.4 Close-up of a straightened gap with $r = \frac{1}{2}\ell_i$

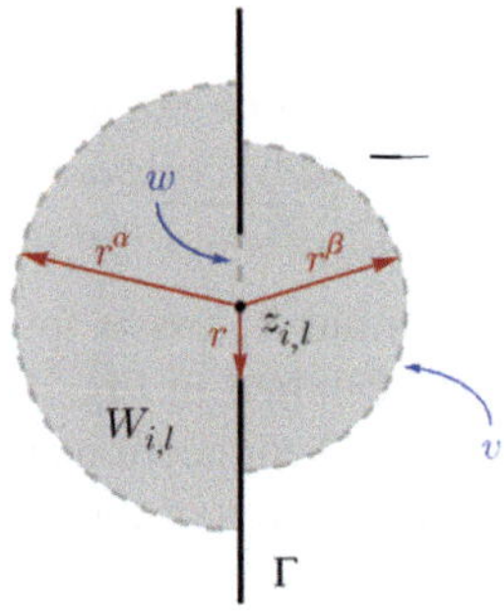

Finally, we define a connecting factor by

$$I_{i,\ell} := \int_{w=(0,w_2)\in\tilde{\Gamma}_{i,\ell}} \left(\int_{v=(v_1,v_2)\in(\partial W_{i,\ell})^+} K_{W_{i,\ell}}(w,v)\ v_1\ d\sigma_v \right) dw_2. \tag{4.31}$$

Remark 4.10 Note that $\ell_i = \left|z_{i,\ell}^- - z_{i,\ell}^+\right| = \left|h_A'(z_i)\right|\ell + \mathcal{O}\left(\ell^2\right)$ and $z_{i,\ell} = h_A(z_i) + \mathcal{O}\left(\ell^2\right)$. Since $\alpha, \beta > 0$ we find for l sufficiently small that $h_A^{-1}\left(W_{i,\ell}\right) \subset \Omega_\ell$.

Lemma 4.2 *Let u_{super} be as in (4.72). Then*

$$I_{i,\ell} = \ell_i^2 \left(\mathcal{I}_\infty + \mathcal{O}\left(\ell_i^{1-\max(\alpha,\beta)}\right)\right).$$

where $\mathcal{I}_\infty = \frac{1}{4}\int_{|w_2|<1} u_{super}(0,w_2)\,dw_2$.

Proof Let u_{R_1,R_2} be as in Proposition 4.2 in the appendix and set $r = \frac{1}{2}\ell_i$. With a rescaling, we find for $w \in W_{i,\ell}$ and hence for $w \in \tilde{\Gamma}_{i,\ell}$ that

$$\int_{v\in(\partial W_{i,\ell})^+} K_{W_{i,\ell}}(w, v)\ v_1\ d\sigma_v = r\ u_{r^{\alpha-1},r^{\beta-1}}\left(r^{-1}\left(w - z_{i,\ell}\right)\right).$$

Setting

$$\mathcal{I}_{R_1,R_2} = \tfrac{1}{4}\int_{|w_2|<1} u_{R_1,R_2}(0, w_2)\,\mathrm{d}w_2$$

we find

$$I_{i,\ell} = \int_{w_2=-r}^{r} r\ u_{r^{\alpha-1},r^{\beta-1}}\left(0, r^{-1}w_2\right)\mathrm{d}w_2 = r^2\int_{-1}^{1} u_{r^{\alpha-1},r^{\beta-1}}(0, x_2)\,dx_2 = \ell_i^2\,\mathcal{I}_{r^{\alpha-1},r^{\beta-1}}.$$

By the same proposition $\mathcal{I}_{r^{\alpha-1},r^{\beta-1}} = \mathcal{I}_\infty + \mathcal{O}\left(r^{1-\max(\alpha,\beta)}\right)$ holds true and the result follows. □

6.2 Comparing G_{Ω_ℓ} by Poisson Kernels

Before we start the proof of (4.29), we derive some asymptotic estimates for the Green function using Poisson kernels. We rewrite the expressions $G_{\Omega_\ell}(z, y)$ and $G_{\Omega_\ell}(x, y)$ with the help of Poisson kernels and then find approximations for these.

Proposition 4.1 *Let $0 < \alpha < \beta < 1$ and let $\mathcal{I}_\infty$ be as in Lemma 4.2. Then there is a constant $C_{\mathrm{VIII}} > 0$ such that :*

1. *For all $\ell \in (0, \ell_0)$, $z \in A$ and $y \in B \backslash U_\rho$:*

$$\left|G_{\Omega_\ell}(z, y) - \ell^2\mathcal{I}_\infty \sum_{j=1}^{k} K_A\left(z, z_j\right) K_B\left(y, z_j\right)\right|$$
$$\leq C_{\mathrm{VIII}}\, d_B(y)\left(\ell^{3-2\alpha}|\log\ell| + \left(\ell^{3-\beta}|\log\ell| + \ell^{1+2\beta}\right)\sum_{j=1}^{k}\left|z - z_j\right|^{-1}\right). \tag{4.32}$$

2. *For all $\ell \in (0, \ell_0)$, $x \in A\backslash U_\rho$ and $y \in B\backslash U_\rho$:*

$$\left|G_{\Omega_\ell}(x, y) - \ell^2\mathcal{I}_\infty \sum_{j=1}^{k} K_A\left(x, z_j\right) K_B\left(y, z_j\right)\right|$$
$$\leq C_{\mathrm{VIII}}\, d_A(x)\, d_B(y)\left(\ell^{3-\beta}|\log\ell| + \ell^{1+2\beta}\right). \tag{4.33}$$

Before we are able to prove this result, we need several lemmas. In those lemmas, the following representation by a Poisson kernel will be used. For $y \in B$, the function $z \mapsto G_{\Omega_\ell}(z, y)$ is harmonic in A and hence we may write

$$G_{\Omega_\ell}(z,y) = \sum_{i=1}^{k} \int_{\Gamma_{i,\ell}} K_A(z,w)\, G_{\Omega_\ell}(w,y)\, d\sigma_w \text{ for all } z \in A \text{ and } y \in B. \tag{4.34}$$

In the first step, we want to approximate $K_A(z,w)$ by $K_A(z,z_i)$. This last term can be put outside of the integral.

Lemma 4.3 *There exists $C_{\mathrm{IX}} > 0$ such that the following holds for $\ell \in (0, \ell_0)$:*

1. For all $\alpha \in (0,1)$, $y \in B \backslash U_\rho$ and $z \in A \backslash U_{\ell^\alpha}$:

$$\left| G_{\Omega_\ell}(z,y) - \sum_{i=1}^{k} K_A(z,z_i) \int_{\Gamma_{i,\ell}} G_{\Omega_\ell}(w,y)\, \mathrm{d}\sigma_w \right| \le C_{\mathrm{IX}}\, d_B(y)\, \ell^{3-2\alpha} \left|\log \ell\right|. \tag{4.35}$$

2. For all $x \in A \backslash U_\rho$ and $y \in B \backslash U_\rho$:

$$\left| G_{\Omega_\ell}(x,y) - \sum_{i=1}^{k} K_A(x,z_i) \int_{\Gamma_{i,\ell}} G_{\Omega_\ell}(w,y)\, \mathrm{d}\sigma_w \right| \le C_{\mathrm{IX}}\, d_A(x)\, d_B(y)\, \ell^{3} \left|\log \ell\right|. \tag{4.36}$$

Proof To start, we go back to (4.34). We want the Poisson kernel expression $K_A(z,w)$ to be independent of w. For this purpose, we use the Lipschitz type estimate of Corollary 4.2, assuming that ℓ is small enough:

$$\begin{aligned}
&\left| G_{\Omega_\ell}(z,y) - \sum_{j=1}^{k} \int_{\Gamma_{j,\ell}} K_A(z,z_j)\, G_{\Omega_\ell}(w,y)\, d\sigma_w \right| \\
&= \left| \sum_{j=1}^{k} \int_{\Gamma_{j,\ell}} \left(K_A(z,w) - K_A(z,z_j)\right) G_{\Omega_\ell}(w,y)\, d\sigma_w \right| \\
&\le c \sum_{j=1}^{k} \int_{\Gamma_{j,\ell}} \left((\ell^\alpha - \ell)^{-2} \left|w - z_j\right|\right) \left(d_B(y)\, \ell \log \ell\right) d\sigma_w \\
&\le c_1\, d_B(y)\, \ell^{1-2\alpha} \left|\log \ell\right| \sum_{j=1}^{k} \int_{\Gamma_{j,\ell}} \left|w - z_j\right| d\sigma_w \le c_2\, d_B(y)\, \ell^{3-2\alpha} \left|\log \ell\right|.
\end{aligned}$$

For the estimate on G_{Ω_ℓ}, we apply Theorem 4.5–3 (and exchange the role of A and B there). For the second estimate, we used the other Lipschitz bound given in Corollary 4.2 to find (4.36). □

Next, we look at $G_{\Omega_\ell}(w, y)$, which appears both in (4.35) and (4.36).

Lemma 4.4 *Fix $j \in \{1, \dots, k\}$, let $y \in B \backslash U_\rho$ and $w \in \tilde{\Gamma}_{j,\ell}$. Then there is some $C_X := C(\Omega, B) > 0$ such that*

$$\left| G_{\Omega_\ell}\left(h_B^{-1}(w), y\right) - \int_{(\partial W_{j,\ell})^+} K_{W_{j,\ell}}(w, v)\, G_B\left(h_B^{-1}(v), y\right) \mathrm{d}\sigma_v \right| \\ \leq C_X\, d_B(y)\, \ell^{2-\max(\alpha,\beta)} |\log \ell| . \tag{4.37}$$

Note that α, β appear through $W_{j,\ell}$ defined in (4.30).

Proof Since $|y - z_j| > \rho > \ell^\beta$ holds for ℓ small enough, one finds that for those ℓ the function $v \mapsto G_{\Omega_\ell}\left(h_B^{-1}(v), y\right)$ is harmonic near the transformed gap $\tilde{\Gamma}_{j,\ell}$. So, for $w \in \tilde{\Gamma}_{j,\ell}$ we may write

$$G_{\Omega_\ell}\left(h_B^{-1}(w), y\right) = \int_{\partial W_{j,\ell}} K_{W_{j,\ell}}(w, v)\, G_{\Omega_\ell}\left(h_B^{-1}(v), y\right) d\sigma_v. \tag{4.38}$$

Since $G_{\Omega_\ell}\left(h_B^{-1}(v), y\right) = 0$ for $v \in \partial W_{j,\ell}$ with $v_1 = 0$, we can rewrite (4.38) as

$$G_{\Omega_\ell}\left(h_B^{-1}(w), y\right) = \int_{(\partial W_{j,\ell})^-} K_{W_{j,\ell}}(w, v)\, G_{\Omega_\ell}\left(h_B^{-1}(v), y\right) d\sigma_v \\ + \int_{(\partial W_{j,\ell})^+} K_{W_{j,\ell}}(w, v)\, G_{\Omega_\ell}\left(h_B^{-1}(v), y\right) d\sigma_v. \tag{4.39}$$

For the first integral in (4.39), we use Theorem 4.5–4 to find

$$0 \leq \int_{(\partial W_{j,\ell})^-} K_{W_{j,\ell}}(w, v)\, G_{\Omega_\ell}\left(h_B^{-1}(v), y\right) d\sigma_v \\ \leq c \int_{(\partial W_{j,\ell})^-} K_{W_{j,\ell}}(w, v)\, d_B(y)\, \ell^{2-\alpha} |\log \ell|\, d\sigma_v \leq c\, d_B(y)\, \ell^{2-\alpha} |\log \ell| . \tag{4.40}$$

For the second integral in (4.39), we use that $G_{\Omega_\ell}\left(h_B^{-1}(v), y\right)$ and $G_B\left(h_B^{-1}(v), y\right)$ are close by Theorem 4.5–2 and find

$$0 \leq \int_{(\partial W_{j,\ell})^+} K_{W_{j,\ell}}(w, v)\, G_{\Omega_\ell}\left(h_B^{-1}(v), y\right) d\sigma_v \\ \leq \int_{(\partial W_{j,\ell})^+} K_{W_{j,\ell}}(w, v)\, G_B\left(h_B^{-1}(v), y\right) d\sigma_v + c\, d_B(y)\, \ell^{2-\beta} |\log \ell| . \tag{4.41}$$

The result follows by combining (4.40) and (4.41). □

Proof [Proof of Proposition 4.1] We start with (4.37) from Lemma 4.4 and approximate $G_B\left(h_B^{-1}(v), y\right)$ for $v \in \left(\partial W_{j,\ell}\right)^+$. Since

$$G_B\left(h_B^{-1}(v), y\right) = G_{h_B(B)}\left(v, h_B(y)\right),$$

we find by Taylor for $v \mapsto G_{h_B}(B)(v, h_B(y))$ and using that $G_{h_B(B)}(v, h_B(y)) = 0$ for $v_1 = 0$, the existence of $\xi_v \in [z_{i,\ell}, v]$ such that

$$G_{h_B(B)}(v, h_B(y)) = v_1 \frac{\partial}{\partial v_1} G_{h_B(B)}(z_{i,\ell}, h_B(y)) + \tfrac{1}{2}(v - z_{i,\ell}) \cdot \mathsf{H}(\xi_v; y)(v - z_{i,\ell}).$$

The Hessian $\mathsf{H}(\cdot; y)$ consists of second derivatives of $v \mapsto G_{h_B(B)}(v, h_B(y))$ and upper bounds for those are given in Theorem 4.4. These are as follows:

$$|\mathsf{H}(\xi_w; y)| \le c\ |h_B(y) - \xi_v|^{-2} \min\left\{1, \frac{d_{h(B)}(h_B(y))}{|h_B(y) - \xi_v|}\right\}.$$

Since $\xi_v \in [z_{i,\ell}, v]$, $|z_{i,\ell} - v| = \ell^\beta$ and $|h_B(y) - z_{i,\ell}| > c\ (\rho - \ell^\beta)$, we find for ℓ small that $|\mathsf{H}(\xi_w; y)| \le c\ d_{h(B)}(h_B(y))$. Since

$$\frac{\partial}{\partial v_1} G_{h_B(B)}(w, h_B(y))_{|w=z_{i,\ell}} = K_{h_B(B)}(h_B(y), z_{i,\ell}),$$

we find

$$\left|G_{h_B(B)}(v, h_B(y)) - v_1 K_{h_B(B)}(h_B(y), z_{i,\ell})\right| \le c\ \ell^{2\beta} d_{h_B(B)}(h_B(y))$$

and we also have

$$\left|\int_{(\partial W_{j,\ell})^+} K_{W_{j,\ell}}(w,v)\left(G_B\left(h_B^{-1}(v), y\right) - v_1 K_{h_B(B)}(h_B(y), z_{i,\ell})\right) d\sigma_v\right| \le c\ \ell^{2\beta} d_B(y). \tag{4.42}$$

By (4.15) and Corollary 4.2 we find

$$K_{h_B(B)}(h_B(y), z_{i,\ell}) = K_B\left(y, h_B^{-1}(z_{i,\ell})\right)\left|\left(h_B^{-1}\right)'(z_{i,\ell})\right| = \frac{K_B(y, z_i)}{|h'_B(z_i)|} + \mathcal{O}(\ell^2)\, d_B(y). \tag{4.43}$$

Combining (4.37) from Lemma 4.4, (4.42), and (4.43) we find

$$\begin{aligned}
&\left|G_{\Omega_\ell}\left(h_B^{-1}(w), y\right) - |h'_B(z_i)|^{-1} \int_{(\partial W_{j,\ell})^+} K_{W_{j,\ell}}(w,v)\, v_1 K_B(y, z_i)\, d\sigma_v\right| \\
&\le \left|\int_{(\partial W_{j,\ell})^+} K_{W_{j,\ell}}(w,v)\left(G_B\left(h_B^{-1}(v), y\right) - v_1 \frac{K_B(y, z_i)}{|h'_B(z_i)|}\right) d\sigma_v\right| + c\ d_B(y)\, \ell^{2-\beta} |\log \ell| \\
&\le \left|\int_{(\partial W_{j,\ell})^+} K_{W_{j,\ell}}(w,v)\left(G_B\left(h_B^{-1}(v), y\right) - v_1 K_{h_B(B)}(h_B(y), z_{i,\ell})\right) d\sigma_v\right| \\
&\qquad + c_1\, d_B(y)\left(\ell^{2-\beta}|\log \ell| + \ell^2\right) \\
&\le c_2\, d_B(y)\left(\ell^{2-\beta}|\log \ell| + \ell^2 + \ell^{2\beta}\right) \le c_3\, d_B(y)\left(\ell^{2-\beta}|\log \ell| + \ell^{2\beta}\right).
\end{aligned} \tag{4.44}$$

We also need

$$\left| \left|h_B'(z_i)\right|^{-1} \int_{\tilde{\Gamma}_{i,\ell}} G_{\Omega_\ell}\left(h_B^{-1}(w), y\right) d\sigma_w - \int_{\Gamma_{i,\ell}} G_{\Omega_\ell}(v, y)\, d\sigma_v \right| \le \int_{\Gamma_{i,\ell}} \left| \frac{|h_B'(w)|}{|h_B'(z_i)|} - 1 \right| G_{\Omega_\ell}(v, y)\, d\sigma_v \le c_4\, \ell^3 d_B(y). \tag{4.45}$$

and, by Lemma 4.2 and Remark 4.10, that

$$\left| \ell^2 \mathcal{I}_\infty - \left|h_B'(z_i)\right|^{-2} I_{i,\ell} \right| \le c_5\, \ell^{1-\beta}. \tag{4.46}$$

By (4.46) and (4.21), we find

$$\left| G_{\Omega_\ell}(z, y) - \ell^2 \mathcal{I}_\infty \sum_{i=1}^k K_A(z, z_i)\, K_B(y, z_i) \right| \le$$
$$\left| G_{\Omega_\ell}(z, y) - \sum_{i=1}^k K_A(z, z_i) K_B(y, z_i) \left|h_B'(z_i)\right|^{-2} I_{i,\ell} \right| + c_6 \ell^{1-\beta} \sum_{i=1}^k K_A(z, z_i)\, d_B(y).$$

By definition, see (4.31), we have

$$\left| G_{\Omega_\ell}(z, y) - \sum_{i=1}^k K_A(z, z_i)\, K_B(y, z_i) \left|h_B'(z_i)\right|^{-2} I_{i,\ell} \right| =$$
$$\left| G_{\Omega_\ell}(z, y) - \sum_{i=1}^k K_A(z, z_i) \left|h_B'(z_i)\right|^{-2} \int_{\tilde{\Gamma}_{i,\ell}} \int_{(\partial W_{j,\ell})^+} K_{W_{j,\ell}}(w, v)\, v_1 K_B(y, z_i)\, d\sigma_v d\sigma_w \right|. \tag{4.47}$$

Next, we split (4.47) as follows:

$$\left| G_{\Omega_\ell}(z, y) - \sum_{i=1}^k K_A(z, z_i) \left|h_B'(z_i)\right|^{-2} \int_{\tilde{\Gamma}_{i,\ell}} \int_{(\partial W_{j,\ell})^+} K_{W_{j,\ell}}(w, v)\, v_1 K_B(y, z_i)\, d\sigma_v d\sigma_w \right|$$
$$\le \left| G_{\Omega_\ell}(z, y) - \sum_{i=1}^k K_A(z, z_i) \int_{\Gamma_{i,\ell}} G_{\Omega_\ell}(v, y)\, d\sigma_v \right| \tag{4.48}$$
$$+ \left| \sum_{i=1}^k K_A(z, z_i) \left(\int_{\Gamma_{i,\ell}} G_{\Omega_\ell}(v, y)\, d\sigma_v - \left|h_B'(z_i)\right|^{-1} \int_{\tilde{\Gamma}_{i,\ell}} G_{\Omega_\ell}\left(h_B^{-1}(w), y\right) d\sigma_w \right) \right| \tag{4.49}$$

$$+ \left| \sum_{i=1}^k \frac{K_A(z, z_i)}{\left|h_B'(z_i)\right|} \int_{\tilde{\Gamma}_{i,\ell}} \left(G_{\Omega_\ell}\left(h_B^{-1}(w), y\right) - \left|h_B'(z_i)\right|^{-1} \times \cdots \right.\right.$$
$$\left.\left. \cdots \times \int_{(\partial W_{j,\ell})^+} K_{W_{j,\ell}}(w, v)\, v_1 K_B(y, z_i)\, d\sigma_v \right) d\sigma_w \right|. \tag{4.50}$$

The item in line (4.48) is estimated by using (4.35) of Lemma 4.3 to find

$$(4.48) \leq \left| G_{\Omega_\ell}(z,y) - \sum_{i=1}^{k} K_A(z,z_i) \int_{\Gamma_{i,\ell}} G_{\Omega_\ell}(v,y)\, d\sigma_v \right| \leq C_{\mathrm{IX}}\, d_B(y)\, \ell^{3-2\alpha} \left|\log \ell\right|.$$

The second item (4.49) uses (4.45) to yield

$$\begin{aligned}(4.49) &\leq \left| G_{\Omega_\ell}(z,y) - \sum_{i=1}^{k} K_A(z,z_i) \textstyle\int_{\Gamma_{i,\ell}} G_{\Omega_\ell}(v,y)\, d\sigma_v \right| \\ &\leq c_7 \sum_{i=1}^{k} K_A(z,z_i)\, d_B(y)\, \ell^3 \leq c_8 \sum_{i=1}^{k} |z-z_i|^{-1}\, d_B(y)\, \ell^3.\end{aligned}$$

Here and in the next estimate, we also use (4.22). The third item (4.50) uses (4.44) to be estimated as follows:

$$\begin{aligned}(4.50) &\leq \sum_{i=1}^{k} \frac{K_A(z,z_i)}{\left|h_B'(z_i)\right|} \int_{\tilde{\Gamma}_{i,\ell}} c_9 d_B(y) \left(\ell^{2-\beta}\left|\log \ell\right| + \ell^{2\beta}\right) d\sigma_w \\ &\leq c_{10} \sum_{i=1}^{k} |z-z_i|^{-1}\, d_B(y) \left(\ell^{3-\beta}\left|\log \ell\right| + \ell^{1+2\beta}\right).\end{aligned}$$

Altogether we obtain that

$$\begin{aligned}&\left| G_{\Omega_\ell}(z,y) - \ell^2 \mathcal{I}_\infty \sum_{i=1}^{k} K_A(z,z_i)\, K_B(y,z_i) \right| \\ &\leq c_{11} d_B(y) \left(\ell^{3-2\alpha}\left|\log l\right| + \left(\ell^{3-\beta}\left|\log \ell\right| + \ell^{1+2\beta}\right) \sum_{i=1}^{k} |z-z_i|^{-1} \right), \qquad (4.51)\end{aligned}$$

which shows our first estimate.

For our second estimate, we change the last steps in (4.51) by using now (4.36) of Lemma 4.3 and $K_A(x,z_i) \leq d_A(x)$ to find

$$\left| G_{\Omega_\ell}(x,y) - \sum_{i=1}^{k} K_A(x,z_i)\, K_B(z_i,y)\, \mathcal{I}_\infty \ell^2 \right|$$

$$\begin{aligned}&\leq \left| G_{\Omega_\ell}(x,y) - \sum_{i=1}^{k} K_A(x,z_i) \left|h_B'(z_i)\right|^{-1} \int_{\tilde{\Gamma}_{i,\ell}} G_{\Omega_\ell}\left(h_B^{-1}(w), y\right) d\sigma_w \right| \\ &\quad + \sum_{i=1}^{k} K_A(x,z_i) \int_{\tilde{\Gamma}_{i,\ell}} c_{12}\, d_B(y) \left(\ell^{2-\beta}\left|\log \ell\right| + \ell^{2\beta}\right) d\sigma_w \\ &\leq c_{13}\, d_A(x)\, d_B(y) \left(\ell^{3-\beta}\left|\log \ell\right| + \ell^{1+2\beta}\right),\end{aligned}$$

which shows the second claim. □

6.3 3G Asymptotics

We still assume that $x \in A$ and $y \in B$ have at least distance ρ to any gap. Our considerations further on only work if z, i.e., the variable of integration, stays away from the gaps. That is why we split the integrals in (4.29) into integrals over those z, which are closer than ℓ^α to the gaps, and over those z, which are farther away. Here, $\alpha \in (0, 1)$ is a fixed exponent, which we determine later in order to get a good rate of convergence. To be precise, in order to show (4.29), we will study the rate of convergence to zero with respect to ℓ for the following expressions:

- The 3G contribution in an ℓ^α-neighborhood of a hole:

$$H^1_{n,\ell,\alpha}(x,y) := \int_{A\cap B_{\ell^\alpha}(z_n)} \frac{G_{\Omega_\ell}(x,z)\, G_{\Omega_\ell}(z,y)}{G_{\Omega_\ell}(x,y)} dz; \tag{4.52}$$

- An asymptotic comparison in an ℓ^α-neighborhood of a hole:

$$H^2_{n,\ell,\alpha,j}(x,y) := \frac{K_A\left(x,z_j\right) K_B\left(y,z_j\right)}{\sum_{m=1}^k K_A(x,z_m)\, K_B(y,z_m)} \int_{A\cap B_{\ell^\alpha}(z_n)} \frac{G_A(x,z)\, K_A\left(z,z_j\right)}{K_A\left(x,z_j\right)} dz; \tag{4.53}$$

- Comparing with asymptotics outside ℓ^α-neighborhoods of the holes:

$$\begin{aligned} H^3_{\ell,\alpha}(x,y) &:= \int_{A\setminus U_{\ell^\alpha}} \frac{G_{\Omega_\ell}(x,z)\, G_{\Omega_\ell}(z,y)}{G_{\Omega_\ell}(x,y)} dz \\ &- \sum_{j=1}^k \frac{K_A\left(x,z_j\right) K_B\left(y,z_j\right)}{\sum_{m=1}^k K_A(x,z_m)\, K_B(y,z_m)} \int_{A\setminus U_{\ell^\alpha}} \frac{G_A(x,z)\, K_A\left(z,z_j\right)}{K_A\left(x,z_j\right)} dz. \end{aligned} \tag{4.54}$$

In (4.52) and (4.53), j and n are fixed coefficients in $\{1, \dots, k\}$. The union of balls U_{ℓ^α} around z_n is defined in (4.17).

Lemma 4.5 *With $H^1_{n,\ell,\alpha}$ as in (4.52), there exists C_{XI} such that for all $\alpha \in (0,1)$, $n \in \{1, \dots, k\}$ and $x \in A$, $y \in B$ with $\left|x - z_j\right| > \rho$ and $\left|y - z_j\right| > \rho$ for all $j \in \{1, \dots . k\}$*

$$H^1_{n,\ell,\alpha}(x,y) \le C_{\mathrm{XI}}\, \ell^\alpha.$$

Proof We start with (4.52) for fixed $n \in \{1, \dots, k\}$ and use the Poisson kernel representation (4.34), the limit behavior of G_{Ω_ℓ} stated in Theorem 4.5–1 and the bounds on G_A and K_A of (4.26) and (4.21) to find:

$$H^1_{n,\ell,\alpha}(x,y) = \int_{A\cap B_{\ell^\alpha}(z_n)} \frac{G_{\Omega_\ell}(x,z) \sum_{j=1}^k \int_{\Gamma_{j,\ell}} K_A(z,w)\, G_{\Omega_\ell}(w,y)\, d\sigma_w}{\sum_{j=1}^k \int_{\Gamma_{j,\ell}} K_A(x,w)\, G_{\Omega_\ell}(w,y)\, d\sigma_w} dz$$

$$\leq \int_{A\cap B_{\ell^\alpha}(z_n)} \frac{\left(c_1\frac{d_A(x)}{|x-z|} + c_2 d_A(x)\,\ell\,|\log \ell|\right)\sum_{j=1}^k \int_{\Gamma_{j,\ell}} c_3 \frac{d_A(z)}{|z-w|^2} G_{\Omega_\ell}(w,y)\,d\sigma_w}{\sum_{j=1}^k \int_{\Gamma_{j,\ell}} c_4 \frac{d_A(x)}{|x-w|^2} G_{\Omega_\ell}(w,y)\,d\sigma_w} dz.$$

Here $c_1, \dots, c_4 > 0$ are suitable constants. As $|x - z_j| > \rho$ for $j = 1, \dots, k$, we can assume $\ell > 0$ to be so small that $|x - z| > \frac{1}{2}\rho$ for all $z \in B_{\ell^\alpha}(z_m)$. Moreover, as A is bounded, there is some upper bound M for $|x - w|$. Finally, $\frac{d_A(z)}{|z-w|} \leq 1$ as before. This yields

$$\begin{aligned} H^1_{n,\ell,\alpha}(x,y) &\leq \int_{A\cap B_{\ell^\alpha}(z_n)} \frac{\left(2c_1\rho^{-1} + c_2\ell\,|\log \ell|\right)\sum_{j=1}^k \int_{\Gamma_{j,\ell}} c_3 \frac{1}{|z-w|} G_{\Omega_\ell}(w,y)\,d\sigma_w}{\sum_{j=1}^k \int_{\Gamma_{j,\ell}} c_4 M^{-2} G_{\Omega_\ell}(w,y)\,d\sigma_w} dz \\ &= \frac{\left(2c_1\rho^{-1} + c_2\ell\,|\log \ell|\right) c_3}{c_4 M^{-2}} \frac{\sum_{j=1}^k \int_{\Gamma_{j,\ell}} \left(\int_{A\cap B_{l^\alpha}(z_n)} \frac{1}{|z-w|} dz\right) G_{\Omega_\ell}(w,y)\,d\sigma_w}{\sum_{j=1}^k \int_{\Gamma_{j,\ell}} G_{\Omega_\ell}(w,y)\,d\sigma_w}. \end{aligned} \tag{4.55}$$

We look at the z-integral, with fixed $w \in \Gamma_{j,\ell}$. We get different upper bounds, depending on whether $n = j$ or not:

$$\int_{A\cap B_{\ell^\alpha}(z_n)} \frac{1}{|z-w|} dz \leq \begin{cases} c_5 \ell^\alpha & \text{if } n = j \\ c_6 \ell^{2\alpha} & \text{if } n \neq j \end{cases} \tag{4.56}$$

The constants c_5 and c_6 can be chosen so large that they do not depend on w. Plugging this into (4.55) and only taking care of the dominating order, we get that there is some constant $c_7 > 0$ such that

$$\int_{A\cap B_{\ell^\alpha}(z_n)} \frac{G_{\Omega_\ell}(x,z)\, G_{\Omega_\ell}(z,y)}{G_{\Omega_\ell}(x,y)} dz \leq c_7 \ell^\alpha \tag{4.57}$$

for all small $\ell > 0$. □

Lemma 4.6 *With $H^2_{n,\ell,\alpha,j}$ as in (4.53), there exists $C_{\mathrm{XII}} > 0$ such that for all $\alpha \in (0,1)$, $n \in \{1, \dots, k\}$ and $x \in A \backslash U_\rho$, $y \in B \backslash U_\rho$:*

$$H^2_{n,\ell,\alpha,j}(x,y) \leq C_{\mathrm{XII}}\,\ell^\alpha. \tag{4.58}$$

Proof The estimates on the Green function (4.26) and the Poisson kernel (4.21) give

$$H^2_{n,\ell,\alpha,j}(x,y) \leq \int_{A\cap B_{\ell^\alpha}(z_n)} \frac{c_1 \frac{d_A(x)}{|x-z|} c_2 \frac{d_A(z)}{|z-z_j|^2}}{c_3 \frac{d_A(x)}{|x-z_j|^2}} dz \leq \int_{A\cap B_{\ell^\alpha}(z_n)} \frac{2c_1\rho^{-1} c_2 \frac{1}{|z-z_j|}}{c_3 M^{-2}} dz,$$

with suitable constants $c_1, c_2, c_3 > 0$. In the last line, we have made the same assumptions on l and used the same upper and lower bounds for the occurring terms as in the preceding lemma. The remaining z-integral has the upper bounds stated in (4.56). We have thus shown that there is a $c_4 > 0$ such that

$$H^2_{n,\ell,\alpha,j}(x,y) \le c_4\,\ell^\alpha \tag{4.59}$$

for all small $\ell > 0$. □

Lemma 4.7 *There exists* $C_{\text{XIII}} = C(\Omega, A, \rho) > 0$ *such that for all* $\alpha \in (0,1)$ *and* $x \in A \backslash U_\rho$, $y \in B \backslash U_\rho$*:*

$$\left| \int_{A\backslash U_{\ell^\alpha}} \frac{G_{\Omega_\ell}(x,z)\,G_{\Omega_\ell}(z,y)}{G_{\Omega_\ell}(x,y)}dz - \int_{A\backslash U_{\ell^\alpha}} \frac{G_A(x,z)\,G_{\Omega_\ell}(z,y)}{G_{\Omega_\ell}(x,y)}dz \right| \\ \le C_{\text{XIII}}\,\ell^{2-2\alpha}\,(1+|\log\ell|)\,.$$

Proof The term $G_{\Omega_\ell}(x,z)$ in the first integral can be approximated by $G_{\Omega_\ell}(x,z)$ according to Theorem 4.5–2:

$$\begin{aligned}
&\left| \int_{A\backslash U_{\ell^\alpha}} \frac{G_{\Omega_\ell}(x,z)\,G_{\Omega_\ell}(z,y)}{G_{\Omega_\ell}(x,y)}dz - \int_{A\backslash U_{\ell^\alpha}} \frac{G_A(x,z)\,G_{\Omega_\ell}(z,y)}{G_{\Omega_\ell}(x,y)}dz \right| \\
&= \int_{A\backslash U_{\ell^\alpha}} \frac{\left(G_{\Omega_\ell}(x,z) - G_A(x,z)\right)G_{\Omega_\ell}(z,y)}{G_{\Omega_\ell}(x,y)}dz \\
&\le \int_{A\backslash U_{\ell^\alpha}} \frac{c_1 d_A(x)\,\ell^{2-\alpha}\,|\log\ell| \sum_{j=1}^k \int_{\Gamma_{j,\ell}} K_A(z,w)\,G_{\Omega_\ell}(w,y)\,d\sigma_w}{\sum_{j=1}^k \int_{\Gamma_{j,\ell}} K_A(x,w)\,G_{\Omega_\ell}(w,y)\,d\sigma_w}dz \\
&\le c_2 \int_{A\backslash U_{\ell^\alpha}} \frac{d_A(x) \sum_{j=1}^k \int_{\Gamma_{j,\ell}} \frac{d_A(z)}{|z-w|^2} G_{\Omega_\ell}(w,y)\,d\sigma_w}{\sum_{j=1}^k \int_{\Gamma_{j,\ell}} \frac{d_A(x)}{|x-w|^2} G_{\Omega_\ell}(w,y)\,d\sigma_w}dz\,\ell^{2-\alpha}\,|\log\ell| \\
&\le c_3 \int_{A\backslash U_{\ell^\alpha}} \frac{\sum_{j=1}^k \int_{\Gamma_{j,\ell}} \ell^{-\alpha} G_{\Omega_\ell}(w,y)\,d\sigma_w}{\sum_{j=1}^k \int_{\Gamma_{j,\ell}} G_{\Omega_\ell}(w,y)\,d\sigma_w}dz\,\ell^{2-\alpha}\,|\log\ell| \\
&\le c_4\,\ell^{2-2\alpha}\,|\log\ell|.
\end{aligned} \tag{4.60}$$

Here, $c_1, \ldots, c_4 > 0$ are suitable constants. We have used the estimates on the Poisson kernel (4.21) and the facts that $|x-w|$ is bounded from above by some constant M and $|z-w| \ge |z-z_j| - |z_j - w| \ge \ell^\alpha - \frac{1}{2}\ell \ge \frac{1}{2}\ell^\alpha$ for small $\ell > 0$. □

Now, we can start to combine the results obtained so far to find a proof of the main theorem.

Proof [Proof of Theorem 4.6] In order to increase readability, we set

$$H_\ell(z,y) := \ell^2 \mathcal{I}_\infty \sum_{j=1}^k K_A(z,z_j)\,K_B(y,z_j)\,,$$

with $\mathcal{I}_\infty$ as in Lemma 4.2.

Step 1. We get that

$$\left| \int_{A\setminus U_{\ell\alpha}} \frac{G_A(x,z)\, G_{\Omega_\ell}(z,y)}{G_{\Omega_\ell}(x,y)} dz - \int_{A\setminus U_{\ell\alpha}} \frac{G_A(x,z)\, H_\ell(z,y)}{H_\ell(x,y)} dz \right|$$
$$\leq \int_{A\setminus U_{\ell\alpha}} \frac{G_A(x,z) \left| G_{\Omega_\ell}(z,y) - H_\ell(z,y) \right|}{G_{\Omega_\ell}(x,y)} dz$$
$$+ \int_{A\setminus U_{\ell\alpha}} \frac{G_A(x,z)\, H_\ell(z,y) \left| G_{\Omega_\ell}(x,y) - H_\ell(x,y) \right|}{G_{\Omega_\ell}(x,y)\, H_\ell(x,y)} dz. \tag{4.61}$$

Before finding an upper bound for this, we take a closer look at H_ℓ and the denominator. By the Poisson kernel estimate (4.21) and using that $d_A(z) \leq \left|z - z_j\right|$, there is some $c_1 > 0$ such that

$$H_\ell(z,y) \leq c_1 \sum_{j=1}^{k} \frac{d_A(z)}{\left|z - z_j\right|^2} \frac{d_B(y)}{\left|y - z_j\right|^2} \ell^2 \leq c_1 \frac{d_B(y)}{\rho^2} \sum_{j=1}^{k} \left|z - z_j\right|^{-1} \ell^2. \tag{4.62}$$

Similarly, a lower bound can be found (with $c_2 > 0$):

$$H_\ell(x,y) \geq c_2 \sum_{j=1}^{k} \frac{d_A(x)}{\left|x - z_j\right|^2} \frac{d_B(y)}{\left|y - z_j\right|^2} \ell^2 \geq c_2\, M^4 d_A(x)\, d_B(y)\, \ell^2. \tag{4.63}$$

Combining this with (4.33) gives a lower bound for $G_{\Omega_\ell}(x,y)$:

$$G_{\Omega_\ell}(x,y) \geq d_A(x)\, d_B(y) \left(c_2 M^4 \ell^2 - c_2 \ell^3 \left|\log \ell\right| - c_3 \ell^{3-\beta} \left|\log \ell\right| - c_4 \ell^{1+2\beta} \right). \tag{4.64}$$

Taking $\beta > \frac{1}{2}$, we find for ℓ small enough that

$$G_{\Omega_\ell}(x,y) \geq c_5\, d_A(x)\, d_B(y)\, \ell^2. \tag{4.65}$$

The Green function estimate (4.26) together with (4.32) and (4.65) give an upper bound for the first integrand in (4.61). Indeed, there is some $c_3 > 0$ such that

$$\frac{G_A(x,z) \left| G_{\Omega_\ell}(z,y) - H_\ell(z,y) \right|}{G_{\Omega_\ell}(x,y)}$$
$$\leq c_3 \frac{\frac{d_A(x)}{|x-z|} d_B(y)\, C_{\mathrm{VIII}} \left(\ell^{3-2\alpha} \left|\log \ell\right| + \left(\ell^{3-\beta} \left|\log \ell\right| + \ell^{1+2\beta} \right) \sum_{j=1}^{k} \left|z - z_j\right|^{-1} \right)}{d_A(x)\, d_B(y)\, \ell^2}$$
$$= c_3 \left|x - z\right|^{-1} C_{\mathrm{VIII}} \left(\ell^{1-2\alpha} \left|\log \ell\right| + \left(\ell^{1-\beta} \left|\log \ell\right| + \ell^{-1+2\beta} \right) \sum_{j=1}^{k} \left|z - z_j\right|^{-1} \right). \tag{4.66}$$

For the second integrand in (4.61), we find by (4.26), (4.62), (4.33), (4.63), and (4.65) that there is some $c_4 > 0$ such that

$$\begin{aligned}
&\frac{G_A(x,z)\, H_\ell(z,y)\, \left|G_{\Omega_\ell}(x,y) - H_\ell(x,y)\right|}{G_{\Omega_\ell}(x,y)\, H_\ell(x,y)} \\
&\leq c_4 \frac{d_A(x)}{|x-z|} \left(d_B(y) \sum_{j=1}^{k} \left|z - z_j\right|^{-1} \ell^2 \right) \frac{d_A(x)\, d_B(y) \left(\ell^{3-\beta} \left|\log \ell\right| + \ell^{1+2\beta}\right)}{d_A(x)\, d_B(y)\, \ell^2\, d_A(x)\, d_B(y)\, \ell^2} \\
&= c_4\, |x-z|^{-1} \sum_{j=1}^{k} \left|z - z_j\right|^{-1} \left(\ell^{1-\beta} \left|\log \ell\right| + \ell^{-1+2\beta}\right). \qquad (4.67)
\end{aligned}$$

Step 2. We integrate the right-hand sides of (4.66) and (4.67) over z. As $\left|x - z_j\right| > \rho$ for all $j \in \{1, \ldots, k\}$, the singularities do not get too close to each other. Hence, there are upper bounds for the integrals which do neither depend neither on x nor on y, and it holds that

$$\begin{aligned}
&\left| \int_{A \setminus U_{\ell^\alpha}} \frac{G_A(x,z)\, G_{\Omega_\ell}(z,y)}{G_{\Omega_\ell}(x,y)} dz - \int_{A \setminus U_{\ell^\alpha}} \frac{G_A(x,z)\, H_\ell(z,y)}{H_\ell(x,y)} dz \right| \\
&\leq c_5 \left(\ell^{1-2\alpha} \left|\log \ell\right| + \ell^{1-\beta} \left|\log \ell\right| + \ell^{-1+2\beta}\right).
\end{aligned}$$

Step 3. Combining the above inequality with (4.60) gives

$$\begin{aligned}
&\left| \int_{A \setminus U_{\ell^\alpha}} \frac{G_{\Omega_\ell}(x,z)\, G_{\Omega_\ell}(z,y)}{G_{\Omega_\ell}(x,y)} dz - \int_{A \setminus U_{\ell^\alpha}} \frac{G_A(x,z)\, H_\ell(z,y)}{H_\ell(x,y)} dz \right| \\
&\leq c_6 \left(\ell^{2-2\alpha} \left|\log \ell\right| + \ell^{1-2\alpha} \left|\log \ell\right| + \ell^{1-\beta} \left|\log \ell\right| + \ell^{-1+2\beta}\right). \qquad (4.68)
\end{aligned}$$

This inequality gives an upper bound for (4.54), as

$$\frac{G_A(x,z)\, H_\ell(z,y)}{H_\ell(x,y)} = \sum_{j=1}^{k} \frac{G_A(x,z)\, K_A\left(z,z_j\right)}{K_A\left(x,z_j\right)} dz \frac{K_A\left(x,z_j\right) K_B\left(y,z_j\right)}{\sum_{m=1}^{k} K_A(x,z_m)\, K_B(y,z_m)}.$$

Step 4. Now it is time to determine α and β . The choice $\alpha = \frac{1}{3}$ and $\beta = \frac{2}{3}$ gives the best rate of convergence. We combine (4.57), (4.58), and (4.68) to get that there is a constant $c_7 > 0$ such that

$$\begin{aligned}
&\left| \int_A \frac{G_{\Omega_\ell}(x,z)\, G_{\Omega_\ell}(z,y)}{G_{\Omega_\ell}(x,y)} dz \right. \\
&\left. - \sum_{j=1}^{k} \mathbb{E}_x^{z_j}(\tau_A) \frac{K_A\left(x,z_j\right) K_B\left(y,z_j\right)}{\sum_{m=1}^{k} K_A(x,z_m)\, K_B(y,z_m)} \right| \leq c_7 \ell^{\frac{1}{3}} \left|\log \ell\right|,
\end{aligned}$$

which is (4.29). □

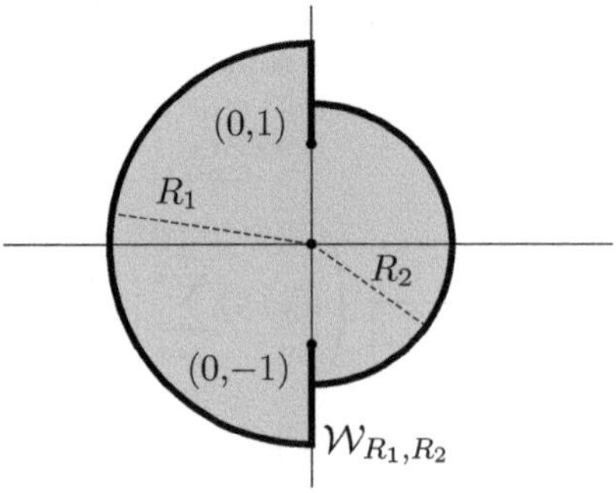

Fig. 4.5 The domain $\mathcal{W}_{R_1,R_2}$ from (4.69)

7 Appendix: Sub- and Supersolution for a Special Dirichlet Problem

For asymptotic estimates on harmonic functions near the gaps, we will need some technical results that we state in this appendix. The key domains are as follows. Setting

$$B_R^+(0)=\left\{x\in\mathbb{R}^2;\,|x|<R\text{ and }x_1>0\right\}\text{ and }B_R^-(0)=\left\{x\in\mathbb{R}^2;\,|x|<R\text{ and }x_1<0\right\},$$

we define for $R_1, R_2 \geq 1$:

$$\mathcal{W}_{R_1,R_2}=B_{R_1}^-(0)\cup B_{R_2}^+(0)\cup B_1(0)\,. \tag{4.69}$$

We are interested in the function u_{R_1,R_2} that satisfies

$$\begin{cases} -\Delta u(x)=0 & \text{for } x\in\mathcal{W}_{R_1,R_2},\\ u(x)=0 & \text{for } x\in\partial\mathcal{W}_{R_1,R_2} \text{ with } x_1\leq 0,\\ u(x)=x_1 & \text{for } x\in\partial\mathcal{W}_{R_1,R_2} \text{ with } x_1>0.\end{cases} \tag{4.70}$$

The scaled limit function for $R_i\to\infty$ governs the local behavior through the gap. We show that this limit exists. We will need two auxiliary functions.

Lemma 4.8 *Two auxiliary functions on $\mathcal{W}_{R_1,R_2}$:*

- *Let $u_{1,1}$ be the solution of (4.70) with $R_1=R_2=1$, then*

$$u_{sub}(x)=\begin{cases} u_{1,1}(x) & \text{for } x\in B_1(0),\\ x_1 & \text{for } x\notin B_1(0) \text{ with } x_1>0,\\ 0 & \text{for } x\notin B_1(0) \text{ with } x_1\leq 0,\end{cases} \tag{4.71}$$

is a subsolution for (4.70) for all $R_1, R_2\geq 1$.

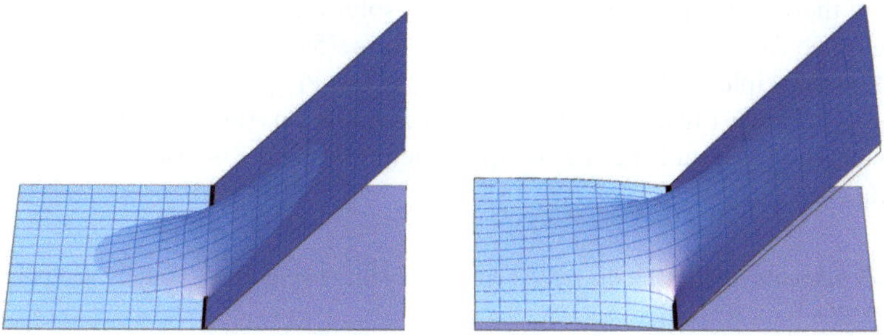

Fig. 4.6 On the *left* the subsolution u_{sub}; on the *right* the supersolution u_{super}

- *Let* $\sqrt{\cdot} : \mathbb{C}\backslash(-\infty, 0] \to \mathbb{R}^+ + i\mathbb{R}$ *be the analytical square root function with the cut on the negative real axis. Then*

$$u_{super}(x) = \tfrac{1}{2}\mathrm{Re}\left(\sqrt{1 - i(x_1 + ix_2)}\sqrt{1 + i(x_1 + ix_2)} + (x_1 + ix_2)\right) \quad (4.72)$$

is a supersolution for (4.70) for all $R_1, R_2 \geq 1$.

Proof Since u_{sub} is harmonic on $B_1(0)$, on $\mathbb{R}^+ \times \mathbb{R}\backslash B_1(0)^+$, on $\mathbb{R}^- \times \mathbb{R}\backslash B_1(0)^-$, is continuous on $\mathbb{R}^2$, and has a jump in the normal derivative on $\partial B_1(0)$ with the right sign, this function is subharmonic. Since also $u_{\text{sub}} = u_{R_1,R_2}$ on $\partial \mathcal{W}_{R_1,R_2}$, it is a subsolution for (4.70) for any $R_1, R_2 \geq 1$.

For the second function, note that the cuts of the square roots coincide with $(-i\infty, -i]$ respectively $[i, i\infty)$. Being the real part of an analytic function, u_{super} is harmonic on $\mathcal{W}_{\infty,\infty}$. It can be continuously extended to $\mathbb{R}^2$ and $u_{\text{super}}(0, x_2) = 0$ for $|x_2| \geq 1$. Moreover, since for $z \in \mathbb{C}$ we have

$$\sqrt{1 - iz}\sqrt{1 + iz} = \mathrm{sign}(\mathrm{Re}(z))\ z\left(1 + \tfrac{1}{2}z^{-2} + \mathcal{O}\left(z^{-4}\right)\right) \text{ for } |z| \to \infty, \quad (4.73)$$

it follows that $u_{\text{super}}(x) = \max(0, x_1) + |x_1|\,\mathcal{O}\left(|x|^{-2}\right)$ for $|x| \to \infty$. The maximum principle yields $u_{\text{super}}(x) \geq \max(0, x_1)$, with even a strict inequality in the interior, and hence that u_{super} is a supersolution for (4.70) for any $R_1, R_2 \geq 1$. □

Proposition 4.2 *Let* $R_1, R_2 \geq 1$ *and let the functions* u_{sub} *and* u_{super} *be defined in (4.71) and (4.72).*

- *Then there is a unique solution* $u_{R_1,R_2} \in C^2\left(\mathcal{W}_{R_1,R_2}\right) \cap C\left(\overline{\mathcal{W}_{R_1,R_2}}\right)$ *of (4.70) with*

$$u_{sub}(x) \leq u_{R_1,R_2}(x) \leq u_{super}(x) \text{ for all } x \in \overline{\mathcal{W}_{R_1,R_2}}. \quad (4.74)$$

- *Moreover, if* $1 \leq r_1 \leq R_1$ *and* $1 \leq r_2 \leq R_2$, *then*

$$u_{r_1,r_2}(x) \leq u_{R_1,R_2}(x). \quad (4.75)$$

- *and for all* $x \in \mathcal{W}_{R_1,R_2}$:

$$0 \leq u_{super}(x) - u_{R_1,R_2}(x) \leq \frac{1}{R_1} + \frac{1}{R_2}.$$

Proof If there is a supersolution above a subsolution then there exists a solution in between. See for example [4]. The estimate in (4.75) is a direct consequence of the maximum principle for first $u_{R_1,R_2} - \max(0, x_1)$ on $\mathcal{W}_{r_1,r_2}$ and next for $u_{R_1,R_2} - u_{r_1,r_2}$ on $\mathcal{W}_{R_1,R_2}$. The last result follows from an estimate on $\partial\mathcal{W}_{r_1,r_2}$ by the power series expansion in (4.73) and the maximum principle for $u_{\text{super}} - u_{R_1,R_2}$ on $\mathcal{W}_{R_1,R_2}$ See Fig. 4.6. □

Acknowledgments We thank the referee for reading the chapter very carefully and pointing out several errors.

References

1. R. Bañuelos, T. Carroll, Extremal problems for conditioned Brownian motion and the hyperbolic metric. Ann. Inst. Fourier (Grenoble). **50**(5), 1507–1532 (2000)
2. M. van den Berg, A. Dall'Acqua, G. Sweers, Estimates for the expected lifetime of conditioned Brownian motion. Proc. R. Soc. Edinburgh Sect. A, **137**(5), 1091–1099 (2007)
3. K.L. Chung, Z.X. Zhao, in *From Brownian Motion to Schrödinger's Equation*. Grundlehren der Mathematischen Wissenschaften, vol. 312 (Springer, Berlin, 1995)
4. P. Clément, G. Sweers, Getting a solution between sub- and supersolutions without monotone iteration. Rend. Istit. Mat. Univ. Trieste. **19**(2), 189–194 (1987)
5. M. Cranston. Lifetime of conditioned Brownian motion in Lipschitz domains. Z. Wahrsch. Verw. Gebiete. **70**(3), 335–340 (1985)
6. M. Cranston, T. R. McConnell, The lifetime of conditioned Brownian motion. Z. Wahrsch. Verw. Gebiete. **65**(1), 1–11 (1983)
7. M. Cranston, E. Fabes, Z. Zhao, Conditional gauge and potential theory for the Schrödinger operator. Trans. Amer. Math. Soc. **307**(1), 171–194 (1988).
8. J.H. Curtiss, *Introduction to Functions of a Complex Variable*. Monographs and Textbooks in Pure and Applied Math, vol. 44 (Marcel Dekker Inc., New York, 1978; with a foreword by E. F. Beckenbach)
9. A. Dall'Acqua, On the lifetime of a conditional Brownian motion in the ball. J. Math. Anal. Appl. **335**(1), 389–405 (2007)
10. A. Dall'Acqua, G. Sweers, Estimates for Green function and Poisson kernels of higher order Dirichlet boundary value problems. J. Differential Equations. **205**(2), 466–487 (2004)
11. A. Dall'Acqua, H.-C. Grunau, G. Sweers, On a conditioned Brownian motion and a maximum principle on the disk. J. Anal. Math. **93**, 309–329 (2004)
12. J.L. Doob, Conditional Brownian motion and the boundary limits of harmonic functions. Bull. Soc. Math. France. **85**, 431–458 (1957)
13. J.L. Doob, *Classical Potential Theory and its Probabilistic Counterpart*. Grundlehren der Mathematischen Wissenschaften, vol. 262 (Springer, New York, 1984)
14. M. Erven, On the lifetime of a conditioned Brownian motion. Ph.D. thesis, University of Cologne, Germany, 2012, http://kups.ub.uni-koeln.de/4565/1/Dissertation_Erven.pdf
15. P.S. Griffin, T.R. McConnell, G. Verchota, Conditioned Brownian motion in simply connected planar domains. Ann. Inst. H. Poincaré Probab. Statist. **29**(2), 229–249 (1993)
16. C. Pommerenke, in *Boundary Behaviour of Conformal Maps*. Grundlehren der Mathematischen Wissenschaften, vol. 299 (Springer, Berlin, 1992)
17. G. Sweers, Positivity for a strongly coupled elliptic system by Green function estimates. J. Geom. Anal. **4**(1), 121–142 (1994)
18. G. Sweers, Positivity for the Navier bilaplace, an anti-eigenvalue and an expected lifetime. DCDS Ser. S **7**(4), 839–855 (2014)

19. J. Xu, The lifetime of conditioned Brownian motion in planar domains of infinite area. Probab. Theory Related Fields. **87**(4), 469–487 (1991)
20. Z.X. Zhao, Uniform boundedness of conditional gauge and Schrödinger equations. Comm. Math. Phys. **93**(1), 19–31 (1984)
21. Z.X. Zhao, Green function for Schrödinger operator and conditioned Feynman-Kac gauge. J. Math. Anal. Appl. **116**(2), 309–334 (1986)

Analyticity of Rotational Water Waves

Joachim Escher and Bogdan-Vasile Matioc

Abstract The aim of this survey is to review some recent results concerning the regularity properties of two-dimensional rotational free-surface flows. It is shown that for large classes of vorticity distributions, the corresponding free water surface together with all streamlines beneath are real-analytic curves. The models considered here include, besides classical periodic water waves of finite depth, solitary waves, waves with infinite depth, capillary waves, and waves over stratified flows. It is also pointed out that the analyticity of the streamlines leads to an intrinsic characterization of symmetric solitary waves with one single crest.

1 Introduction

The regularity theory for nonlinear water waves has both a long tradition and represents an active field of mathematical research. Various (partly very) recent studies on this topic are located between two landmarks: On the one hand, there is Lewy's famous result [34] ensuring that two-dimensional waves traveling over an *irrotational* flow and which do have a C^1-profile are in fact represented by real-analytic curves. On the other hand, the existence of the so-called Stokes's wave of greatest height (conjectured by Stokes in [52] and rigorously established in [3, 55]) implies that there are steady waves with a stagnation point at their crest where the free surface is only Lipschitz continuous but not C^1.

The picture for *rotational* water waves appears with a much deeper complexity, the study of rotational flows being vital in the physical context, particularly from the point of view of modeling wave–current interactions [7, 54]. It is the aim of this survey chapter to review some recent results on the regularity theory for the classical water wave problem for rotational flows. We first expose that Lewy's result holds true for periodic and solitary rotational water waves without stagnation points, even if the

J. Escher (✉) · B.-V. Matioc
Institute for Applied Mathematics, Leibniz University of Hannover,
30167 Hannover, Germany
e-mail: escher@ifam.uni-hannover.de

B.-V. Matioc
e-mail: matioc@ifam.uni-hannover.de

J. Escher et al., (eds.), *Elliptic and Parabolic Equations,* Springer Proceedings in Mathematics & Statistics 119, DOI 10.1007/978-3-319-12547-3_5

vorticity field is merely integrable. This is not only mathematically intriguing but also, in view of Helmholtz's law[1], quite natural and experimentally substantiated, since discontinuous or unbounded vorticity functions are rather common in applications, cf. [4, 50, 53].

It is a matter of common knowledge that flows with stagnation points may loose regularity. We already mentioned the prominent Stokes's wave which is a paradigmatic example for this phenomenon. In contrast to irrotational flows, in which stagnation points can only appear on the free surface, cf. [7, 11], rotational flows allow for internal stagnation points. Even more, there are situations in which rotational flows with internal stagnation points possess nevertheless a real-analytic wave profile.

Finally, let us remark that for solitary water waves, the analyticity of the free surface is the main ingredient to prove an intrinsic characterization of symmetric waves: A solitary wave is symmetric and has precisely one crest if and only if all fluid particles maximize their distance to the fluid bed on a vertical line, cf. Theorem 6. This result is new, even for irrotational flows.

2 Two-Dimensional Periodic Flows on Water of Finite Depth

This section is dedicated to periodic water waves propagating over a flat fluid bed. The basis of our investigation is the inviscid isothermal fluid mechanics. We assume that gravity is the only body force, but we also include capillary forces on the free surface. This is a common physical setting to study water waves, cf. [33, 35].

2.1 The Governing Equations and Equivalent Formulations

We shall consider only water waves propagating at the surface of a two-dimensional fluid body. To present the model equations, we choose the Cartesian coordinates (x, y, z) such that the x-axis is the direction of wave propagation and the y-axis points vertically upwards, the fluid's motion being independent of the z-coordinate. Moreover, the free surface of the fluid is assumed to be the graph $y = \eta(t, x)$, with t denoting the time variable. The fluid domain is denoted by Ω_η and it is bounded from below by the flat bed $y = -d$. The constant d is chosen such that $-d < \eta(t, x)$ for all $(t, x) \in \mathbb{R}^2$. Because of our assumptions, the dynamic in the fluid is modeled

[1] Helmholtz's law for two-dimensional flows ensures that the vorticity is constant along C^1-trajectories. This implies particularly that once the vorticity does not vanish in a single point irrotationality is lost forever.

by Euler's equations:

$$\begin{cases} u_t + uu_x + vu_y = -P_x \\ v_t + uv_x + vv_y = -P_y - g \end{cases} \qquad \text{in } \Omega_\eta. \tag{1a}$$

Hereby we use P to denote the dynamic pressure, u is the horizontal velocity component, v is the vertical velocity component, g stands for the gravity constant, and the water' s density is taken to be $\rho = 1$. Additionally, we assume that the fluid is incompressible, meaning that the divergence of the velocity field vanishes:

$$u_x + v_y = 0 \qquad \text{in } \Omega_\eta. \tag{1b}$$

We still need to impose boundary conditions on the bed and at the wave surface. First, the flat bed is taken to be impermeable, expressed by

$$v = 0 \qquad \text{on } y = -d. \tag{1c}$$

The Laplace–Young equation states that the pressure jump at the wave surface is proportional to the mean curvature of the surface, the constant of proportionality being the surface tension coefficient σ, which leads us to the *dynamic boundary condition:*

$$P = P_0 - \sigma \frac{\eta_{xx}}{(1+\eta_x^2)^{3/2}} \qquad \text{on } y = \eta(t,x). \tag{1d}$$

The constant P_0 stands for the uniform pressure in the air. The second boundary condition on the wave surface expresses the fact that the free surface of the water consists at all times of the same particles, and reads as

$$v = \eta_t + u\eta_x \qquad \text{on } y = \eta(t,x), \tag{1e}$$

the so-called *kinematic boundary condition*. We refer to [7, 33] for a detailed derivation of the above boundary conditions for surface free water waves.

We consider herein only traveling wave solutions of the problem (1). To be more precise, we presuppose that

$$\eta(t,x) = \tilde{\eta}(x - ct), \qquad (u,v)(t,x,y) = (\tilde{u},\tilde{v})(x - ct, y),$$

with $c > 0$, denoting the wave speed. This assumption eliminates the time from the problem and we are left with the following set of equations, where we use again the notation η and (u, v) as a substitute for $\tilde{\eta}$ and $(\tilde{u}, \tilde{v})$:

$$\begin{cases} (u-c)u_x + vu_y = -P_x & \text{in } \Omega_\eta, \\ (u-c)v_x + vv_y = -P_y - g & \text{in } \Omega_\eta, \\ u_x + v_y = 0 & \text{in } \Omega_\eta, \\ P = P_0 - \sigma\eta''/(1+\eta'^2)^{3/2} & \text{on } y = \eta(x), \\ v = (u-c)\eta' & \text{on } y = \eta(x), \\ v = 0 & \text{on } y = -d. \end{cases} \tag{2}$$

Classical solutions (u, v, P, η) of (2) belong to the following regularity class:

$$(u, v, P, \eta) \in (C^1_{per}(\overline{\Omega}_\eta))^3 \times C^2_{per}(\mathbb{R}), \tag{3}$$

the subscript *per* being used to express the λ-periodicity of (u, v, P, η) with respect to the variable x, and where the constant $\lambda > 0$ is the wavelength. We take η to have zero integral mean over a period, so that d is the average mean depth of the flow. A last assumption that we make is that the wave speed c exceeds the horizontal velocity of each particle within the fluid, that is,

$$\sup_{\Omega_\eta} u < c. \tag{4}$$

Recall that *stagnation points* are fluid particles that travel horizontally with the wave speed c. Particularly, condition (4) thus rules out the existence of stagnation points in the flow.

The structure of system (2) can be simplified by introducing the so-called *stream function* ψ. More precisely, let

$$\psi(x, y) := -p_0 + \int_{-d}^{y} (u(x, s) - c)\,\mathrm{d}s \qquad \text{for } (x, y) \in \overline{\Omega}_\eta. \tag{5}$$

Let us observe that $\psi_y = u - c$, while (1b) and the last two equations of (2) yield

$$\psi_x(x, y) = \int_{-d}^{y} u_x(x, s)\,\mathrm{d}s = -\int_{-d}^{y} v_y(x, s)\,\mathrm{d}s = -v(x, y).$$

Hence, $\nabla\psi = (-v, u - c)$ is orthogonal to the velocity field $(u - c, v)$ of the fluid motion observed from a reference frame that moves with the wave at the same speed. Moreover, $\psi \in C^2_{per}(\overline{\Omega}_\eta)$ and

$$\frac{\mathrm{d}}{\mathrm{d}x}\psi(x, \eta(x)) = \nabla\psi(x, \eta(x)) \cdot (1, \eta'(x)) = 0 \qquad \text{for } x \in \mathbb{R}, \tag{6}$$

meaning that we can choose $\psi = 0$ on $y = \eta(x)$ if we define the constant p_0 by

$$p_0 := \int_{-d}^{\eta(x)} (u(x, s) - c)\,\mathrm{d}s$$

for some $x \in \mathbb{R}$. The constant $p_0 < 0$ is independent of x, cf. (6). It represents the relative mass flux.

No restriction is made here on the vorticity:

$$\omega := u_y - v_x$$

of the water flows that we consider. Clearly, we have that $\omega = \Delta\psi$. Moreover, under the assumption (4), the vorticity possesses the following useful representation.

Lemma 1 (The vorticity function) *There is a function* $\gamma \in C([p_0, 0])$ *— the so-called* vorticity function *of the flow — such that*

$$\omega(x, y) = \gamma(-\psi(x, y)) \qquad \text{for all } (x, y) \in \bar{\Omega}_\eta. \tag{7}$$

Proof We first introduce new variables via the (partial) *hodograph transformation*[2] $\mathcal{H} : \bar{\Omega}_\eta \to \bar{\Omega}$, with $\Omega := \mathbb{R} \times (p_0, 0)$, by the relation

$$\mathcal{H}(x, y) := (q, p)(x, y) = (x, -\psi(x, y)) \qquad \text{for } (x, y) \in \bar{\Omega}_\eta. \tag{8}$$

This transformation appears already in [17]. We note that the assumption (4) ensures that $\mathcal{H} \in \mathrm{Diff}^2(\Omega_\eta, \Omega)$. Observing that

$$\frac{\partial(q, p)}{\partial(x, y)} = \begin{pmatrix} 1 & 0 \\ v & c - u \end{pmatrix} \qquad \text{and} \qquad \frac{\partial(x, y)}{\partial(q, p)} \circ \mathcal{H} = \begin{pmatrix} 1 & 0 \\ -\frac{v}{c-u} & \frac{1}{c-u} \end{pmatrix}, \tag{9}$$

we compute by using integration by parts and Euler's equations that for each test function $\varphi \in \mathcal{D}(\Omega)$ we have

$$\begin{aligned}
\left\langle \frac{\mathrm{d}}{\mathrm{d}q}(\omega \circ \mathcal{H}^{-1}) \Big| \varphi \right\rangle &= \int_{\Omega_\eta} \omega(u - c)\varphi_q \circ \mathcal{H} \, \mathrm{d}(x, y) \\
&= \int_{\Omega_\eta} (u_y - v_x)[(u - c)(\varphi \circ \mathcal{H})_x + v(\varphi \circ \mathcal{H})_y] \, \mathrm{d}(x, y) \\
&= - \int_{\Omega_\eta} [(u - c)v_x + vv_y](\varphi \circ \mathcal{H})_x \, \mathrm{d}(x, y) \\
&\quad + \int_{\Omega_\eta} [(u - c)u_x + vu_y](\varphi \circ \mathcal{H})_y \, \mathrm{d}(x, y) \\
&= \int_{\Omega_\eta} [(P_y + g)(\varphi \circ \mathcal{H})_x - P_x(\varphi \circ \mathcal{H})_y] \, \mathrm{d}(x, y) = 0.
\end{aligned}$$

This shows that $\omega \circ \mathcal{H}^{-1}$ depends only on the variable p. Hence, $\gamma := \omega \circ \mathcal{H}^{-1} \in C([p_0, 0])$ and $\omega = \gamma \circ \mathcal{H}$, the latter relation being equivalent to (7). □

To complete the stream function formulation, we recall Bernoulli's law which states that the *total energy*

$$E := \frac{(u - c)^2 + v^2}{2} + g(y + d) + P - \int_0^\psi \gamma(-s) \, \mathrm{d}s \tag{10}$$

[2] The mapping $\mathcal{H}$ is called a *partial* hodograph transformation, because its first component is trivial. The full hodograph transformation involves the velocity potential ϕ – the holomorphic conjugate to the stream function ψ, that is $\varphi + i\psi$ is holomorphic. Note that φ is only well defined for irrotational flows. Furthermore, it is worthwhile to mention that the full hodograph transformation is a conformal mapping, in contrast to the partial hodograph transformation $\mathcal{H}$.

is constant in the fluid domain of a flow with vorticity function γ. Indeed, this follows from (2) in combination with (7). Concluding, $(\eta, \psi) \in C^2_{per}(\mathbb{R}) \times C^2_{per}(\bar{\Omega}_\eta)$ solves the following free boundary problem:

$$\begin{cases} \Delta\psi = \gamma(-\psi) & \text{in } \Omega_\eta, \\ |\nabla\psi|^2 + 2g(y+d) - 2\sigma\eta''/(1+\eta'^2)^{3/2} = Q & \text{on } y = \eta(x), \\ \psi = 0 & \text{on } y = \eta(x), \\ \psi = -p_0 & \text{on } y = -d, \end{cases} \tag{11}$$

with the positive constant $Q := 2(E - P_0)$. In terms of ψ, the condition (4) appears as

$$\sup_{\Omega_\eta} \psi_y < 0. \tag{12}$$

Proposition 1 *The Euler and the stream function formulation are equivalent, that is:*

(i) The Eqs. (2) and (4) with (u, v, P, η) satisfying (3);

are equivalent to

(ii) The Eqs. (11) and (12) with $(\eta, \psi) \in C^2_{per}(\mathbb{R}) \times C^2_{per}(\bar{\Omega}_\eta)$ and $\gamma \in C([p_0, 0])$

Proof We are left to show that if $(\eta, \psi) \in C^2_{per}(\mathbb{R}) \times C^2_{per}(\bar{\Omega}_\eta)$ is a solution of (11) and (12) for some $\gamma \in C([p_0, 0])$, then we can associate to it a tuple (u, v, P, η) satisfying (2)–(4). It is natural to define the velocity field via $\nabla\psi = (-v, u - c)$, and P via Bernoulli's law (10):

$$P := P_0 + \frac{Q}{2} - \frac{|\nabla\psi|}{2} - g(y+d) + \int_0^\psi \gamma(-s)\,ds.$$

The rest of the claim follows by direct computations. □

For the existence theory of solutions to (11), cf. [10], it is useful to express the hydrodynamical problem in terms of the *height function* $h : \bar{\Omega} \to \mathbb{R}$, defined by

$$h := y \circ \mathcal{H}^{-1} + d. \tag{13}$$

The function $h \in C^2_{per}(\bar{\Omega})$ associates to each pair $(q, p) \in \Omega$, the height of the fluid particle $(x, y) := \mathcal{H}^{-1}(q, p)$ above the bed. It follows easily from (9) that h solves the following quasilinear problem:

$$\begin{cases} (1+h_q^2)h_{pp} - 2h_p h_q h_{pq} + h_p^2 h_{qq} - \gamma h_p^3 = 0 & \text{in } \Omega, \\ 1 + h_q^2 + (2gh - Q)h_p^2 - 2\sigma \dfrac{h_p^2 h_{qq}}{(1+h_q^2)^{3/2}} = 0 & \text{on } p = 0, \\ h = 0 & \text{on } p = p_0, \end{cases} \tag{14}$$

while the assumption (4) is equivalent to

$$\inf_{\Omega} h_p > 0. \tag{15}$$

Proposition 2 *The stream function and the height function formulation are equivalent, that is:*

(i) The Eqs. (11) and (12) with $(\eta, \psi) \in C^2_{per}(\mathbb{R}) \times C^2_{per}(\bar{\Omega}_\eta)$ and $\gamma \in C([p_0, 0])$;

are equivalent to

(ii) The Eqs. (14) and (15) with $h \in C^2_{per}(\bar{\Omega})$ and $\gamma \in C([p_0, 0])$

Proof Given $\gamma \in C([p_0, 0])$, we only need to associate to each solution $h \in C^2_{per}(\bar{\Omega})$ of (14) and (15) a constant $d > 0$ and a pair $(\eta, \psi) \in C^2_{per}(\mathbb{R}) \times C^2_{per}(\bar{\Omega}_\eta)$ solving (11) and (12). We set

$$d := \frac{1}{\lambda} \int_0^\lambda h(q, 0)\, dq \qquad \text{and} \qquad \eta := h(\cdot, 0) - d. \tag{16}$$

Because $h > 0$ in Ω, the constant d is positive, $\eta \in C^2_{per}(\mathbb{R})$ has integral mean zero, and $\eta + d > 0$. In view of (13), it is natural to define, for fixed $x \in \mathbb{R}$, the function $\psi(x, \cdot)$ as the solution of the initial value problem:

$$\psi_y(x, y) = -1/h_p(x, -\psi(x, y)) \quad \text{for } y \le \eta(x), \qquad \psi(x, \eta(x)) = 0. \tag{17}$$

For each $x \in \mathbb{R}$, the solution of (5) exists on a maximal interval $[y(x), \eta(x)]$ with $\psi(x, y(x)) = -p_0$. Recalling (16), we obtain by integrating (5) on $[y, \eta(x)]$, with $y \in [y(x), \eta(x))$, that $h(x, -\psi(x, y)) = y + d$ in $[y(x), \eta(x)]$. It follows now from the last equation of (14) and (15) that $y(x) = -d$ for all $x \in \mathbb{R}$, and then it is easy to see that $\psi \in C^2_{per}(\bar{\Omega}_\eta)$. That the pair (η, ψ) solves (11) and (12) follows via (9). □

The equivalent formulations of the water wave problem (2) presented in the Propositions 1 and 2 are classical [7, 10]. The ideas used in the proofs of these results are essential also to derive the equivalence of the three formulations in the setting of strong solutions[3] (see (ii) of the discussion following Theorem 2).

We close this subsection by observing that the height function formulation is not only appealing from the mathematical point of view – as it leads to a uniformly quasilinear elliptic boundary value problem for only one unknown, and the problem is posed in the fixed domain Ω, cf. (15), – but it also gives information about the flow associated to each h. As defined in (16), the function $[q \mapsto h(q, 0) - d]$ is the free surface of the wave, while $h(\cdot, p_0) - d$ parametrizes the flat bed. Furthermore, for each $p \in (p_0, 0)$, it holds that

$$\psi(q, h(q, p) - d) = -p,$$

[3] By a *strong solution* of any of the above formulations, we mean a set of functions that satisfies the corresponding system of equations almost everywhere and which fulfils (4), (12), or (15), respectively.

meaning that $[q \mapsto h(q,p) - d]$ is, for each $p \in [p_0, 0]$, a level curve of ψ. Since $\nabla\psi$ is orthogonal on $(u-c, v)$, the level curves of ψ are the *streamlines* of the flow (2) in the moving frame, that is fluid lines which are tangent to the velocity field $(u-c, v)$. Recalling (12), we conclude that each streamline is parametrized by the function $[q \mapsto h(q,p) - d]$ for a unique $p \in [p_0, 0]$.

2.2 Weak Interpretations of the Height Function Formulation

Recently, Constantin and Strauss [12] considered the problem (2) and (4) in the context of gravity water waves with merely bounded vorticity. They established the existence of strong solutions of the hydrodynamical problem corresponding to water waves with small and large amplitude by using a weak formulation of the Eqs. (14). Small amplitude capillary and capillary-gravity periodic waves with a rough – that is discontinuous or unbounded – vorticity, when merely $\gamma \in L_r((p_0, 0))$ for some $r \in (1, \infty]$, were constructed in [39, 48], based also on a suitable weak interpretation of the height function formulation (14).

We also present these weak interpretations here, since they have been successfully exploited in [20, 48] to establish the real-analyticity of the wave surface together with all other streamlines for various types of waves (even solitary waves) over water of finite depth under fairly mild regularity assumptions on the vorticity function.

These formulations originated from the observation made in [12] that the first equation of (14) can be written in divergence form:

$$\left(\frac{h_q}{h_p}\right)_q - \left(\Gamma + \frac{1+h_q^2}{2h_p^2}\right)_p = 0 \qquad \text{in } \Omega, \tag{18}$$

with $\Gamma : [p_0, 0] \to \mathbb{R}$ given by

$$\Gamma(p) = \int_0^p \gamma(s)\, \mathrm{d}s \qquad \text{for } p \in [p_0, 0].$$

Hence, given $\gamma \in L_1((p_0, 0))$, we may aspire to realize (18) in the distributional sense. Following [12], we introduce now the notion of weak solution to (14).

Definition 1 Assume that $\sigma = 0$ and $\gamma \in L_1((p_0, 0))$. A function $h \in C^1_{per}(\overline{\Omega})$ which satisfies (15) is called a *weak solution* of (14) if (18) holds in the distributional sense and if h satisfies the boundary conditions

$$\begin{cases} 1 + h_q^2 + (2gh - Q)h_p^2 = 0 & \text{on } p = 0, \\ h = 0 & \text{on } p = p_0 \end{cases} \tag{19}$$

pointwise.

Given $u, v \in W^1_{1,loc}$ with $uv \in L_{1,loc}$ and $u\partial v + v\partial u \in L_{1,\mathrm{loc}}$, we have

$$\partial(uv) = u\partial v + v\partial u, \tag{20}$$

where ∂ means (weak) partial differentiation, cf. Sect. 7 in [24]. Based on this product formula it is not difficult to see that any strong solution $h \in W_1^2(\Omega) \cap C^1(\bar{\Omega})$ of (14) and (15) is a weak solution of (14) too.[4] Particularly, each classical solution $h \in C^2_{per}(\bar{\Omega})$ of (14) and (15) is a weak solution of (14).

Let us emphasize that when considering surface tension effects, that is in the case $\sigma > 0$, the boundary condition of (14) on $p = 0$ is not well defined even in the setting of strong solutions $h \in W_r^2(\Omega) \cap C^{1+\alpha}(\bar{\Omega})$ of (14) and (15), with $r \in [1, \infty]$ and $\alpha := (r-1)/r \in [0, 1]$.[5] Therefore, a weak interpretation of this boundary condition has to be defined. As observed in [48], the operator

$$1 - \partial_q^2 : C^2_{per}(\mathbb{R}) \to C_{per}(\mathbb{R})$$

is an isomorphism. Hence, if h is a classical solution of (14) and (15), the boundary condition on $p = 0$ can be reexpressed as the following nonlinear and nonlocal equation:

$$h + (1 - \partial_q^2)^{-1} \operatorname{tr}_0 \left(\frac{\left(1 + h_q^2 + (2gh - Q)h_p^2\right)(1 + h_q^2)^{3/2}}{2\sigma h_p^2} - h \right) = 0 \qquad \text{on } p = 0, \tag{21}$$

with tr_0 denoting the trace operator on with respect to the boundary component $p = 0$ of $\bar{\Omega}$, that is $\operatorname{tr}_0 h = h(\cdot, 0)$ for all $h \in C_{per}(\bar{\Omega})$. This fact enables us to introduce a suitable notion of weak solution to (14) in the case when capillarity effects are present.

Definition 2 Assume that $\sigma > 0$ and $\gamma \in L_1((p_0, 0))$. A function $h \in C^1_{per}(\bar{\Omega})$ with $h = 0$ on $p = p_0$ and which satisfies (15) is called a *weak solution* of (14) if (18) holds in the distributional sense and if h satisfies (21) in $C_{per}(\mathbb{R})$.

2.3 Regularity Results for Periodic Flows on Water of Finite Depth

As we shall see, the notion of weak solutions is quite useful since it enables us to consider the hydrodynamical problem (14) under weak assumptions on the vorticity function in suitable abstract functional analytic settings, cf. [12, 39, 48]. The benefit of this approach is twofold: Firstly, it allows the development of a rigorous existence theory for periodic waves of finite depth with rough vorticity [12, 39, 48]; and secondly, it is the basis of the derivation of a priori regularity properties for such waves [20, 48]. We discuss here the regularity results obtained in [20, Theorem 1.1] and [48, Theorem 5.1] and present some recent implications of them.

---

[4] In the Sects. 2.2 and 2.3, the Sobolev spaces are defined in the context of λ-periodic functions with respect to the variable x.

[5] If $r = \infty$, then $\alpha = 1$. In this case, we use the notation $C^{1+1}(\bar{\Omega}) := W^2_\infty(\Omega)$.

Theorem 1 (Regularity of periodic waves) *Assume that* $\gamma \in L_1((p_0, 0))$ *and let* $\sigma \geq 0$ *and* $\alpha \in (0,1)$. *Given a weak solution* $h \in C^{1+\alpha}_{per}(\bar{\Omega})$ *of 14, we have that the distributional derivatives* $\partial_q^m h \in C^{1+\alpha}_{per}(\bar{\Omega})$ *for all* $m \in \mathbb{N}$. *Moreover, there exists a constant* $L > 1$ *with the property that*

$$\|\partial_q^m h\|_{1+\alpha} \leq L^{m-2}(m-3)! \tag{22}$$

for all integers $m \geq 3$.

Particularly, all the streamlines (including the wave profile) are real-analytic curves.

The real-analyticity of the streamlines and of the wave profile is a direct consequence of (22), because this relation ensures the locally uniform convergence in q of the Taylor series associated to $[q \mapsto h(q, p)]$ towards the corresponding function for each $p \in [p_0, 0]$. The convergence takes place on open intervals with uniform length $1/L$. The proof of Theorem 1 exploits the invariance of the Eqs. (18), (19), and (21) with respect to horizontal translations and an iteration procedure which is based on the following global Schauder estimate for weak solutions of elliptic boundary value problems. We note that no periodicity assumption is made.[6]

Theorem 2 ([2, Theorem 9.3] and [24, Theorem 8.33]) *Let* $w \in BUC^{1+\alpha}(\Omega)$ *be a weak solution of the elliptic boundary value problem:*

$$\begin{cases} (a_{11}w_q)_q + (a_{12}w_p)_q + (a_{21}w_q)_p + (a_{22}w_p)_p = f_q + g_p & \text{in } \Omega, \\ \delta w + (1-\delta)(aw_q + bw_p) = \varphi & \text{on } p = p_0, \\ w = 0 & \text{on } p = p_0, \end{cases}$$

where $\delta \in \{0,1\}$, $\varphi \in BUC^{\delta+\alpha}(\Omega)$, *and* $f, g, a, b, a_{ij} \in BUC^{\alpha}(\Omega)$, $1 \leq i, j \leq 2$. *If the first equation is uniformly elliptic in* Ω *and* $\inf_\Omega |b| > 0$, *then*

$$\|w\|_{1+\alpha} \leq K(\|w\|_0 + \|f\|_\alpha + \|g\|_\alpha + \|\varphi\|_{\delta+\alpha}).$$

with a constant K *independent of* w.

It is worthwhile to add some comments concerning the statements of Theorem 1.

(*i*) Theorem 1 establishes the real-analyticity of the wave surface and of the streamlines for any classical solution:

$$(u, v, P, \eta) \in C^1_{per}(\bar{\Omega}_\eta) \times C^1_{per}(\bar{\Omega}_\eta) \times C^1_{per}(\bar{\Omega}_\eta) \times C^2_{per}(\mathbb{R})$$

of (2) and (4). The statement is true for periodic gravity, capillary-gravity, and pure capillary water waves.

---

[6] Given an open subset $U \subset \mathbb{R}^m$, $m \geq 1$, the space $BUC^k(U)$, $k \in \mathbb{N}$, consists of all functions which possess bounded uniformly continuous derivatives up to order k. Given $\alpha \in (0,1)$, the space $BUC^{k+\alpha}(U)$ contains all functions of $BUC^k(U)$ which have uniformly Hölder continuous derivatives of order k.

For gravity water waves, the real-analyticity of the free surface when $\gamma = 0$, that is for irrotational flows, has been established in [34] by using Schwarz's reflection principle, the property holding also for irrotational water waves with capillarity effects [40]. The real-analyticity of the free surface of rotational flows with additional Hölder regularity was established for general real-analytic vorticity functions in [9] for gravity waves, and in [26–29] for waves with surface tension effects, the streamlines being real-analytic even when requiring only Hölder continuity from the vorticity function [9, 26–29]. The method of Constantin and Escher [9] uses a novel weighted translation which is an adaption of an approach which had been used when analyzing the regularity of solutions of parabolic problems [1, 21]. Subsequently, it was shown with the help of a new iteration technique [5] that the Hölder continuity of the vorticity function ensures also the real-analyticity of the free surface of the wave.
All the above mentioned results are contained and generalized in Theorem 1.

(ii) Based on Theorem 1 it was recently shown in [48, Lemma 4.7] and [39, Theorem 2.1] that if $\gamma \in L_r((p_0, 0))$ and $r \in (1, \infty]$, then any weak solution $h \in C^{1+\alpha}_{per}(\bar{\Omega})$, with $\alpha := (r-1)/r$, of (14) possesses the additional regularity $h \in W^2_r(\Omega)$. Because weak solutions satisfy $\mathrm{tr}_0\, h \in C^{\omega}_{per}(\mathbb{R})$, it follows that any such solution belongs to $W^2_r(\Omega) \cap C^{1+\alpha}_{per}(\bar{\Omega})$ and is a strong solution of (14) and (15). It turns out that the three formulations are equivalent also in the setting of strong Sobolev solutions with weak Hölder regularity. Indeed, [39, Theorem 2.1] establishes the equivalence of:

(a) The height function formulation (14) and (15) with $h \in W^2_r(\Omega) \cap C^{1+\alpha}_{per}(\bar{\Omega})$ such that $\mathrm{tr}_0\, h \in W^2_r(\mathbb{R})$ and $\gamma \in L_r((p_0, 0))$.
(b) The formulation (11) and (12) with the stream function $\psi \in W^2_r(\Omega) \cap C^{1+\alpha}_{per}(\bar{\Omega})$, $\eta \in W^2_r(\mathbb{R})$, and $\gamma \in L_r((p_0, 0))$.
(c) The velocity (Euler) formulation (2) and (4) with $u, v, P \in W^1_r(\Omega_\eta) \cap C^{\alpha}_{per}(\bar{\Omega}_\eta)$ and $\eta \in W^2_r(\mathbb{R})$.

The only restriction on r is that $r > 1$, the surface tension coefficient being arbitrary $\sigma \geq 0$. The assumption that $\mathrm{tr}_0\, h \in W^2_r(\mathbb{R})$ is in fact an implication of $h \in W^2_r(\Omega) \cap C^{1+\alpha}_{per}(\bar{\Omega})$ solving (14) and (15). This equivalence result is fundamental when establishing the existence of small amplitude waves with surface tension and unbounded vorticity [39, Theorem 2.3].
For periodic gravity waves, there are former equivalence results established in the particular case of W^2_∞-solutions in [41] when $\gamma \in L_\infty((p_0, 0))$, or for strong W^2_r-solutions when $r > 2$, cf. [12, 19] (in this case one has due to Sobolev's embedding $W^2_r(\Omega) \hookrightarrow C^{1+\beta}_{per}(\bar{\Omega})$, but with $\beta := (r-2)/r$).

(iii) The real-analyticity of the wave surface and of the profile for periodic gravity waves of finite depth combined with the principle of analytic continuation were the crucial arguments in the proof of the following intrinsic characterization [45, Theorem 2.1] (see also [47]) of the symmetric waves:
Consider a periodic gravity water wave solution of (2)–(4) with a non-flat free surface. Moreover, assume that the vorticity function is Lipschitz continuous

in $[p_1, 0]$ for some $p_1 \in (p_0, 0)$. Then, the wave profile is symmetric and has only one crest and trough per period if and only if there exists a vertical line within the fluid domain such that all the fluid particles located on that line minimize there their distance to the fluid bed.

The above result is new even for irrotational flows. A similar result is true also for solitary waves of finite depth, cf. Theorem 6. We emphasize that a different characterization of the symmetric gravity waves was obtained also quite recently [8, 14, 49], the symmetry being implied by the existence of a unique crest per period.

(iv) Analyticity results in the limit case when assuming only that $\gamma \in L_1((p_0, 0))$ and $h \in C^1(\bar{\Omega})$ are not known so far. Consequently, the equivalence of the three formulations mentioned in (ii) is an open problem when $r = 1$ (and $\alpha = 0$).

2.4 Regularity of Periodic Waves with Stagnation Points

When allowing for stagnation points, the height function cannot be defined in general because the determinant of the Jacobian of $\mathcal{H}$ vanishes at these points. However, given $\gamma \in C(\mathbb{R})$ and any classical solution $(\eta, \psi) \in C^2_{per}(\mathbb{R}) \times C^2_{per}(\bar{\Omega}_\eta)$ of (11) which satisfies $\nabla\psi = 0$ at some point P in $\bar{\Omega}_\eta$, one can associate a classical solution of (2) with the corresponding flow containing a stagnation point at P (see the proof of Proposition 1). Notice that $\nabla\psi(P) = 0$ is equivalent to saying that P is a stagnation point.

The presence of stagnation points is intriguing and it has an impact on the regularity of the streamlines and of the wave profile as well. As mentioned in the introduction, irrotational water waves can possess stagnation points only at the wave surface, the most prominent example being the Stokes wave of extreme form possessing a stagnation point at each crest, with the wave profile forming an angle of $2\pi/3$ there. On the other hand, when allowing for surface tension effects the regularity of the free surface appears not to be influenced by the presence of stagnation points.

Theorem 3 ([42, Theorem 2.1 and Remark 2.1]) *Let $\sigma > 0$ and assume that the vorticity function $\gamma \in C^\omega(\mathbb{R})$ is bounded and satisfies $\gamma' \leq 0$. Then, the free surface of the flow associated to any classical solution $(\eta, \psi) \in C^2_{per}(\mathbb{R}) \times C^2_{per}(\bar{\Omega}_\eta)$ of (11) is real-analytic, that is, $\eta \in C^\omega(\mathbb{R})$.*

The proof of this result uses a parameter trick which exploits the parabolicity of the curvature operator and the regularity theory for nonlinear parabolic problems.

The Case of Constant Vorticity In contrast to Stokes waves, rotational water waves with constant vorticity can posses stagnation points within the flow [13, 37, 38, 56]. Nevertheless, even if the profile is overhanging, one can show that the free surface of such flows is smooth. The analysis of flows with constant vorticity, stagnation points, and possessing a possibly overhanging profile has been initiated in [13] (see [38] for the case when $\sigma > 0$), where a new formulation of the mathematical problem has been derived:

$$\left(a+\gamma\left(\frac{[w^2]}{2h}-w+\mathcal{C}_h(ww')-w\mathcal{C}_h(w')\right)\right)^2 \tag{23}$$

$$=\left(a^2+b+2\sigma\frac{w''+w''\mathcal{C}_h(w')-w'\mathcal{C}_h(w'')}{\left(w'^2+(1+\mathcal{C}_h(w'))^2\right)^{3/2}}-2gw\right)\left(w'^2+\left(1+\mathcal{C}_h(w')\right)^2\right). \tag{24}$$

The positive constant h is the *conformal mean depth,* γ is the vorticity constant, a and b are two real parameters, $[w]$ denotes the average of w over $[0, 2\pi]$, and $\mathcal{C}_h$ is the Fourier multiplier

$$\sum_{n\in\mathbb{Z}\setminus\{0\}} a_n e^{inx} \mapsto \sum_{n\in\mathbb{Z}\setminus\{0\}} -i\coth(nh)a_n e^{inx}. \tag{25}$$

It is shown in [13, 38] that if $w \in C^{2+\alpha}(\mathbb{R})$ is a 2π-periodic solution of the Eq. (23) which verifies

$$[w] = 0, \tag{26}$$

$$w > -h, \tag{27}$$

$$[\mathbb{R} \ni x \mapsto (x+\mathcal{C}_h(w)(x), w(x)+h)] \quad \text{is a one-to-one regular curve,} \tag{28}$$

then (28) describes the profile of a wave traveling above the flat bed located at $y = 0$. The next theorem collects results established in [13, Theorem 4.1], [37, Theorem 2.1], and [38, Theorem 10].

Theorem 4 *Let $h > 0$ and $\alpha \in (0, 1)$ be given.*

(i) Let $\sigma = 0$ and assume that $w \in C^{1+\alpha}(\mathbb{R})$ is a 2π-periodic solution of (23)–(28) which satisfies $a^2 + b - 2gw > 0$ in $\mathbb{R}$. Then $w \in C^\infty(\mathbb{R})$.
(ii) If $\sigma \neq 0$, then any 2π-periodic solution $w \in C^{2+\alpha}(\mathbb{R})$ of (23)–(28) is smooth $w \in C^\infty(\mathbb{R})$.

3 Solitary Waves on Water of Finite Depth

We consider now fluid motions confined to the influence of gravity only, that is we choose $\sigma = 0$. We present next the appropriately modified equations that describe the dynamics of solitary waves on water of finite depth.

3.1 Regularity of Solitary Waves

In the context of the hydrodynamical problem for solitary traveling waves, the assumption of periodicity of (u, v, P, η) is dropped, the system (2) being supplemented instead by the far-field boundary conditions:

$$\begin{cases} \eta(x) \to 0 & \text{as } |x| \to \infty, \\ v(x,y) \to 0 & \text{as } |x| \to \infty \text{ uniformly in } y, \end{cases} \tag{29}$$

expressing the fact that the flow is almost laminar far away, cf. Lemma 4. Classical solutions of the problem (2), (4), and (29) belong now to the class

$$(u, v, P, \eta) \in (BUC^1(\Omega_\eta))^3 \times BUC^2(\mathbb{R}). \tag{30}$$

The positive constant d is now the asymptotic depth of the fluid. In terms of the stream function ψ, which is defined by the same relation (5), the problem (2), (4), and (29) is expressed by (11) and (12), complemented with

$$\begin{cases} \eta(x) \to 0 & \text{for } |x| \to \infty, \\ \psi_x(x,y) \to 0 & \text{for } |x| \to \infty \text{ uniformly in } y. \end{cases} \tag{31}$$

Because of Bernoulli's principle and of (31), the positive constant

$$\mu := Q - 2gd \tag{32}$$

is the square of the relative upstream horizontal velocity in the far field. Correspondingly, in terms of the height function h, which is defined again by (13), the problem (2), (4), and (29) is described by (14), (15), and the far-field conditions:

$$\begin{cases} h(q,0) \to d & \text{as } |q| \to \infty, \\ h_q(q,p) \to 0 & \text{for } |q| \to \infty \text{ uniformly in } p. \end{cases} \tag{33}$$

The arguments used in the Propositions 1 and 2 show that the three formulations of the solitary wave problem are equivalent. In a similar manner to Definition 1, we define the following notion of weak solution to (14), (15), and (33).

Definition 3 Assume that $\sigma = 0$ and $\gamma \in L_1((p_0, 0))$. A function $h \in BUC^1_{per}(\Omega)$ which satisfies (15) and (33) is called a *weak solitary wave solution* if (18) holds in the distributional sense and if h satisfies the boundary conditions (19) pointwise.

Clearly, any classical solution of (14), (15), and (33) is a weak solitary wave solution. On the other hand, we have the following result.

Lemma 2 *Let $r \in (1, \infty]$ and $\alpha := (r-1)/r$. We assume that $\gamma \in L_r((p_0, 0))$ and that $h \in BUC^{1+\alpha}(\Omega)$ is a weak solitary wave such that $h_q \in BUC^{1+\alpha}(\Omega)$. Then, $h \in W^2_{r,loc}(\Omega)$ solves (14) almost everywhere in Ω.*

Proof Because $\partial_p(h_q) \in BUC^\alpha(\Omega)$, it follows that h_p is continuously differentiable with respect to q and that $\partial_q h_p = \partial_p h_q \in BUC^\alpha(\Omega)$. This property and the relation (18) imply that

$$\Gamma + \frac{1 + h_q^2}{2h_p^2} \in BUC^{1+\alpha}(\Omega). \tag{34}$$

Particularly, since $\Gamma \in W^1_{r,\mathrm{loc}}(\Omega)$, we deduce that

$$\frac{1+h_q^2}{2h_p^2} \in W^1_{r,loc}(\Omega).$$

Using the chain rule [24, Lemma 7.5] and the property (20), we get $h_p \in W^1_{r,loc}(\Omega)$, and the claimed assertions follow at once. □

As noted in Theorem 2, the global Schauder estimates can be applied to the Eqs. (14), (15), and (33), yielding the same regularity result as for periodic waves, cf. [20, Remark 1.3], and revealing that the assumption $h_q \in BUC^{1+\alpha}(\Omega)$ in Lemma 2 is not needed.

Theorem 5 (Regularity of solitary waves) *Assume that $\gamma \in L_1((p_0, 0))$. Given a weak solitary wave solution $h \in BUC^{1+\alpha}(\Omega), \alpha \in (0,1)$, of (14), (15), and (33), the distributional derivatives $\partial_q^m h \in BUC^{1+\alpha}(\Omega)$ for all $m \in \mathbb{N}$. Moreover, there exists a constant $L > 1$ with the property that*

$$\|\partial_q^m h\|_{1+\alpha} \leq L^{m-2}(m-3)!$$

for all integers $m \geq 3$.

Particularly, the streamlines and the wave profile of each weak solitary wave solution are real-analytic curves.

Theorem 5 generalizes previous results [5, 32, 46] where the regularity of the streamlines is established for $BUC^{2+\alpha}(\Omega)$, $\alpha \in (0,1)$, solutions with a Hölder continuous vorticity function under certain restriction on the parameter μ [32], or for arbitrary μ [46]. While in [32, 46] the real-analyticity of the profile is obtained only for real-analytic vorticity functions, the authors of [5] established this property for arbitrary Hölder continuous vorticity functions.

3.2 An Intrinsic Characterization of the Symmetric Solitary Waves

In the light of Theorem 5, the symmetry result established in [46, Theorem 3.1] can be improved in the sense that several conditions may be dropped: the assumption (*b*) of [46, Theorem 3.1], the Hölder continuity of the second derivatives of the solution, and the assumption that the difference to a laminar flow solution belongs to a small Hölder space. The statement can be reformulated as a characterization of the symmetric solitary waves in terms of the underlying flow.

Theorem 6 *Let $h \in BUC^2(\Omega)$ be a classical solution of (14), (15), and (33) with a non-flat free surface. Additionally, assume that the vorticity function $\gamma \in C([p_0, 0])$ is Lipschitz continuous in $[p_1, 0]$ for some $p_1 \in (p_0, 0)$.*

Then, the wave profile is symmetric and has only one crest [respectively trough] if and only if there exists a vertical line within the fluid domain such that all the fluid particles located on that line maximize [respectively minimize] there their distance to the fluid bed.

Moreover, the profile of a symmetric wave with one crest (respectively trough) is strictly monotone on either side of the crest line.

Remark 1 The Lipschitz continuity of γ is naturally satisfied if we consider more regular classical solutions, that is for $h \in W^3_\infty(\Omega)$. In particular, this property is satisfied in the setting of irrotational flows.

Remark 2 We note that the claim that the wave profile of a symmetric wave with only one crest [respectively trough] is strictly monotone on each side of the crest [respectively trough] line is a simple consequence of the real analyticity of the wave profile, cf. Theorem 5. In fact, the arguments presented in the study of the *Case* 3 in the proof of Proposition 4 below show that $\eta'(x) = 0$ if and only if $x = x_0$, whereby $x = x_0$ is the crest [respectively trough] line of the symmetric wave.

While the symmetry problem for periodic gravity waves has been considered first in [23], the symmetry of solitary gravity waves has been addressed later: In the irrotational context, Craig and Sternberg [15] have shown that if the constant μ, defined in (32) is larger than a certain critical value, then solitary waves possess a unique crest and are symmetric with respect to the crest line, being strictly monotone on each side of it. A similar result is true also for sufficiently regular rotational waves [31]. On the other hand, our result makes no restriction on $\mu > 0$ and the regularity assumptions are weaker than in [31]. The main tools for establishing Theorem 6 are sharp elliptic maximum principles (see Lemma 3 below) combined with the moving plane method and the regularity results of Theorem 5.

Lemma 3 (Serrin's corner point lemma) *Let*

$$R = \{(q, p) \in \mathbb{R}^2 : a < q < b,\ p_0 < p < f(q)\}$$

where $a < b$, $f \in C^2([a, b], (p_0, \infty))$ *and* $f'(a) = 0$ *[respectively* $f'(b) = 0$*]. Let further* $H \in C^2(\overline{R})$ *satisfy* $\mathcal{L}H \geq 0$ *in* R *for a uniformly elliptic operator* $\mathcal{L} = a_{ij}\partial_{ij} + b_i\partial_i$ *with continuous coefficients in* $\overline{R}$*. Additionally, assume that there exists a positive constant* K *such that*

$$|a_{12}(q, p)| \leq K(q - a) \qquad \textit{[respectively } |a_{12}(q, p)| \leq K(b - q)\textit{]} \tag{35}$$

for all $(q, p) \in R$*. If the corner point* $P = (a, f(a))$ *[respectively* $P = (b, f(b))$*] satisfies* $H(P) = 0$ *and if* $H < 0$ *in* R*, then either*

$$\frac{\partial H}{\partial s} < 0 \quad \textit{or} \quad \frac{\partial^2 H}{\partial s^2} < 0 \quad \textit{at } P,$$

where $s \in \mathbb{R}^2$ *is any direction at* P *that enters* R *non-tangentially.*

Proof See [51, Lemma 2]. □

Before proceeding to the proof of Theorem 6, we remark that because the flows that we consider are steady, the particle paths in the moving frame, that is the solutions of the system

$$\begin{cases} x' = u(x,y) - c, \\ y' = v(x,y), \end{cases}$$

coincide with the level curves of ψ. The latter are parametrized by the mappings $[q \mapsto h(q,p) - d]$ with $p \in [p_0, 0]$. Hence, the condition in Theorem 6 on the particle trajectories is equivalent to saying that all the streamlines attain their global maximum [respectively minimum] on the same line. This global maximum [respectively minimum] is a priori not unique though.

We establish herein the characterization in Theorem 6 only for symmetric waves which possess a single crest line, the other case being similar. In the first step, we prove the sufficiency. To this end, we observe first that the conditions (33) determine the flow in the far field.

We emphasize that merely the existence of a classical solution of (14), (15), and (33) imposes restrictions on the parameters μ and d and ensures the existence of a laminar flow solution $H = H(p)$. We note that the existence of such a laminar flow solution appears as a requirement in [31, 46].

Lemma 4 *Let $\gamma \in C([p_0, 0])$ and let $h \in BUC^2(\Omega)$ be a solution of (14), (15), and (33). Then, the constant μ defined by (32) satisfies $\mu > 2\max_{[p_0,0]}\Gamma$ and the asymptotic depth is*

$$d = \int_{p_0}^{0} \frac{dp}{\sqrt{\mu - 2\Gamma(p)}}. \tag{36}$$

Moreover, the function $H \in C^2([p_0, 0])$, defined by

$$H(p) := \int_{p_0}^{p} \frac{ds}{\sqrt{\mu - 2\Gamma(s)}} \qquad \textit{for } p \in [p_0, 0] \tag{37}$$

is a laminar flow solution of (14), (15), and (33), and we have

$$h(q,p) - H(p) \to 0 \qquad \textit{for } |q| \to \infty \textit{ uniformly in } p. \tag{38}$$

Proof We first claim that if $q_n \to \infty$, then we may extract a subsequence $(q_{n_k})_k$ such that

$$\lim_{k\to\infty} \max_{\substack{p \in [p_0,0] \\ q \in [-1,1]}} \left| \frac{1}{2h_p^2(q_{n_k} + q, p)} - \left(\frac{\mu}{2} - \Gamma(p)\right) \right| = 0. \tag{39}$$

The validity of (39) is shown in the second part of this proof.

We then note that (15) together with (39) yields that $\mu > 2\max_{[p_0,0]}\Gamma$. Hence, the function H given by (37) is well defined. In order to prove (38), we assume by contradiction that there is $\varepsilon > 0$ and sequences $q_n \to \infty$ and $(p_n)_n \subset [p_0, 0]$ with

$|h(q_n, p_n) - H(p_n)| \geq \varepsilon$ for all $n \in \mathbb{N}$. Even more, because $[p_0, 0]$ is compact, we may choose $p_n = \widetilde{p}$ for some $\widetilde{p} \in [p_0, 0]$ and all $n \in \mathbb{N}$, so that

$$|h(q_n, \widetilde{p}) - H(\widetilde{p})| \geq \varepsilon \qquad \text{for all } n \in \mathbb{N}. \tag{40}$$

We define now $f_n : \overline{D} \to \mathbb{R}$ by setting $f_n(q, p) := h(q_n + q, p) - H(p)$ for $(q, p) \in \overline{D}$, whereby we set $D := (-1, 1) \times (p_0, 0)$. Because $(f_n)_n$ is bounded in $C^2(\overline{D})$, there exists a function $f \in C^1(\overline{D})$ and a subsequence $(n_k)_k$ such that

$$f_{n_k} \to f \qquad \text{in } C^1(\overline{D}).$$

We may also assume that the sequence (q_{n_k}) satisfies (39). Recalling (33), it is easy to see that $\partial_q f = 0$ in D. Additionally, it follows readily from (39) that we have $h_p(q_{n_k} + q, p) \to H'(p)$ for all $(q, p) \in \overline{D}$, so that also $\partial_p f = 0$ in D. Particularly, f is a constant function. Evaluating f_{n_k} at $(0, 0)$ and $(0, p_0)$ and using the fact that $h = 0$ at $p = p_0$ we find, together with the relations (33) and (37), that

$$f = d - \int_{p_0}^{0} \frac{1}{\sqrt{\mu - 2\Gamma(p)}} \, \mathrm{d}p = 0.$$

This proves (36) and, since $f_{n_k}(0, \widetilde{p}) = h(q_{n_k}, \widetilde{p}) - H(\widetilde{p}) \to_{k \to \infty} 0$, we also see that our assumption (40) is false.

Since it is clear from (36) that H is a solution of (14), (15), and (33), we are left to establish (39). To this end, we define on the same rectangle $\overline{D} = [-1, 1] \times [p_0, 0]$ the functions $g_n : \overline{D} \to \mathbb{R}$ by

$$g_n(q, p) := \frac{1}{2h_p^2(q_n + q, p)} + \Gamma(p) \qquad \text{for } (q, p) \in \overline{D}.$$

Because $(g_n)_n$ is bounded in $C^1(\overline{D})$, we may extract a subsequence $(g_{n_k})_k$ which converges in $C(\overline{D})$ towards a function G. To identify G, we pick $\phi \in C_0^\infty((-1, 1))$ and $\varphi \in C_0^\infty((p_0, 0))$, and we let $\xi_k \in C_0^\infty(\Omega)$, $k \in \mathbb{N}$, be given by

$$\xi_k(q, p) := \phi(q - q_{n_k}) \, \varphi(p) \qquad \text{for } (q, p) \in \Omega.$$

Because h satisfies (18) and $\operatorname{supp} \xi_k \subset q_{n_k} + D$, we find that

$$\int_{q_{n_k}+D} \frac{h_q}{h_p} \partial_q \xi_k \, \mathrm{d}(q, p) = \int_{q_{n_k}+D} \left(\Gamma(p) + \frac{1 + h_q^2}{2h_p^2} \right) \partial_p \xi_k \, \mathrm{d}(q, p).$$

Changing variables and rearranging the terms we get

$$\begin{aligned} &\int_D \left(\frac{h_q}{h_p}(s + q_{n_k}, p)\phi'(s)\varphi(p) - \frac{h_q^2}{2h_p^2}(s + q_{n_k}, p)\phi(s)\varphi'(p) \right) \mathrm{d}(s, p) \\ &= \int_D g_{n_k}(s, p)\phi(s)\varphi'(p) \, \mathrm{d}(s, p) \end{aligned}$$

for all $k \in \mathbb{N}$. For $k \to \infty$, (33) yields

$$\int_{p_0}^{0} \Big(\int_{-1}^{1} G(s,p)\phi(s)\,\mathrm{d}s \Big) \varphi'(p)\,\mathrm{d}p = 0,$$

meaning that

$$\int_{-1}^{1} (G(s,p) - G(s,0))\phi(s)\,\mathrm{d}s = 0 \qquad \text{for all } \phi \in C_0^\infty((-1,1)) \text{ and } p \in [p_0, 0].$$

Since Bernoulli's principle and the relations (33) imply that $h_p^2(q_{n_k} + q, 0) \to 1/\mu$ for all $q \in (-1,1)$, the previous integral identity yields that

$$G(q,p) = G(q,0) = \frac{\mu}{2} \qquad \text{for all } (q,p) \in \overline{D}.$$

This relation leads us to (39). □

We are now prepared to prove Theorem 6. As the first result, we establish the necessity of the condition presented in Theorem 6.

Proposition 3 (The necessity) *Let the assumptions of Theorem 6 hold.*

If the wave profile is symmetric and has only one crest, then there exists a vertical line within the fluid domain such that all the fluid particles located on that line maximize there their distance to the fluid bed.

Proof Because of the invariance of the Eqs. (14), (15), and (33) with respect to horizontal translations, we may assume that $x_0 = 0$. The wave profile being symmetric, we have that $h(q,0) = h(-q,0)$ for all $q \in \mathbb{R}$. We need to show that $h(\cdot, p)$ attains its global maximum at $q = 0$ for each $p \in [p_0, 0]$. To do so, let $H : \overline{\Omega} \to \mathbb{R}$ be the mapping given by

$$H(q,p) := h(-q,p) - h(q,p) \qquad \text{for all } (q,p) \in \overline{\Omega}.$$

Since the function $\widetilde{h} \in C^2(\overline{\Omega})$ given by $\widetilde{h}(q,p) := h(-q,p)$ is also a solution of (14), (15), and (33), we find that H solves the elliptic equation $\mathcal{L}H = 0$ in $\overline{\Omega}$, whereby

$$\mathcal{L} := (1 + \widetilde{h}_q^2)\partial_p^2 - 2\widetilde{h}_p\widetilde{h}_q\partial_{pq} + \widetilde{h}_p^2\partial_q^2 + b_1\partial_p + b_2\partial_q \tag{41}$$

and with

$$\begin{aligned} b_1 &:= (h_q + \widetilde{h}_q)h_{pp} - 2h_p h_{pq}, \\ b_2 &:= (h_p + \widetilde{h}_p)h_{qq} - 2\widetilde{h}_q h_{pq} - \gamma(h_p^2 + h_p\widetilde{h}_p + \widetilde{h}_p^2). \end{aligned} \tag{42}$$

Since $H = 0$ on $\partial\Omega$ and $H(q,p) \to 0$ for $|q| \to \infty$, cf. Lemma 4, the strong elliptic maximum principle yields that $H \equiv 0$, or equivalently $h(q,p) = h(-q,p)$ for all $(q,p) \in \overline{\Omega}$. Particularly, $h_q(0,p) = 0$ for all $p \in [p_0, 0]$.

Our assumptions imply that in fact $h_q \le 0$ on $\partial\Omega_+$, with $\Omega_+ := (0,\infty) \times (p_0, 0)$. Because $h \in BUC^2(\Omega)$, we infer from Theorem 5 that $\partial_q^2 h \in BUC^1(\Omega)$. Therewith, we get $h_q \in BUC^2(\Omega)$. Differentiating now the first equation of (14) with respect to q shows that h_q solves the elliptic equation:

$$(1+h_q^2)w_{pp} - 2h_ph_qw_{pq} + h_p^2w_{qq} + 2h_qh_{pp}w_q \\ - 2h_ph_{pq}w_p - 3\gamma h_p^2w_p = 0 \quad \text{in } \overline{\Omega}_+. \tag{43}$$

Since by (33) we also have $h_q(q,p) \to 0$ for $q \to \infty$ and $h_q(\cdot,0)$ is not constant, we obtain that $h_q < 0$ in Ω_+. This property and the symmetry of h with respect to the line $q = 0$ prove the claim. □

Finally, we prove the sufficiency part.

Proposition 4 (The sufficiency) *Let the assumptions of Theorem 6 hold.*

If there exists a vertical line within the fluid domain such that all the fluid particles located on that line maximize there their distance to the fluid bed, then the wave profile has a unique crest and it is symmetric with respect to the crest line.

Proof Because of the invariance of the Eqs. (14), (15), and (33) with respect to horizontal translations, we may assume that $h(\cdot,p)$ attains its global maximum at $q = 0$ for each $p \in [p_0, 0]$. Then, $h_q(0,p) = h_{qp}(0,p) = 0$ for all $p \in [p_0, 0]$. This property ensures that the function $w : \overline{\Omega} \to \mathbb{R}$ defined by

$$w(q,p) := \begin{cases} h(q,p) & \text{for } q \geq 0, \\ h(-q,p) & \text{for } q < 0 \end{cases}$$

belongs to the class $BUC^2(\Omega)$. Because w is a solution of (14), (15), and (33) which coincides with h in the set $[q \geq 0]$, the principle of analytic continuation together with Theorem 5 yields that $h = w$ in $\overline{\Omega}$. Particularly, the wave profile is symmetric with respect to the crest line $x = 0$.

It remains to show that $h(\cdot,0)$ does not possess a local maximum to the right of $q = 0$. Because $h(\cdot,0)$ is real-analytic, nonconstant, and attains the global maximum at $q = 0$, there exists an $\delta > 0$ such that $h_q(\cdot,0) < 0$ on $(0,\delta)$. Hence, if $\delta > 0$ is small, the reflection Ω_δ^r of the set

$$\Omega_\delta := \{(x,y) : 0 < x < \delta, -d < y < \eta(x)\},$$

with respect to $x = \delta$ contains the set $\{(x,y) \in \Omega_\eta : \delta < x < 2\delta\}$. We define now

$$\delta_* := \sup\{\delta : \{(x,y) \in \Omega_\eta : \delta < x < 2\delta\} \subset \Omega_\delta^r\}.$$

Proving by contradiction, we assume that $h(\cdot,0)$ has a local maximum to the right of $q = 0$, then the supremum δ_* is finite and the reflection of the wave surface with respect to $x = \delta_*$ intersects the wave profile tangentially. Let $(x_1, \eta(x_1))$, with $x_1 \in [\delta_*, 2\delta_*]$, be a point of intersection. Then, we have

$$h(x_1,0) = h(2\delta_* - x_1, 0) \quad \text{and} \quad h_q(x_1,0) = -h_q(2\delta_* - x_1, 0). \tag{44}$$

We distinguish the following three cases.

Case 1 : Assume that $x_1 \in (\delta_*, 2\delta_*)$. Defining $R := (\delta_*, 2\delta_*) \times (p_0, 0)$ and introducing the function $H \in C^2(\overline{R})$ by

$$H(q,p) = h(q,p) - h(2\delta_* - q, p) \qquad \text{for } (q,p) \in \overline{R}, \tag{45}$$

we see that $\mathcal{L}H = 0$ in $\overline{R}$. Hereby, $\mathcal{L}$ denotes the uniformly elliptic operator defined by (41) and (42), but with $\widetilde{h}(q,p) := h(2\delta_* - q, p)$. Our assumption and the construction of δ_* ensure that $H \le 0$ on ∂R. The weak elliptic maximum principle yields then $\max_R H = 0 = H(P)$, where $P := (x_1, 0)$. If $H \not\equiv 0$, then by Hopf's lemma, $H_p(P) > 0$. On the other hand, the boundary condition on $p = 0$ in (14), (15), and (44) imply that $h_p(P) = h_p(2\delta_* - x_1, 0)$, thus $H_p(P) = 0$. Consequently, $H \equiv 0$, meaning that $h(q,p) = h(2\delta_* - q, p)$ for all $(q,p) \in [0, 2\delta_*] \times [p_0, 0]$. The latter property implies that $h_q = h_{qp} = 0$ for $q \in \{0, 2\delta_*\}$. This allows us to extend the restriction $h|_{[0,2\delta_*]\times[p_0,0]}$ to a $2\delta_*$−periodic function h_0 in $C^2(\overline{\Omega})$. Since h_0 is a solution of (14) and (15), we know from Theorem 1 that $h_0(\cdot, 0)$ is also real-analytic. But, $h(\cdot, 0) = h_0(\cdot, 0)$ on $[0, 2\delta_*]$, which contradicts in virtue of the principle of analytic continuation the far-field conditions (33). This case is therefore excluded.

Case 2 : Assume that $x_1 = 2\delta_*$. Letting $H \in C^2(\overline{R})$ be the function defined in *Case* 1, we see again that $H \le 0$ on ∂R. Clearly, $\mathcal{L}H = 0$ in $\overline{R}$, and therewith we get $\max_R H = 0 \le H(P)$ if we set $P := (2\delta_*, 0)$. Our goal is to show that $H \equiv 0$ in $\overline{R}$, property which excludes this case, cf. *Case* 1. So, if $H \not\equiv 0$, then by the strong elliptic maximum principle $H < 0$ in R. We apply Serrin's corner point lemma in the rectangle R at the corner P. To verify the assumptions of this lemma, we observe that the coefficient a_{12} of $\mathcal{L}$ is $a_{12}(q,p) = -\widetilde{h}_q(q,p)\widetilde{h}_p(q,p)$, $(q,p) \in \overline{R}$, and, since $\widetilde{h}_q(2\delta_*, p) = 0$ for all $p \in [p_0, 0]$, we have

$$\begin{aligned}|a_{12}(q,p)| &\le (\max_{\overline{\Omega}} h_p)|\widetilde{h}_q(q,p)| \le (\max_{\overline{\Omega}} h_p)|\widetilde{h}_q(q,p) - \widetilde{h}_q(2\delta_*, p)| \\ &\le (\max_{\overline{\Omega}} h_p)(\max_{\overline{\Omega}} |h_{qq}|)(2\delta_* - q).\end{aligned}$$

Hence, all the assumptions of Lemma 3 are satisfied, and therewith

$$\frac{\partial H}{\partial s}(P) < 0 \qquad \text{or} \qquad \frac{\partial^2 H}{\partial s^2} H(P) < 0$$

for any direction $s \in \mathbb{R}^2$ at P that enters R non-tangentially. Because of (44), we have $h_q(2\delta_*, 0) = h_q(0,0) = 0$ and, with (14), $h_p(2\delta_*, 0) = h_p(0,0)$, meaning that $H_q(P) = H_p(P) = 0$. Therefore, we get that the second directional derivative

$$\frac{\partial^2 H}{\partial s^2}(P) = s_1^2 H_{qq}(P) + 2s_1 s_2 H_{pq}(P) + s_2^2 H_{pp}(P) < 0$$

for the choice $s = (s_1, s_2) := (-\widetilde{h}_p(P), -1)$, that is

$$H_{pp}(P) + 2\widetilde{h}_p(P) H_{pq}(P) + \widetilde{h}_p^2(P) H_{qq}(P) < 0. \tag{46}$$

Differentiating the boundary condition on $p = 0$ of (14) with respect to q yields that

$$h_q h_{qq} + (2gh - Q)h_p h_{pq} + gh_q h_p^2 = 0 \qquad \text{on } p = 0.$$

Therewith, it is easy to see that $h_{pq}(P) = h_{pq}(0,0) = 0$, which implies $H_{pq}(P) = 0$. Since $\mathcal{L}H = 0$ and $H_q(P) = H_p(P) = 0$, we also have

$$H_{pp}(P) + \widetilde{h}_p^2(P)H_{qq}(P) = 0,$$

which contradicts (46). Hence, we can also rule out *Case* 2.

Case 3 : Assume that $x_1 = \delta_*$. Since the function H introduced at *Case* 1 satisfies $H \leq 0$ on ∂R and $\mathcal{L}H \leq 0$ in $\overline{R}$, we have that $H \leq 0$. Assume that $H < 0$ in $\overline{R}$. In order to apply Serrin's lemma, we use the stream function formulation. To this end, we define the function $\Xi : \overline{R_\eta^*} \to \mathbb{R}$, whereby $R_\eta^* := \{(x, y) \in \bar{\Omega}_\eta : \delta_* < x < 2\delta_*\}$, by setting

$$\Xi(x, y) := \psi(x, y) - \psi(2\delta_* - x, y) \qquad \text{for } (x, y) \in \overline{R_\eta^*}.$$

Letting $P := (\delta_*, \eta(\delta_*))$, we see that $\Xi(P) = 0$. Moreover, given $(x, y) \in R_\eta^*$, let $p_1 := -\psi(2\delta_* - x, y)$ and $p_2 := -\psi(x, y)$. Then, $h(2\delta_* - x, p_1) = y + d = h(x, p_2)$, and, since $H < 0$ in R and $(x, p_1) \in R$, we have $h(x, p_1) < h(x, p_2)$. We infer now form (15) that $p_1 < p_2$. To conclude, we have shown so far that

$$\Xi < 0 = \Xi(P) \qquad \text{in} \quad R_\eta^*. \tag{47}$$

In order to derive an elliptic equation for Ξ, we use the assumption that the vorticity function is Lipschitz continuous in $[p_1, 0]$ for some $p_1 \in (p_0, 0)$. To be more specific, we fix a ball $\mathbb{B}$ with center P such that for all $(x, y) \in \mathbb{B} \cap R_\eta^*$, we have $\psi(x, y), \psi(2\delta_* - x, y) \in [0, -p_1]$. Then, we find

$$\Delta\Xi(x, y) = \gamma(-\psi(x, y)) - \gamma(-\psi(2\delta_* - x, y)) = c(x, y)\Xi(x, y), \tag{48}$$

whereby $c : \overline{\mathbb{B} \cap R_\eta^*} \to \mathbb{R}$ is the bounded function defined by

$$c(x, y) = \begin{cases} \dfrac{\gamma(-\psi(x, y)) - \gamma(-\psi(2\delta_* - x, y))}{\psi(x, y) - \psi(2\delta_* - x, y)}, & \text{if } \psi(x, y) \neq \psi(2\delta_* - x, y)), \\ 0, & \text{if } \psi(x, y) = \psi(2\delta_* - x, y)). \end{cases}$$

Finally, we consider the function $\Phi : \overline{\mathbb{B} \cap R_\eta^*} \to \mathbb{R}$ by setting $\Phi(x, y) := \Xi(x, y)e^{Mx}$ for $(x, y) \in \overline{\mathbb{B} \cap R_\eta^*}$, with M chosen such that $M^2 \geq \sup_{\mathbb{B} \cap R_\eta^*} c$. Then, $\Phi < 0 = \Phi(P)$ in $\mathbb{B} \cap R_\eta^*$, cf. (47), and it is easy to see from (48) that

$$\Delta\Phi - 2M\Phi_x = (c - M^2)\Phi \geq 0 \qquad \text{in } \overline{\mathbb{B} \cap R_\eta^*}.$$

We are in the position to apply Serrin's lemma at the corner P of $\mathbb{B} \cap R_\eta^*$. Thus, we have

$$\frac{\partial\Phi}{\partial s}(P) < 0 \qquad \text{or} \qquad \frac{\partial^2\Phi}{\partial s^2}(P) < 0 \tag{49}$$

for any direction $s \in \mathbb{R}^2$ at P that enters $\mathbb{B} \cap R_\eta^*$ non-tangentially.

Recalling (44), we have that $\eta'(\delta_*) = 0$. Because $\psi = 0$ at the wave surface, we get $\psi_x(P) = 0$, and therefore $\Xi_x(P) = \Xi_y(P) = 0$. Moreover, differentiation of the second equation of (11) with respect to x yields

$$\psi_x \psi_{xx} + \psi_y \psi_{xy} + \eta'(\psi_x \psi_{xy} + \psi_y \psi_{yy}) + 2g\eta' = 0 \qquad \text{on } y = \eta(x),$$

hence $\psi_{xy}(P) = 0$. Consequently, all second-order derivatives of Ξ vanish at P. Summarizing, all first- and second-order derivatives of Φ together with Φ itself, vanish at P which contradicts (49). This excludes also the *Case* 3.

To finish the proof, we observe that since none of the *Cases* $1-3$ occurs, we must have $\delta_* = \infty$. Particularly, the wave surface has no local maximum to the right of the crest line $x = 0$. This completes the proof of the proposition. □

4 Deep-water Waves

For fluid motions in water of infinite depth, one has to replace in the system of Euler's Eq. (2), the condition of having an impermeable bottom, that is the last equation of the system, by the far-field condition

$$(u, v) \to 0 \qquad \text{for } y \to -\infty \text{ uniformly in } x, \tag{50}$$

when considering periodic fluid motions. This expresses the fact that at great depths there is actually no flow. For solitary deep-water waves instead, the physically relevant far field condition is

$$\begin{cases} \eta \to 0 & \text{for } |x| \to \infty, \\ (u, v) \to 0 & \text{for } |(x, y)| \to \infty. \end{cases} \tag{51}$$

Classical solutions of (2) (with (50) or (51)) and (4), belong to the following class:

$$(u, v, P, \eta) \in (BUC^1(\Omega_\eta))^3 \times BUC^2(\mathbb{R}).$$

This problem can also be reformulated equivalently in terms of the stream function and of the height function, cf. [43] for details. We note that due to the condition (15) which excludes the presence of stagnation points in the flow, the height function is no longer bounded. We should also mention that the existing results concerning the regularity of deep-water waves are in the setting of fluid motions for which gravity is the sole force of restoration.

The height function formulation was first employed in [43] to establish the real-analyticity of the streamlines of periodic deep-water waves with a Hölder continuous function vorticity which is additionally integrable against a cubic weight, the wave surface being shown to be real analytic only for real-analytic vorticity functions.

Based upon the results established in [43] it was shown in [44, Theorem 2.1], by using the iteration procedure introduced in [5], that the streamlines, including the wave surface, of both solitary and periodic deep-water gravity waves have the optimal regularity.

Theorem 7 (Regularity of deep-water waves) *Let* $\alpha \in (0,1)$ *and assume* $\sigma = 0$. *If* (u, v, P, η) *is a classical solution of (2) (with (50) or (51)) and (4) within the class*

$$(BUC^{1+\alpha}(\Omega_\eta))^3 \times BUC^{2+\alpha}(\mathbb{R}),$$

then, the wave profile together with all other streamlines are real-analytic graphs.

This result confirms that the regularity properties of Gerstner's solution [6, 22, 25] – which is the only known explicit solution that describes gravity water waves traveling at the surface of a fluid with infinite depth, with the streamlines being all real-analytic trochoids – are inherited by all classical solutions of the deepwater wave problem.

5 Periodic Stratified Flows in Water of Finite Depth

Stratification is a phenomenon that is common in ocean flows where in the presence of salinity and under the influence of the gravity force a heterogeneity in the fluid is produced. Stratification corresponds to the formation of fluid layers, normally arranged horizontally with the less dense layers sitting on top of the more dense ones. A particular feature of stratified water flows is that the density varies, but only as a function of the streamlines. Periodic steady stratified flows are described by a system which is a modification of system (2) in the following way. Due to the fact that the density of the fluid is nonconstant, one has to replace the first two equations in (2) (that is Euler's equations) by

$$\begin{cases} \rho(u-c)u_x + \rho v u_y = -P_x & \text{in } \Omega_\eta, \\ \rho(u-c)v_x + \rho v v_y = -P_y - g\rho & \text{in } \Omega_\eta, \\ \rho_x(u-c) + \rho_y v = 0 & \text{in } \Omega_\eta. \end{cases} \tag{52}$$

In the absence of stagnation points, that is when assuming (4), an equivalent formulations of the problem in terms of the so-called pseudo-stream function and of the height function are due to Long and Yih [36, 58] and Dubreil-Jacotin [16, 18], respectively. Regularity properties for such type of flows were first established in [30], some of the results being restricted to the context of flows which are not stably stratified, that is the density decreases with depth. Later on, by employing the iteration technique introduced in [5], Wang [57, Theorems 2.2 and 5.1] improved the results obtained in [30]. The theorem presented below is an equivalent statement of [57, Theorems 2.2 and 5.1].

Theorem 8 (Regularity of periodic stratified waves) *Let* $\alpha \in (0, 1)$ *and assume that* $\sigma \geq 0$. *If* (ρ, u, v, P, η) *is a classical solution of (2) (with 52) and (4) within the class*

$$(C^{1+\alpha}_{per}(\overline{\Omega}_\eta))^4 \times C^{2+\alpha}_{per}(\mathbb{R}),$$

then, the wave profile together with all other streamlines are real-analytic curves.

References

1. S. Angenent, Parabolic equations for curves on surfaces. I. Curves with $p-$integrable curvature. Ann. Math. **132**(3), 451–483 (1990)
2. S. Agmon, A. Douglis, L. Nirenberg, Estimates near the boundary for solutions of elliptic partial differential equations satisfying general boundary conditions. I. Comm. Pure Appl. Math. **12**, 623–727 (1959)
3. C.J. Amick, L.E. Fraenkel, J.F. Toland, On the Stokes conjecture for the wave of extreme form. Acta Math. **148**, 193–214 (1982)
4. B.T. Benjamin, The solitary wave on a stream with an arbitrary distribution of vorticity. J. Fluid Mech. **12**, 97–116 (1962)
5. H. Chen, W.-X. Li, L.-J. Wang, Regularity of traveling free surface water waves with vorticity. J. Nonlinear Sci. **23**, 1111–1142 (2013)
6. A. Constantin, On the deep water wave motion. J. Phys. A. **34**, 1405–1417 (2001)
7. A. Constantin, *Nonlinear Water Waves with Applications to Wave-current Interactions and Tsunamis*. CBMS-NSF Conference Series in Applied Mathematics, vol. 81 (SIAM, Philadelphia, 2011)
8. A. Constantin, J. Escher, Symmetry of steady periodic surface water waves with vorticity. J. Fluid Mech. **498**, 171–181 (2004)
9. A. Constantin, J. Escher, Analyticity of periodic travelling free surface water waves with vorticity, Ann. Math. **173**, 559–568 (2011)
10. A. Constantin, W. Strauss, Exact steady periodic water waves with vorticity. Comm. Pure Appl. Math. **57**, 481–527 (2004)
11. A. Constantin, W. Strauss, Pressure beneath a Stokes wave. Comm. Pure Appl. Math. **63**, 533–557 (2010)
12. A. Constantin, W. Strauss, Periodic traveling gravity water waves with discontinuous vorticity. Arch. Ration. Mech. Anal. **202**, 133–175 (2011)
13. A. Constantin, E. Varvaruca, Steady periodic water waves with constant vorticity: Regularity and local bifurcation. Arch. Ration. Mech. Anal. **199**, 33–67 (2011)
14. A. Constantin, M. Ehrnström, E. Wahlén, Symmetry of steady periodic gravity water waves with vorticity. Duke Math J. **140**, 591–603 (2007)
15. W. Craig, P. Sternberg, Symmetry of solitary waves. Comm. Partial Differential Equations. **13**, 603–633 (1988)
16. M.-L. Dubreil-Jacotin, Sur les ondes de type permanent dans les liquides hétérogènes. Atti Accad. Naz. Lincei Rend. **6**, 814–819 (1932)
17. M.-L. Dubreil-Jacotin, Sur la détermination rigoureuse des ondes permanentes périodiques d'ampleur finie. J. Math. Pures Appl. **13**, 217–291 (1934)
18. M.-L. Dubreil-Jacotin, Sur les théorèmes d'existence relatifs aux ondes permanentes périodiques a deux dimensions dans les liquides hétérogènes. J. Math. Pures et Appl. **9**, 43–67 (1937)
19. J. Escher, Regularity of rotational travelling water waves. Philos. Trans. R. Soc. Lond. A. **370**, 1602–1615 (2011)
20. J. Escher, B.-V. Matioc, On the analyticity of periodic gravity water waves with integrable vorticity function. Differential Integral Equations. **27**(3–4), 217–232 (2014)

21. J. Escher, G. Simonett, Analyticity of the interface in a free boundary problem. Math. Ann. **305** (3–4), 439–459 (1996)
22. F. Gerstner, Theorie der Wellen samt einer daraus abgeleiteten Theorie der Deichprofile. Ann. Phys. **2**, 412–445 (1809)
23. P.R. Garabedian, Surface waves of finite depth. J. Anal. Math. **14**, 161–169 (1965)
24. D. Gilbarg, N.S. Trudinger, *Elliptic Partial Differential Equations of Second Order* (Springer, Berlin, 2001)
25. D. Henry, On Gerstner's water wave. J. Nonlinear Math. Phys. **15**, 87–95 (2008)
26. D. Henry, Analyticity of the streamlines for periodic travelling free surface capillary-gravity water waves with vorticity. SIAM J. Math. Anal. **42**(6), 3103–3111 (2010)
27. D. Henry, On the regularity of capillary water waves with vorticity. C. R. Acad. Sci. Paris Ser. I. **349** (3–4), 171–173 (2011)
28. D. Henry, Analyticity of the free surface for periodic travelling capillary-gravity water waves with vorticity. J. Math. Fluid Mech. **14**(2), 249–254 (2012)
29. D. Henry, Regularity for steady periodic capillary water waves with vorticity. Philos. Trans. R. Soc. Lond. A. **370**, 1616–1628 (2012)
30. D. Henry, B.-V. Matioc, On the regularity of steady periodic stratified water waves. Comm. Pure Appl. Anal. **11** (4), 1453–1464 (2012)
31. V.M. Hur, Symmetry of solitary water waves with vorticity. Math. Res. Lett. **15**, 491–509 (2008)
32. V.M. Hur, Analyticity of rotational flows beneath solitary water waves. Int. Math. Res. Not. **11**, 2550–2570 (2012)
33. R.S. Johnson, *A Modern Introduction to the Mathematical Theory of Water Waves* (Cambridge University Press Cambridge, 1997)
34. H. Lewy, A note on harmonic functions and a hydrodynamical application. Proc. Amer. Math. Soc. **3**, 111–113 (1952)
35. J. Lighthill, *Waves in Fluids* (Cambridge University Press, Cambridge, 1978)
36. R.R. Long, Some aspects of the flow of stratified fluids. Part I : A theoretical investigation. Tellus. **5**, 42–57 (1953)
37. C.I. Martin, Regularity of steady periodic capillary water waves with constant vorticity. J. Nonlinear Math. Phys. **19**, 1240006(7 p) (2012)
38. C.I. Martin, Local bifurcation and regularity for steady periodic capillary-gravity water waves with constant vorticity. Nonlinear Anal. Real World Appl. **14**, 131–149 (2013)
39. C.I. Martin, B.-V. Matioc, Steady periodic water waves with unbounded vorticity: Equivalent formulations and existence results. J. Nonlinear Sci. **26**, 633–659 (2014)
40. A.-M. Matei, The Neumann problem for free boundaries in two dimensions. C. R. Acad. Sci. Paris Ser. I. **335**, 597–602 (2002)
41. B.-V. Matioc, Analyticity of the streamlines for periodic traveling water waves with bounded vorticity. Int. Math. Res. Not. **17**, 3858–3871 (2011)
42. A.-V. Matioc, Steady internal water waves with a critical layer bounded by the wave surface. J. Nonlinear Math. Phys. **19**, 1250008(21p) (2012)
43. B.-V. Matioc, On the regularity of deep-water waves with general vorticity distributions. Quart. Appl. Math. **LXX**(1), 393–405 (2012)
44. B.-V. Matioc, Regularity results for deep-water waves with Hölder continuous vorticity. Appl. Anal. **92**, 2144–2151 (2013)
45. B.-V. Matioc, A characterization of the symmetric steady water waves in terms of the underlying flow. Discrete Contin. Dyn. Syst. Ser. A. **34**(8), 3125–3133 (2014)
46. A.-V. Matioc, B.-V. Matioc, Regularity and symmetry properties of rotational solitary water waves. J. Evol. Equ. **12**, 481–494 (2012)
47. A.-V. Matioc, B.-V. Matioc, On the symmetry of periodic gravity water waves with vorticity. Differential Integral Equations **26**, 129–140 (2013)
48. A.-V. Matioc, B.-V. Matioc, Capillary-gravity water waves with discontinuous vorticity: Existence and regularity results. Comm. Math. Phys. **330**, 859–886 (2014)

49. H. Okamoto, M. Shōji, in *The Mathematical Theory of Permanent Progressive Water-Waves*. Adv. Ser. Nonlinear Dynam., vol. 20 (World Scientific, River Edge, 2001)
50. O.M. Philllips, Banner M.L., Wave breaking in presence of wind drift and swell. J. Fluid Mech. **66**, 625–640 (1974)
51. J. Serrin, A symmetry problem in potential theory. Arch. Rational Mech. Anal. **43**, 304–318 (1971)
52. G.G. Stokes, in *Considerations relative to the greatest height of oscillatory irrational waves which can be propagated without change of form*. Mathematical and Physical Papers, vol. I (Cambridge University Press, Cambridge, 1880), pp. 225–228
53. C. Swan, I.P. Cummins, R.L. James, An experimental study of two-dimensional surface water waves propagating on depth-varying currents. Part 1. Regular waves. J. Fluid Mech. **428**, 273–304 (2001)
54. G. Thomas, G. Klopman, *Wave-current Interactions in the Near-shore Region* (WIT, Southampton, (1997)
55. J.F. Toland, Stokes waves. Topol. Methods Nonlinear Anal. **1**, 1–48 (1996)
56. E. Wahlén, Steady water waves with a critical layer. J. Differential Equations. **246**, 1468–2483 (2009)
57. L.-J. Wang, Regularity of traveling periodic stratified water waves with vorticity. Nonlinear Anal. **81**, 247–263 (2013)
58. C.S. Yih, The role of drift mass in the kinetic energy and momentum of periodic water waves and sound waves. J. Fluid Mech. **331**, 429–438 (1997)

Degenerate and Singular Porous Medium Type Equations with Measure Data

Ugo Gianazza

Abstract We consider the inhomogeneous porous medium equation

$$\partial_t u - \Delta u^m = \mu, \qquad m > \tfrac{(N-2)_+}{N},$$

and more general equations of porous medium type with a non-negative Radon measure μ on the right-hand side. In a first step, we prove a priori estimates for weak solutions in terms of a linear Riesz potential of the right-hand side measure, which takes exactly the same form as the one for the classical heat equation. Then, we give an optimal criterium for the continuity of weak solutions, again in terms of a Riesz potential. Finally, we prove the existence of non-negative, very weak solutions and show that these constructed very weak solutions satisfy the same estimates. We deal with both the degenerate case $m > 1$ and the singular case $\frac{(N-2)_+}{N} < m < 1$.

1 Introduction

In this chapter, we present some recent results obtained in collaboration with Verena Boegelein and Frank Duzaar (see [9, 10, 11, 12]). Here, we collect the main results and remarks, and refer to the original papers for all the proofs.

A first goal is to develop a certain kind of *linear potential theory* for porous medium type equations, in both the degenerate ($m > 1$) and the singular ($\frac{(N-2)_+}{N} < m < 1$) range, providing sharp regularity results for the solutions u, such as boundedness and continuity, in terms of a linear Riesz potential of the right-hand side measure μ. It will be shown, that the same results, valid for solutions of the heat equation

$$u_t - \Delta u = \mu,$$

hold true also for parabolic equations as defined in (1), with the vector field **A** satisfying the structure conditions (2) below. A second aim is to prove the existence of the so-called very weak solutions, and extend to them the previous regularity results.

U. Gianazza (✉)
"F. Casorati" Department of Mathematics, University of Pavia,
27100, Pavia, Italy
e-mail: gianazza@imati.cnr.it

J. Escher et al., (eds.), *Elliptic and Parabolic Equations*, Springer Proceedings in Mathematics & Statistics 119, DOI 10.1007/978-3-319-12547-3_6

We deal with nonhomogeneous porous medium type equations and the associated Cauchy–Dirichlet problems in a space–time cylinder $E_T := E \times (0,T)$, where $E \subset \mathbb{R}^N$ is a bounded open domain, $N \geq 2$, and $T > 0$. The equations are of the type

$$u_t - \operatorname{div} \mathbf{A}(x,t,u,Du) = \mu \qquad \text{in } E_T, \tag{1}$$

where μ is a non-negative Radon measure on E_T with finite total mass $\mu(E_T) < \infty$. Without loss of generality, we may assume that the measure μ is defined on $\mathbb{R}^{N+1}$, by letting $\mu\lfloor(\mathbb{R}^{N+1} \setminus E_T) = 0$. For the vector field $\mathbf{A} : E_T \times \mathbb{R} \times \mathbb{R}^N \to \mathbb{R}^N$, we assume that it is measurable with respect to $(x,t) \in E_T$ for all $(u,\xi) \in \mathbb{R} \times \mathbb{R}^N$ and continuous with respect to (u,ξ) for a.e. $(x,t) \in E_T$; moreover, it satisfies the following growth and ellipticity conditions:

$$\begin{cases} \mathbf{A}(x,t,u,\xi) \cdot \xi \geq mC_o|u|^{m-1}|\xi|^2, \\ |\mathbf{A}(x,t,u,\xi)| \leq mC_1|u|^{m-1}|\xi|, \end{cases} \tag{2}$$

whenever $(x,t) \in E_T$, $u \in \mathbb{R}$ and $\xi \in \mathbb{R}^N$, for some $0 < C_o \leq C_1 < \infty$ and $m > 0$.

The most prominent example for equations treated in the sequel is given by the classical porous medium equation

$$u_t - \operatorname{div}(\mathbf{a}(x,t)Du^m) = \mu \qquad \text{in } E_T, \tag{3}$$

where the matrix $\mathbf{a}$ is only measurable and positive definite in E_T.

The dynamics of fluids in porous media is extremely important in a lot of different applications, such as gas and oil extraction, distribution and treatment of underground waters, dam hydraulics, pollution of rivers, etc.

Since in all these examples, the porous matrix presents a wide range of behaviors according to the different physical situations, a flow analysis performed through the classical fluid dynamics laws, coupled with the solid–liquid interaction, turns out to be definitely impractical. Therefore, the description of the motion of a fluid in a porous medium is usually done at a so-called mesoscopic level, in terms of the fluid density u, its velocity $\mathbf{v}$, and the pressure p. When these quantities are linked by the mass conservation law, the Darcy diffusion law, and the ideal gas law, it is well known that the resulting equation is indeed (3), where $\mathbf{a}(x,t) = \mathbb{I}_N$ and $\mu = 0$.

Besides their intrinsic mathematical interest, Radon measures on the right-hand side (e.g., instead of properly integrable functions) play a role in the description of phenomena with concentrated sources.

There exists an enormous amount of mathematical literature concerning the porous medium equation, both for the degenerate ($m > 1$) and the singular ($0 < m < 1$, also known as *fast diffusion equation*) case. In order to get at least a feeling of how the understanding of the equation has developed over the past 30 years, the interested reader might first refer to the *old* but still very interesting surveys [3, 32], and then to the much more *recent* monographs [22, 26, 50, 49] and the lists of references therein. Indeed, despite all the progresses achieved so far, the subject is far from being completely settled.

As far as the regularity for equations with the same structure considered here and $m > 1$ is concerned, when $\mu \equiv 0$ the local boundedness of weak energy solutions was proved in [2]. An important step forward was the proof that locally bounded solutions are locally Hölder continuous, due to DiBenedetto and Friedman [25]. Hölder continuity for solutions of the Cauchy problem for the prototype equations (3), with $\mathbf{a}(x,t) = \mathbb{I}_N$ and $\mu = 0$ had been established before by Caffarelli and Friedman [15]; their approach relies on the special property (of such *global* solutions)

$$u_t \geq -\frac{k}{t}\,u, \qquad \frac{1}{k} = m - 1 + \frac{2}{N}.$$

Continuity of solutions of degenerate parabolic equations $u_t = \Delta(|u|^{m-1}u)$ was proved by Caffarelli and Evans [14], but the modulus of continuity implicit in their proof is essentially of logarithmic kind.

Coming to the singular case, that is $0 < m < 1$, to our knowledge, the boundedness of solutions when $\mu = 0$, was first proved in [29; Theorem 2.2] for the prototype fast diffusion equation with $\frac{(N-2)_+}{N} < m < 1$, and later extended to a more general equation in [17; Corollary 3.20]. In [13; Theorem 2.1], a unified treatment to the boundedness of solutions of the prototype equation in the whole range $0 < m < 1$ is given; in [26; Appendix B] the same result is proved for solutions of general quasilinear parabolic equations of the form (1) with coefficients satisfying (2), once more in the full range $0 < m < 1$.

As for the continuity, Chen and DiBenedetto [16] proved that locally bounded solutions of the parabolic p-Laplacian with $1 < p < 2$ are locally Hölder continuous, and their methods extend to singular porous medium equations of the form (1)–(2). Such a result can also be obtained from the estimates proved in [51] for doubly nonlinear singular equations, setting $p = 2$. In [26; Appendix B], a complete proof of the Hölder continuity of solutions of general quasilinear parabolic equations of the form (1) with coefficients satisfying (2) is given.

Finally, few comments on potential estimates for degenerate, respectively singular parabolic equations of p-Laplacian type, are due. Nonlinear Wolff potential estimates for the *solution* u were first established in the degenerate case $p \geq 2$ for time independent measures in [42]. Later on, the result was extended to the singular case $\frac{2N}{N+2} < p < 2$ in [40], while the case of general measures was treated in the degenerate case in [44]. These results can be viewed as the natural extension of the Kilpelainen and Maly results from [33] to parabolic p-Laplacian type equations. Results concerning *gradient estimates* for degenerate quasilinear parabolic equations in terms of nonlinear potentials can be found in [43] for the right-hand side given by functions in Kato classes, whereas the full result for a right-hand side measure is in [35]. In these papers local boundedness of the gradient of weak solutions in terms of nonlinear Wolff potentials of the right-hand side was established for the degenerate case $p \geq 2$. Starting with [34, 36] pointwise estimates for the gradient in terms of nonlinear Wolff potentials in the degenerate case, and linear Riesz potentials in the singular case has been achieved; see [37] for the final step and the Riesz potential estimates for the gradient of the solution in the degenerate case. As already mentioned before, the role of linear Riesz potentials for gradient continuity results has been observed in [38].

2 A Priori Estimates for Weak Solutions: The Degenerate Case

The theory heavily depends on the chosen notion of solution, and indeed we will give two different notions of solutions, depending on the value of m. We start considering $m > 1$.

Definition 1 [Weak energy solution, $m \geq 1$] A non-negative function $u : E_T \to \mathbb{R}$ satisfying

$$u \in C^0_{\rm loc}(0,T;L^2_{\rm loc}(E)), \quad u^{\frac{m+1}{2}} \in L^2_{\rm loc}(0,T;W^{1,2}_{\rm loc}(E)) \tag{4}$$

is termed *weak energy solution of the (degenerate) porous medium type equation* (1) if and only if for every subset $U \Subset E$ and every subinterval $[t_1,t_2] \subset (0,T)$ the following identity

$$\int_U u\varphi\, dx\Big|_{t_1}^{t_2} + \int_{t_1}^{t_2}\int_U [-u\varphi_t + \mathbf{A}(x,t,u,Du)\cdot D\varphi]\, dxdt = \iint_{U\times(t_1,t_2)} \varphi\, d\mu \tag{5}$$

holds true for any testing function $\varphi \in C^\infty(\overline{U} \times [t_1,t_2])$ vanishing on $\partial U \times (t_1,t_2)$. In (5), the symbol Du has to be understood in the sense of the following definition:

$$Du := \tfrac{2}{m+1}\mathbf{1}_{\{u>0\}} u^{\frac{1-m}{2}} Du^{\frac{m+1}{2}}\,. \tag{6}$$

The assumption on the testing function φ can be weakened, e.g., if one already knows that u is bounded, then it is enough to assume that

$$\varphi \in W^{1,2}_{\rm loc}(0,T;L^2(U)) \cap L^2_{\rm loc}(0,T;W^{1,2}_0(U)).$$

In the general case, the additional hypothesis that the testing function φ must be bounded has to be imposed, in order to guarantee that the right-hand side of (5) is well defined. All other integrals appearing there are finite, due to the other assumptions on u and φ. For bounded solutions the above notion of a weak energy solution can for instance be retrieved from [26, 31] for the homogeneous, respectively inhomogeneous porous medium equation with a right-hand side $\mu \in L^\infty(E_T)$. The notion differs from the most common one, where the regularity condition on $u^{\frac{m+1}{2}}$ is replaced by the assumption $u^m \in L^2_{\rm loc}(0,T;W^{1,2}_{\rm loc}(E))$. The requirements in (4) allow the testing of the homogeneous equation only by the solution u itself (and not by u^m) and lead to a natural energy estimate for u in terms of $Du^{\frac{m+1}{2}}$. For the homogeneous, respectively inhomogeneous equation with a bounded right-hand side $\mu \in L^\infty(E_T)$, this notion seems to be the weakest one which allows natural energy estimates.

To formulate our results, we define the localized (or truncated) parabolic Riesz potential by

$$\mathbf{I}^\mu_\beta(z_o,r,\theta) := \int_0^r \frac{\mu(Q_{\rho,\rho^2\theta/r^2}(z_o))}{\rho^{N+2-\beta}}\frac{d\rho}{\rho}, \qquad \beta \in (0,N+2], \tag{7}$$

for $z_o = (x_o, t_o) \in E_T$ and $r, \theta > 0$ such that $Q_{r,\theta}(z_o) \Subset E_T$, where

$$Q_{r,\theta}(z_o) := B_r(x_o) \times (t_o - \theta, t_o).$$

Whenever writing $2Q$ for a cylinder $Q \equiv Q_{r,\theta}(z_o)$, we mean $2Q = Q_{2r,4\theta}(z_o)$.

In the case that $\theta = r^2$, the potential $\mathbf{I}^{\mu}_{\beta}$ reduces to the standard localized parabolic Riesz potential. With the definition of $\mathbf{I}^{\mu}_{\beta}$ at hand, we can state our first main result, a linear potential estimate for weak energy solutions of the porous medium equation.

Theorem 1 (Riesz potential bound for weak energy solutions, $m \geq 1$, [9]) *Let u be a non-negative weak energy solution of the porous medium Eq.* (1) *in the sense of Definition 1, where the vector field* $\mathbf{A}$ *fulfils the growth and ellipticity conditions* (2) *and $m \geq 1$. Then, for any $0 < \lambda \leq \frac{1}{N}$ and almost every $z_o \in E_T$ and every parabolic cylinder $Q_{r,\theta}(z_o) \Subset E_T$, the following potential estimate*

$$u(z_o) \leq 2\Big(\frac{r^2}{\theta}\Big)^{\frac{1}{m-1}} + \boldsymbol{\gamma}\Big[\frac{1}{r^{N+2}}\iint_{Q_{r,\theta}(z_o)} u^{m+\lambda}\,dxdt\Big]^{\frac{1}{1+\lambda}} + \boldsymbol{\gamma}\,\mathbf{I}^{\mu}_{2}(z_o, r, \theta) \qquad (8)$$

holds true with a universal constant $\boldsymbol{\gamma}$ depending only on N, C_o, C_1, m, and λ.

Some comments on Theorem 1 and the history of the problem are in order. First, when $\mu \equiv 0$ the pointwise estimate (8) reduces to the well-known sup-bound for weak energy solutions of the porous medium equation from [2]. Second, in the nonhomogeneous case $m = 1$, i.e., when the equation does not show a degenerate behavior anymore and $\mu \neq 0$, we arrive at most at the potential estimate from [28]. The only difference in this case stems from the integral

$$\boldsymbol{\gamma}\Big[\frac{1}{r^{N+2}}\iint_{Q_{r,r^2}(z_o)} u^{1+\lambda}\,dxdt\Big]^{\frac{1}{1+\lambda}} \quad \text{with } \lambda > 0.$$

In this term, we are not allowed to pass to the limit $\lambda \downarrow 0$, since the constant $\boldsymbol{\gamma} \uparrow \infty$ blows up in the limit $\lambda \downarrow 0$. If such an operation were possible, this would yield the zero order linear potential estimate from [28] for parabolic equations with linear growth in the gradient variable. However, this is in perfect accordance with the zero order potential estimate for the elliptic p-Laplacian equation from the seminal paper [33], where the right-hand side of the potential estimate contains an integral of the form

$$\boldsymbol{\gamma}\Big[\frac{1}{r^{N}}\int_{B_\rho(x_o)} u^{p-1+\lambda}\,dx\Big]^{\frac{1}{1+\lambda}}$$

with $\lambda > 0$ sufficiently small. Completely analogous estimates for the parabolic p-Laplacian are given in [[42], [44]]. Even here the classical case $p = 2$ cannot be completely recovered by letting $\lambda \downarrow 0$. With this respect, our result is in perfect accordance with the known results. Furthermore, it generalizes a previously obtained potential estimate for weak solutions of the porous medium equation from [41]. This result was formulated for weak solutions satisfying $u^m \in L^2(0, T; W^{1,2}(E))$, but the proof works with minor modifications also for weak energy solutions as above. The potential estimate [41; Theorem 1.5], however, is weaker than ours, since its right-hand side differs from the right-hand side in (8) by a sup-term of the form

$$\gamma \sup_{t\in(t_o-\theta,t_o)} \frac{1}{\rho^N}\int_{B_\rho(x_o)} u(x,t)\,dx.$$

Hence, the classical sup-bound from [2] will not follow in the case $\mu = 0$. With this respect, the potential estimate (8) seems to be the natural one, since it reduces in the cases $m = 1$, respectively $\mu = 0$, to the known results from [2, 28, 42, 44].

Finally, it is well-known that there exists a strong connection between estimates for solutions of the porous medium equation and estimates for *gradients* of solutions of the parabolic p-Laplacian (see, e.g., the introduction of [24]).

Therefore, under this point of view, Theorem 1 can be regarded as some sort of porous medium analogue of [36; Theorems 1.1, 1.2] (see also [37]).

We can now give a sufficient criterion guaranteeing the continuity of solutions.

Theorem 2 (Continuity of weak energy solutions via linear potentials, $m > 1$, [10]) *Let u be a non-negative, locally bounded, weak energy solution of the porous medium Eq.* (1) *in the sense of Definition* 1, *where the vector field* $\mathbf{A}$ *fulfils the growth and ellipticity conditions* (2). *Furthermore, consider $E_o \Subset E_T$ and assume that*

$$\lim_{r\downarrow 0}\sup_{z_o\in E_o} \mathbf{I}_2^\mu(z_o,r,r^2) = 0 \tag{9}$$

holds true. Then, u is continuous in E_o.

The subtle point here is, that the local boundedness of $\mathbf{I}_2^\mu(\cdot,r,r^2)$ for some $r > 0$, ensures the local boundedness of u, and moreover implies $\lim_{\rho\downarrow 0}\mathbf{I}_2^\mu(z,\rho,\rho^2) = 0$ for a.e. $z \in E_T$, while the information of locally uniform convergence to zero of the Riesz potential from (9) implies the local continuity of the weak energy solution. With that respect, our results, i.e., the local boundedness and the continuity of weak solutions via Riesz potentials, are of borderline type. It is somewhat surprising that the Riesz $\mathbf{I}_2^\mu$ potential plays the same role as in the linear setting. It would be interesting to consider measures μ for which the Riesz potential $\mathbf{I}_2{}^\mu(\cdot,r,r^2)$ is locally bounded, and moreover satisfy

$$\lim_{\rho\downarrow 0}\frac{\mu(Q_{\rho,\rho^2}(z))}{\rho^N} = 0,$$

locally uniformly in E_T with respect to z. By our potential estimate (8), weak energy solutions would be locally bounded, and one might conjecture that they are also locally VMO on E_T. Such a result would be between local boundedness and continuity, but we do not dwell on it here.

As an application of Theorem 2, one can consider measures given by measurable functions $\mu \in L^1(E_T)$. In [10; Chap. 5] we establish the following important assertion:

$$\mu \in L(\tfrac{N+2}{2},1) \quad\Longrightarrow\quad u \text{ is locally continuous in } E_T.$$

For the definition of the Lorentz space $L(\frac{N+2}{2},1)$, we refer to [30]. One can see how sharp this result actually is, by the classical theory for parabolic equations of the form

$$u_t - \operatorname{div} \mathbf{A}(x,t,Du) = \mu \quad \text{in} \quad E_T$$

with coefficients satisfying (2) with $m = 1$. Here, it is known that the assumption $\mu \in L^{\frac{N+2}{2}+\varepsilon}(E_T)$, for some arbitrary small $\varepsilon > 0$, implies the continuity of u. This can be retrieved for example from [24; Sect. IV]. On the other hand, for $\mu \in L^{\frac{N+2}{2}}(E_T)$ solutions might be even unbounded. We note that $L^{\frac{N+2}{2}+\varepsilon} \subset L(\frac{N+2}{2}, 1)$ for any $\varepsilon > 0$. It is worth pointing out that the assumption $\mu \in L(\frac{N+2}{2}, 1)$ is independent of $m \geq 1$.

3 A Priori Estimates for Weak Solutions: The Singular Case

We can now consider the singular case, namely when $0 < m < 1$.

Definition 2 [Weak energy solution, $0 < m < 1$] A non-negative function $u : E_T \to \mathbb{R}$ satisfying

$$u \in C^0_{\mathrm{loc}}(0,T; L^{1+m}_{\mathrm{loc}}(E)), \quad u^m \in L^2_{\mathrm{loc}}(0,T; W^{1,2}_{\mathrm{loc}}(E)) \tag{10}$$

is termed *weak energy solution of the singular porous medium type Eq.* (1) if and only if for every subset $U \Subset E$ and every subinterval $[t_1, t_2] \subset (0,T)$ the following equation

$$\int_U u\varphi \, dx \Big|_{t_1}^{t_2} + \int_{t_1}^{t_2} \int_U [-u\varphi_t + \mathbf{A}(x,t,u,Du) \cdot D\varphi] \, dxdt = \int_{U\times(t_1,t_2)} \varphi \, d\mu \tag{11}$$

holds true for any testing function

$$\varphi \in L^2_{\mathrm{loc}}(0,T; W^{1,2}_0(U)) \quad \text{with} \quad \varphi, \varphi_t \in L^\infty(U_T).$$

In (11) the symbol Du has to be understood in the sense of the following definition:

$$Du := \tfrac{1}{m} u^{1-m} Du^m . \tag{12}$$

The assumption that the testing function φ and its time derivative φ_t must be bounded has to be imposed in order to guarantee that the terms involving the time derivative and the right-hand side of (11) are well defined. All other integrals appearing there are finite, due to the other assumptions on u and φ.

Even though in the literature (1) with $0 < m < 1$ is frequently referred to as *fast diffusion equation*, as we have already mentioned above, in the following we will prefer the term *singular porous medium equation*.

With the definition of $\mathbf{I}^\mu_\beta$ given in (7), we can state the first result of this section, a linear potential estimate for weak energy solutions of the singular porous medium equation, of the same kind as Theorem 1.

Theorem 3 (Riesz potential bound for weak energy solutions, $\frac{(N-2)_+}{N} < m < 1$, [11]) *Let u be a non-negative weak energy solution of the porous medium equation* (1) *in the sense of Definition* 2*, where the vector field* $\mathbf{A}$ *fulfils the growth and ellipticity conditions* (2) *with*

$$\frac{(N-2)_+}{N} < m < 1. \tag{13}$$

Then, for any given $\lambda \in (0, \frac{\kappa}{N(1+m)}]$, *where* $\kappa := 2-N(1-m)$, *almost every* $z_o \in E_T$, *and every parabolic cylinder* $Q_{4r,\theta}(z_o) \subset E_T$, *the following potential estimate*

$$u(z_o) \le \boldsymbol{\gamma}\Big(\frac{1}{\theta^{1+\frac{N}{2}}}\iint_{Q_{r,\theta}} u^{1+\lambda}\,dxdt\Big)^{\frac{2}{\kappa+2\lambda}} + \boldsymbol{\gamma}\big(\frac{\theta}{r^2}\big)^{\frac{1}{1-m}} + \boldsymbol{\gamma}\,\big(\frac{r^2}{\theta}\big)^{\frac{N}{\kappa}}\mathbf{I}_2(z_o, 4r, \theta)^{\frac{2}{\kappa}} \tag{14}$$

holds true with a universal constant $\boldsymbol{\gamma}$ *depending only on* N, m, C_o, C_1, *and* λ.

Some words on Theorem 3 are in order. First of all, in the case of the homogeneous porous medium equation, i.e. when $\mu \equiv 0$, from (14) one almost recovers the well known sup-estimate for weak energy solutions, as given for example in [26; Proposition B.4.1] (with $r = 1$ and $\lambda_1 = \kappa \equiv 2 - N(1-m)$), i.e.,

$$\sup_{\frac{1}{2}Q_{r,\theta}(z_o)} u \le \boldsymbol{\gamma}\big(\frac{r^2}{\theta}\big)^{\frac{N}{\kappa}}\Big(\frac{1}{|Q_{r,\theta}|}\iint_{Q_{r,\theta}} u\,dxdt\Big)^{\frac{2}{\kappa}} + \boldsymbol{\gamma}\big(\frac{\theta}{r^2}\big)^{\frac{1}{1-m}}.$$

We should mention that our estimate differs from the one in [26; Proposition B.4.1] due to the presence of the parameter λ, which in our case can be chosen arbitrarily small, but not equal to zero, since the constant $\boldsymbol{\gamma}$ blows up in the limit $\lambda \downarrow 0$. On the other hand, in the nonhomogeneous case $m = 1$, i.e., when the equation does not admit a singular structure anymore and $\mu \neq 0$, we arrive almost at the zero order linear Riesz potential estimate from [28] for nonlinear parabolic equations with linear growth in the gradient variable. Again, the only difference appears in the integral

$$\boldsymbol{\gamma}\Big[\frac{1}{r^{N+2}}\iint_{Q_{r,r^2}(z_o)} u^{1+\lambda}\,dxdt\Big]^{\frac{1}{1+\lambda}}.$$

Exactly as before, here the parameter λ can be chosen arbitrarily small, but not equal to zero, since the constant $\boldsymbol{\gamma}$ might blow up as $\lambda \downarrow 0$.

Finally, we are considering $m \in (\frac{(N-2)_+}{N}, 1)$, and not the whole range $(0, 1)$: this is no surprise, since it is well-known that already in the homogeneous case, i.e., when $\mu = 0$, unbounded solutions exist if $0 < m \le \frac{(N-2)_+}{N}$. In order to have bounded solutions when $m \in (0, \frac{(N-2)_+}{N}]$, one needs to impose further conditions (see [24; Chap. 5], and [13]).

As an immediate consequence of the potential estimate (14) we obtain:

Corollary 1 *Additionally to the hypotheses of Theorem 3, assume that*

$$z \mapsto \mathbf{I}_2(z, r)$$

is locally bounded in E_T for some $r > 0$. Then, $u \in L^\infty_{\mathrm{loc}}(E_T)$.

Note that the local boundedness of the truncated Riesz potential $\mathbf{I}_2^\mu(\cdot, r)$ for some $r > 0$, guarantees the local boundedness of u, and moreover yields

$$\lim_{\rho \downarrow 0} \mathbf{I}_2^\mu(z, \rho) = 0 \qquad \text{for a.e. } z \in E_T.$$

Now, the second main result of this chapter is a sufficient criterion ensuring the continuity of solutions.

Theorem 4 (Continuity of weak energy solutions via linear potentials, $\frac{(N-2)_+}{N} < m < 1$, [11]) . *Let u be a non-negative, locally bounded, weak energy solution of the porous medium equation* (1) *in the sense of Definition* 2, *where the vector-field* $\mathbf{A}$ *fulfils the growth and ellipticity conditions* (2) *and m is as in* (13). *Furthermore, consider $E_o \Subset E_T$ and assume that*

$$\lim_{r \downarrow 0} \sup_{z_o \in E_o} \mathbf{I}_2^\mu(z_o, r) = 0 \tag{15}$$

holds true. Then, u is continuous in E_o.

As in the degenerate case, the notable point in the previous results is the assertion that local boundedness of the truncated Riesz potential implies the local boundedness of weak energy solutions, while locally uniform convergence to zero of the truncated Riesz potential implies the local continuity of weak energy solutions.

By now, it is well known that the Riesz potential $\mathbf{I}_2^\mu$ plays the same role as in the linear setting; in [27] a related result for stationary p-Laplacian type systems concerning the characterization of the continuity of the gradient in terms of a nonlinear Wolff potential can be found. The results specialize in the case $p = 2$ to a characterization in terms of a linear Riesz potential, although the equations considered are nonlinear. A similar phenomenon, again in terms of linear Riesz potentials, concerning evolutionary p-Laplacian type equations with respect to gradient continuity has been observed in [38].

From Theorem 4, one can easily derive a simple consequence, which has already been discussed when $m > 1$ in the previous section; we have the following important assertion:

$$\mu \in L(\tfrac{N+2}{2}, 1) \quad \Longrightarrow \quad u \text{ is locally continuous in } E_T.$$

As before, $L(\frac{N+2}{2}, 1)$ denotes the Lorentz space for the parameters $(\frac{N+2}{2}, 1)$. Note that the assumption $\mu \in L(\frac{N+2}{2}, 1)$ is independent of $m \leq 1$, exactly as in the degenerate range. The implication

$$\mu \text{ is in a proper Lorentz space} \quad \Longrightarrow \quad u \text{ is locally continuous in } E_T,$$

has been widely discussed in [27], and then in [35, 37, 38].

4 Existence of Very Weak Solutions when $m > 1$

Let us now assume that, besides (2), for any $u \in \mathbb{R}$, $\xi_1, \xi_2 \in \mathbb{R}^n$, and a.e. $(x,t) \in E_T$ the vector field $\mathbf{A}$ satisfies the following monotonicity assumption

$$(\mathbf{A}(x,t,u,\xi_1) - \mathbf{A}(x,t,u,\xi_2)) \cdot (\xi_1 - \xi_2) \geq \nu |u|^{m-1} |\xi_1 - \xi_2|^2. \tag{16}$$

The aim is then to establish under conditions (2) and (16) the existence of a solution of the Cauchy–Dirichlet problem

$$\begin{cases} u_t - \operatorname{div} \mathbf{A}(x,t,u,Du) = \mu, & \text{in } E_T, \\ u = 0 \quad \text{on } \partial_{\mathrm{par}} E_T, \end{cases} \tag{17}$$

in the sense of distributions, when μ is a bounded, non-negative Radon measure. As before, we first deal with $m > 1$. In this context we use the notion of very weak solutions, which are defined as follows:

Definition 3 [Very weak solution, $m > 1$] A non-negative function $u : E_T \to \mathbb{R}$ satisfying

$$u \in L^\infty(0,T; L^1(E)), \quad u^{\frac{m+1}{2}} \in L^1(0,T; W_0^{1,1}(E)) \tag{18}$$

is termed *very weak solution* of the Cauchy–Dirichlet problem (17) for the (degenerate) porous medium type equation if and only if $\mathbf{A}(x,t,u,Du) \in L^1(E_T, \mathbb{R}^N)$ and the following identity

$$\iint_{E_T} \Big[- u\varphi_t + \mathbf{A}(x,t,u,Du) \cdot D\varphi \Big] \, dxdt = \int_{E_T} \varphi \, d\mu \tag{19}$$

holds true for any testing function $\varphi \in C^\infty(\overline{E}_T)$ vanishing on $[\partial E \times (0,T)] \cup [\overline{E} \times \{T\}]$.

As for the weak energy solutions from Definition 1, the symbol Du has to be interpreted in the sense of (6). In terms of this notion we have the following existence result for very weak solutions.

Theorem 5 (Existence of very weak solutions, $m > 1$, [9]). *Assume that hypotheses* (2) *and* (16) *hold, and μ is a non-negative Radon measure in E_T. Then, there exists at least one non-negative very weak solution u of* (17) *satisfying*

$$u \in L^\infty(0,T; L^1(E)), \quad u^{\frac{m+1}{2}} \in L^q(0,T; W_0^{1,q}(E)) \tag{20}$$

for any q such that

$$1 \leq q < 2 - \frac{2N}{N(m+1)+2}. \tag{21}$$

Note that in the case $m = 1$, we recover the existence result for measure data problems for nonlinear parabolic equations with linear growth of the coefficients in the gradient variable from [8; Theorem 1.2]. Indeed, with the choices $r = q$ and $p = 2$ conditions [8; (1.7), (1.8)] reduce to the hypothesis $1 \le q < 2 - \frac{N}{N+1}$, which is exactly condition (21) from above when $m = 1$.

The very weak solution constructed in Theorem 5 is a *solution obtained by limit of approximations* (SOLA) to (17). The approximating sequence is given by weak energy solutions

$$u_j \in C^0([0,T]; L^2(E)), \quad u_j^{\frac{m+1}{2}} \in L^2(0,T; W^{1,2}(E))$$

of Cauchy–Dirichlet problems

$$\begin{cases} \partial_t u_j - \operatorname{div} \mathbf{A}(x,t,u_j,Du_j) = \mu_j, & \text{in } E_T, \\ u_j = 0 \quad \text{on } \partial_{\text{par}} E_T, \end{cases}$$

where $\mu_j \in L^\infty(E_T)$ is a regularization of μ such that $\mu_j \, dxdt \to \mu$ in the limit $j \to \infty$ in the sense of Radon measures, and $\mu_j(E_T) \le \mu(E_T)$ for $j \ge 1$.

The existence of such weak energy solutions in the sense of Definition 1 follows for example from [31]. The construction of very weak solutions of measure data problems related to elliptic and parabolic p-Laplacian equations is a well established procedure; we refer to [4–7, 18, 19, 20]. Measure data problems for the porous medium equation in its simplest form (3) with $\mathbf{a}(x,t) \equiv \mathbb{I}_N$ have been studied in [45]. There, the stronger assumption $u^m \in L^2(0,T; W_0^{1,2}(E))$ has been used to establish the existence of very weak solutions. Finally, in [21] a somewhat related existence problem for inhomogeneous porous medium equations of the type

$$u_t - \operatorname{div}(\mathbf{a}(x,t,u)Du) = \mathbf{b}(x,t,u,Du) \quad \text{in } E_T$$

has been studied.

To our knowledge, the general case of nonsplitting coefficients $\mathbf{A}(x,t,u,Du)$ with a degenerate behavior as described in (2) was not studied up to now.

By a further truncation procedure in the approximation procedure described above, we obtain from Theorem 1 the following regularity result.

Theorem 6 (Potential estimates for general measure data problems, $m > 1$, [9]) *Suppose that u is the very weak solution built in Theorem 5. Then, u satisfies the potential estimate* (8).

Indeed, Theorem 6 continues to hold true for any very weak solution u satisfying the additional regularity requirement that $\min\{u,k\}^{\frac{m+1}{2}} \in L^2(0,T; W^{1,2}(E))$ for any $k > 0$.

Now, we test our results versus the *Barenblatt fundamental solution*

$$\mathcal{B}_m(x,t) := \begin{cases} \frac{1}{t^{\frac{N}{k}}} \left[1 - b \left(\frac{|x|}{t^{\frac{1}{k}}} \right)^2 \right]_+^{\frac{1}{m-1}}, & t > 0 \\ 0 & t \le 0 \end{cases} \tag{22}$$

where $m > 1$,

$$\kappa = N(m-1)+2, \quad \text{and} \quad b = b(N,m) = \frac{N(m-1)}{2Nm\kappa}. \tag{23}$$

It is well-known, that the Barenblatt fundamental solution is a very weak solution of the porous medium equation

$$\partial_t u - \Delta u^m = \delta_{(0,0)},$$

in $\mathbb{R}^N \times [-\varepsilon, +\infty)$, $\varepsilon > 0$, where $\delta_{(0,0)}$ is the delta-function at the origin.

The uniqueness of such a solution when δ is seen as *initial datum*, has been proved in [47]. In order to establish that the restriction (21) on q in Theorem 5 is sharp, we need to show that the Barenblatt solution admits the following integrability properties:

$$\int_0^1 \int_{\mathbb{R}^N} |D\mathcal{B}_m{}^{\frac{m+1}{2}}|^q \, dxdt < \infty \qquad \forall\, 1 \le q < 2 - \frac{N}{N(m+1)+2},$$

$$\int_0^1 \int_{\mathbb{R}^N} |D\mathcal{B}_m{}^{\frac{m+1}{2}}|^q \, dxdt = \infty \qquad \forall\, q \ge 2 - \frac{N}{N(m+1)+2}.$$

By a straightforward computation and a change of variable we obtain that

$$\begin{aligned}
&\int_0^1 \int_{\mathbb{R}^N} \mathcal{B}_m^{\frac{(m-1)q}{2}} |D\mathcal{B}_m|^q \, dxdt \\
&= \left(\frac{2b}{m-1}\right)^q \int_0^1 \int_{\mathbb{R}^N} \frac{1}{t^{\frac{N(m-1)q}{2\kappa}+\frac{Nq}{\kappa}}} \left(\frac{|x|}{t^{\frac{2}{\kappa}}}\right)^q \left[1 - b\left(\frac{|x|}{t^{\frac{1}{\kappa}}}\right)^2\right]_+^{\frac{q(2-m)}{m-1}+\frac{q}{2}} dxdt \\
&= \left(\frac{2b}{m-1}\right)^q \omega_N \int_0^1 \frac{t^{\frac{N}{\kappa}}}{t^{\frac{N(m-1)q}{2\kappa}+\frac{Nq}{\kappa}+\frac{q}{\kappa}}} \, dt \int_0^{\rho_*} s^{q+N-1} \left(1 - bs^2\right)^{\frac{q(2-m)}{m-1}+\frac{q}{2}} ds
\end{aligned}$$

holds true, where we defined $\rho_* := \sqrt{\frac{1}{b}}$ and $\omega_N := \mathcal{H}^{N-1}(S^{N-1})$ for short in the last line. For the integral to be finite, the following conditions must hold:

$$\frac{q(2-m)}{m-1} + \frac{q}{2} > -1 \quad \text{and} \quad \frac{N}{\kappa} - \frac{N(m-1)q}{2\kappa} - \frac{Nq}{\kappa} - \frac{q}{\kappa} > -1.$$

The first condition is void in the range $1 < m \le 3$, and otherwise yields the restriction $q < 2(1 + \frac{2}{m-3})$; as for the second condition, taking into account the value of κ, we find the restriction (21) on q as expected. Since, obviously $2 - \frac{N}{N(m+1)+2} < 2(1 + \frac{2}{m-3})$, the claimed integrability property of the Barenblatt solution follows, and proves the sharpness of the restriction (21).

We claim that also the potential estimate (8) shows the correct decay at the origin. To this end, we rewrite (8) in the form

$$u(z_o) \le 2\left(\frac{r^2}{\theta}\right)^{\frac{1}{m-1}} + \gamma \left(\frac{\theta}{r^2}\right)^{\frac{1}{1+\lambda}} \left[\iint_{Q_{r,\theta}(z_o)} u^{m+\lambda} dxdt\right]^{\frac{1}{1+\lambda}} + \gamma\, \mathbf{I}_2^{\mu}(z_o, r, \theta),$$

where λ is a real number in $(0, \frac{1}{N}]$. By its very definition, the Riesz potential can be bounded from above by

$$\mathbf{I}_2^{\mu}(z_o, r, \theta) \le \int_0^{\infty} \frac{\mu(Q_{\rho,\rho^2\theta/r^2}(z_o))}{\rho^N} \frac{d\rho}{\rho}.$$

Now, let $\mu = \delta_{(0,0)}$, and for the sake of simplicity, let us write u instead of $\mathcal{B}_m$. Consider $z_o = (0, t_o)$ with $t_o > 0$ (note that $u(0, t_o) > 0$), and let $\theta := [u(0, t_o)]^{1-m} r^2$, which is the correct *intrinsic* scaling for the porous medium equation. For the cylinder $Q_{\rho,\rho^2\theta/r^2}$ to contain the origin, we need to have that

$$\rho^2 \frac{[u(0,t_o)]^{1-m} r^2}{r^2} \ge t_o \qquad \Longleftrightarrow \qquad \rho \ge \sqrt{t_o [u(0,t_o)]^{m-1}}.$$

Consequently,

$$\begin{aligned} \mathbf{I}_2^{\delta_o}(z_o, r, \theta) &\le \int_{\sqrt{t_o[u(0,t_o)]^{m-1}}}^{\infty} \frac{\mu(Q_{\rho,\rho^2\theta/r^2}(z_o))}{\rho^N} \frac{d\rho}{\rho} \\ &= \int_{\sqrt{t_o[u(0,t_o)]^{m-1}}}^{\infty} \frac{1}{\rho^{N+1}} d\rho = \frac{1}{N t_o^{\frac{N}{2}} [u(0,t_o)]^{\frac{(m-1)N}{2}}}. \end{aligned}$$

Now, if we limit ourselves to consider only the bound from above that comes from the Riesz potential, we obtain

$$u(0, t_o) \le \frac{\gamma}{t_o^{\frac{N}{2}} [u(0,t_o)]^{\frac{(m-1)N}{2}}} \qquad \Longleftrightarrow \qquad u(0, t_o) \le \gamma\, t_o^{-\frac{N}{\kappa}}.$$

This corresponds exactly to the time decay of the Barenblatt solution $\mathcal{B}_m$, and shows that (8) is optimal under this point of view. A similar decay estimate is given in [36][§ 1.4] for the gradient of the Barenblatt fundamental solution of the parabolic p–Laplacian, when $p \ge 2$.

We stated Theorem 2 as a result for weak energy solutions, but it also applies to the very weak solutions u built in Theorem 5. Indeed, due to the boundedness of the weak energy solutions u_j making up the approximating sequence of the very weak solution u, it is possible to build the starting cylinder (and consequently, the whole approximating sequence of shrinking cylinders $Q_{r_n,\theta_n}(z_o)$) in a way that is independent of j. A similar argument is discussed, e.g., in [23; Chap. 6].

Finally, for more regular data, we can improve in our existence Theorem 5 the integrability properties of the very weak solution.

Theorem 7 [[9]] *Assume that hypotheses* (2) *and* (16) *hold and that the Radon measure is given by a non-negative function* $\mu \in L^s(E_T)$ *for some* s *in the range*

$$1 < s < 1 + \frac{N}{Nm + 4}. \tag{24}$$

Then, there exists at least one non-negative very weak solution u of the measure data problem (17) *satisfying*

$$u \in L^\infty(0,T;L^1(E)), \quad u^{\frac{m+1}{2}} \in L^q(0,T;W_0^{1,q}(E)) \tag{25}$$

where q is given by

$$q = \frac{2s(Nm+2)}{N(m+1)+4-2s}\,. \tag{26}$$

We note that in the case $m = 1$ the restriction on s and the value of q coincide with the corresponding values from [8; Theorem 1.9] (take $p = 2$ in (1.9) and (1.10) in Theorem 1.9 of [8]). With this respect, our result perfectly matches with the classical existence result for very weak solutions of the measure data problems.

5 Existence of Very Weak Solutions when $\frac{(N-2)_+}{N} < m < 1$

In the singular range $0 < m < 1$, the situation is much more complicated, and we cannot prove the existence of very weak solutions with the same generality, as in § 4. We have to limit ourselves to equations of the type

$$u_t - D_j[\mathbf{a}_{ij}(x,t,u)D_i u^m] = \mu \qquad \text{in } E_T. \tag{27}$$

For the coefficients

$$\mathbf{a}_{ij} : E_T \times \mathbb{R} \to \mathbb{R}^{N\times N}, \qquad 1 \le i,j \le N,$$

we assume measurability with respect to $(x,t) \in E_T$ for all $u \in \mathbb{R}$, and continuity with respect to u for a.e. $(x,t) \in E_T$. Moreover, we assume the following growth and ellipticity conditions

$$\mathbf{a}_{ij}(x,t,u)\xi_i\xi_j \ge C_o|\xi|^2 \qquad \text{and} \qquad \mathbf{a}_{ij}(x,t,u)\xi_i\eta_j \le C_1|\xi||\eta|, \tag{28}$$

whenever $(x,t) \in E_T$, $u \in \mathbb{R}$ and $\xi,\eta \in \mathbb{R}^N$, for some $0 < C_o \le C_1 < \infty$; they substitute the more general conditions given in (2). The most prominent example for equations treated in the sequel is given by the classical *singular* porous medium equation (also called *fast diffusion equation*), in which case $\mathbf{a}_{ij}(x,t,u) \equiv \mathbf{a}(x,t)\delta_{ij}$ and therefore

$$u_t - \operatorname{div}(\mathbf{a}(x,t)Du^m) = \mu \qquad \text{in } E_T. \tag{29}$$

The main aim of this final section is to establish the existence of a solution of the Cauchy–Dirichlet problem

$$\begin{cases} u_t - D_j[\mathbf{a}_{ij}(x,t,u)D_i u^m] = \mu, & \text{in } E_T, \\ u = 0 & \text{on } \partial_{\text{par}} E_T, \end{cases} \tag{30}$$

in the sense of distributions, when μ is a bounded non-negative Radon measure, and $m \in (0, 1)$. In this context, we use the notion of very weak solutions, which is defined as follows:

Definition 4 [Very weak solution, $0 < m < 1$] A non-negative function $u \in L^1(E_T)$ satisfying

$$u^m \in L^1(0, T; W_0^{1,1}(E)) \tag{31}$$

is termed *very weak solution* of the Cauchy–Dirichlet problem (30) for the singular porous medium type equation, if and only if the following identity

$$\iint_{E_T} [-u\varphi_t + \mathbf{a}_{ij}(x, t, u) D_i u^m D_j \varphi]\, dxdt = \int_{E_T} \varphi\, d\mu \tag{32}$$

holds true for any testing function $\varphi \in C^\infty(\overline{E_T})$ vanishing in a neighborhood of $[\partial E \times (0, T)] \cup [\overline{E} \times \{T\}]$. □

In terms of this notion, we have the following existence result for very weak solutions.

Theorem 8 (Existence of very weak solutions, $\frac{(N-2)_+}{N} < m < 1$, [12]). *Assume that hypotheses* (28) *hold, m is in the range given by* (13)*, and* μ *is a non-negative Radon measure in* E_T*. Then, there exists at least one non-negative very weak solution u of* (30) *satisfying*

$$u^m \in L^q(0, T; W_0^{1,q}(E)) \tag{33}$$

for any q such that

$$1 \le q < 1 + \frac{1}{Nm + 1}. \tag{34}$$

Note that in the case $m = 1$, once more we recover the existence result for measure data problems for nonlinear parabolic equations with linear growth of the coefficients in the gradient variable from [8; Theorem 1.2]. Indeed, with the choices $r = q$ and $p = 2$ the conditions [8; (1.7), (1.8)] reduce to the hypothesis $1 \le q < 1 + \frac{1}{N+1}$, which is exactly condition (34) from above.

The very weak solution constructed in Theorem 8 is a SOLA to (30). The approximating sequence is given by weak energy solutions

$$u_j \in C^0([0, T]; L^{1+m}(E)), \quad u_j^m \in L^2(0, T; W_0^{1,2}(E)) \tag{35}$$

of Cauchy–Dirichlet problems

$$\begin{cases} \partial_t u_j - D_k[\mathbf{a}_{ik}(x, t, u) D_i u^m] = \mu_j, & \text{in } E_T, \\ u_j = 0 \quad \text{on } \partial_{\text{par}} E_T, \end{cases} \tag{36}$$

where $\mu_j \in L^\infty(E_T)$ is a regularization of μ such that $\mu_j\, dxdt \to \mu$ in the limit $j \to \infty$ in the sense of Radon measures, and $\mu_j(E_T) \le \mu(E_T)$ for $j \ge 1$.

The existence of such weak energy solutions follows for example from [1]. The construction of very weak solutions of the measure data problems is a well-established procedure; for more details, we refer to the references given in § 4. Measure data problems for the singular porous medium equation in its simplest form (29) with $\mathbf{a}(x,t) \equiv 1$ have been studied in [46].

To our knowledge, till now, the general case of multiplicative coefficients $\mathbf{a}(x,t,u)Du^m$ with a behavior as described in (28) has not been studied.

By a further truncation procedure in the approximation scheme described above, we obtain the following pointwise bound for very weak solutions.

Theorem 9 (Potential estimates for general measure data problems, $\frac{(N-2)_+}{N} < m < 1$, [12]). *Suppose that u is the very weak solution built in Theorem* 8, *with m in the range* (13). *Then, u satisfies the potential estimates* (14).

Indeed, Theorem 9 continues to hold true for any very weak solution u satisfying the additional regularity requirement that $\min\{u,k\}^m \in L^2(0,T;W^{1,2}(E))$ for any $k > 0$.

We want to test the sharpness of our results: under this point of view, the *Barenblatt fundamental solution*

$$\mathcal{B}_m(x,t) := \begin{cases} \frac{1}{t^{\frac{N}{\kappa}}}\left[1+b\left(\frac{|x|}{t^{\frac{1}{\kappa}}}\right)^2\right]_+^{\frac{1}{m-1}}, & t>0 \\ 0, & t \le 0 \end{cases} \tag{37}$$

where $\frac{(N-2)_+}{N} < m < 1$,

$$\kappa = N(m-1)+2, \quad \text{and} \quad b = b(N,m) = \frac{N(1-m)}{2Nm\kappa}, \tag{38}$$

turns out to be very useful. It is well-known, that the Barenblatt fundamental solution is a very weak solution of

$$\partial_t u - \Delta u^m = \delta_{(0,0)},$$

in $\mathbb{R}^N \times [-\varepsilon,+\infty)$, $\varepsilon > 0$, where $\delta_{(0,0)}$ is the delta function at the origin. The uniqueness of the solution of the porous medium equation, when the *initial datum* is a locally integrable function has been proved in [29], and it has been extended to measures in [17] (see also [22; Chap. 3]).

Even though we have stated our results in E_T, where E is a bounded domain and not the whole $\mathbb{R}^N$, as it is the case for the Barenblatt fundamental solution, nevertheless it can be shown that we can apply the potential estimate (14) in this context too.

We claim that (14) shows the correct decay at the origin. By its very definition, the Riesz potential can be bounded from above by

$$\mathbf{I}_2^{\mu}(z_o,r,\theta) \le \int_0^{\infty} \frac{\mu(Q_{\rho,\rho^2\theta/r^2}(z_o))}{\rho^N}\frac{d\rho}{\rho}.$$

For the sake of simplicity, in the following we will write u instead of $\mathcal{B}_m$. Consider a point $z_o = (0,t_o)$ with $t_o > 0$ (note that $u(0,t_o) > 0$), and let $\theta := [u(0,t_o)]^{1-m}r^2$, which is the correct *intrinsic* scaling for the porous medium equation. For the cylinder $Q_{\rho,\rho^2\theta/16r^2}(z_o)$ to contain the origin, we need to have that

$$\rho^2\frac{[u(0,t_o)]^{1-m}r^2}{16r^2}\ge t_o \qquad \Longleftrightarrow \qquad \rho\ge 4\sqrt{t_o[u(0,t_o)]^{m-1}}.$$

Consequently, we have

$$\begin{aligned}\mathbf{I}_2^{\delta_o}(z_o,4r,\theta)&\le\int_{4\sqrt{t_o[u(0,t_o)]^{m-1}}}^{\infty}\frac{\mu(Q_{\rho,\rho^2\theta/16r^2}(z_o))}{\rho^N}\frac{d\rho}{\rho}\\&=\int_{4\sqrt{t_o[u(0,t_o)]^{m-1}}}^{\infty}\frac{d\rho}{\rho^{N+1}}=\frac{1}{4^N N t_o^{\frac{N}{2}}[u(0,t_o)]^{\frac{(m-1)N}{2}}}.\end{aligned}$$

Now, if we limit ourselves to consider only the bound from above that comes from the Riesz potential, namely

$$u(z_o)\le\gamma\left(\frac{r^2}{\theta}\right)^{\frac{N}{\kappa}}\mathbf{I}_2(z_o,4r,\theta)^{\frac{2}{\kappa}},$$

we obtain that

$$u(0,t_o)\le\gamma\left(\frac{r^2}{[u(0,t_o)]^{1-m}r^2}\right)^{\frac{N}{\kappa}}\left(\frac{1}{t_o^{\frac{N}{2}}[u(0,t_o)]^{\frac{(m-1)N}{2}}}\right)^{\frac{2}{\kappa}}=\gamma\, t_o^{-\frac{N}{\kappa}}$$

holds true. This corresponds exactly to the time decay of $\mathcal{B}_m$, and shows that (14) is optimal under this point of view. A similar decay estimate is given in [35; § 1.3] for the gradient of the Barenblatt fundamental solution of the parabolic p–Laplacian, when $p<2$.

It would be nice, if we could test also (34) versus the Barenblatt fundamental solution, as we did in § 4. Unfortunately, as mentioned above, $\mathcal{B}_m$ is the very weak solution in $\mathbb{R}^N\times[-\varepsilon,+\infty)$, where $\varepsilon>0$, and therefore, it cannot be a solution of (30), which is stated in E_T, E being a bounded domain, with *homogeneous* conditions on the parabolic boundary. We do not have this difficulty when $m>1$, since in such a case $\mathcal{B}_m$ has compact support, and provided we choose E large enough, it is always possible to satisfy homogeneous conditions on the parabolic boundary of E_T.

Finally, for more regular data, we can improve in Theorem 8.

Theorem 10 [[12]]. *Assume that hypothesis* (28) *holds and that the Radon measure is given by a non-negative function* $\mu\in L^s(E_T)$ *for some* s *in the range*

$$1<s<1+\frac{Nm}{Nm+2+2m}. \tag{39}$$

Then, there exists at least one non-negative very weak solution u *of the measure data problem* (30) *satisfying*

$$u^m\in L^q(0,T;W_0^{1,q}(E)) \tag{40}$$

where q is given by

$$q = \frac{s(Nm+2)}{Nm+m+1-ms}. \tag{41}$$

As in § 4, in the case $m = 1$ the restriction on s and the value of q coincide with the corresponding values from [8; Theorem 1.9] (take $p = 2$ in (1.9) and (1.10) in Theorem 1.9 of [8]). With this respect, once more, our result perfectly matches with the classical existence result for very weak solutions of the measure data problems.

References

1. H.W. Alt, S. Luckhaus, Quasilinear elliptic-parabolic differential equations. Math. Z. **183**, 311–341 (1983)
2. D. Andreucci, $L^\infty_{\rm loc}$-estimates for local solutions of degenerate parabolic equations. SIAM J. Math. Anal. **22**, 138–145 (1991)
3. D.G. Aronson: The porous medium equation. Nonlinear diffusion problems (Montecatini Terme, 1985). Lecture Notes in Math. **1224**, 1–46 Springer, Berlin (1986)
4. L. Boccardo, Problemi differenziali ellittici e parabolici con dati misure. Boll. Un. Mat. Ital. A. **11**(7), 439–461 (1997)
5. L. Boccardo, T. Gallouët, Nonlinear elliptic and parabolic equations involving measure data. J. Funct. Anal. **87**, 149–169 (1989)
6. L. Boccardo, T. Gallouët, Nonlinear elliptic equations with right-hand side measures. Comm. Partial Differ. Equ. **17**, 641–655 (1992)
7. L. Boccardo, T. Gallouët, L. Orsina, Existence and uniqueness of entropy solutions for nonlinear elliptic equations with measure data. Ann. Inst. H. Poincaré Anal. Non Linéaire **13**, 539–551 (1996)
8. L. Boccardo, A. Dall'Aglio, T. Gallouët, L. Orsina, Nonlinear parabolic equations with measure data. J. Funct. Anal. **147**, 237–258 (1997)
9. V. Bögelein, F. Duzaar, U. Gianazza, Porous medium type equations with measure data and potential estimates. SIAM J. Math. Anal. **45**, 3283–3330 (2013)
10. V. Bögelein, F. Duzaar, U. Gianazza, Continuity estimates for porous medium type equations with measure data. J. Funct. Anal. **267**(9), 3351–3396 (2014)
11. V. Bögelein, F. Duzaar, U. Gianazza, Sharp boundedness and continuity results for the singular porous medium equation. Preprint Mittag-Leffler Institute, ISRN IML-R- -31-13/14- -SE+fall, to appear in Israel J. Math.
12. V. Bögelein, F. Duzaar, U. Gianazza, Very weak solutions of singular porous medium equations with measure data. Commun. Pure Appl. Anal. **14**(1) 23–49 (2015)
13. M. Bonforte, J.L. Vázquez, Positivity, local smoothing, and Harnack inequalities for very fast diffusion equations. Adv. Math. **223**, 529–578 (2010)
14. L.A. Caffarelli, C.L. Evans, Continuity of the temperature in the two phase Stefan problem. Archive Ration. Mech. Anal. **81**, 199–220 (1983)
15. L.A. Caffarelli, A. Friedman, Regularity of the free boundary of a gas flow in an n-dimensional porous medium. Indiana Univ. Math. J. **29**, 361–391 (1980)
16. Y.-Z. Chen, E. DiBenedetto, On the local behavior of solutions of singular parabolic equations. Arch. Ration. Mech. Anal. 103, 319–345 (1988)
17. B.E. Dahlberg, C.E. Kenig, Non-negative solutions to fast diffusions. Rev.Mat. Iberoam. **4**, 11–29 (1988)
18. A. Dall'Aglio, Approximated solutions of equations with L^1 data. Application to the H-convergence of quasi-linear parabolic equations. Ann. Mat. Pura Appl. **170**(4), 207–240 (1996)

19. A. Dall'Aglio, L. Orsina, Existence results for some nonlinear parabolic equations with nonregular data. Differ. Integral Equ. **5**, 1335–1354 (1992)
20. A. Dall'Aglio, L. Orsina, Nonlinear parabolic equations with natural growth conditions and L^1 data. Nonlinear Anal. **27**, 59–73 (1996)
21. A. Dall'Aglio, D. Giachetti, C. Leone, S. Segura de León, Quasi-linear parabolic equations with degenerate coercivity having a quadratic gradient term. Ann. Inst. H. Poincar Anal. Non Linéaire **23**, 97–126 (2006)
22. P. Daskalopoulos, C.E. Kenig, Degenerate diffusions. EMS Tracts in Mathematics, 1, European Mathematical Society (EMS), Zürich (2007)
23. E. DiBenedetto, Continuity of weak solutions to certain singular parabolic equations. Ann. Mat. Pura. Appl. **130**, 131–176 (1982)
24. E. DiBenedetto, *Degenerate Parabolic Equations. Springer Universitext*. (Springer, New York, 1993)
25. E. DiBenedetto, A. Friedman, Hölder estimates for nonlinear degenerate parabolic systems. J. Reine Angew. Math. **357**, 1–22 (1985)
26. E. DiBenedetto, U. Gianazza, V. Vespri, *Harnack's inequality for degenerate and singular parabolic equations. Springer Monographs in Mathematics*. (Springer, New York, 2012)
27. F. Duzaar, G. Mingione, Gradient continuity estimates. Calc. Var. Partial Differ. Equ. **39**, 379–418 (2010)
28. F. Duzaar, G. Mingione, Gradient estimates via non-linear potentials. Am. J. Math. **133**, 1093–1149 (2011)
29. M.A. Herrero, M. Pierre, The Cauchy problem for $u_t = \Delta u^m$ when $0 < m < 1$. Trans. Amer. Math. Soc. **291**, 145–158 (1985)
30. L.I. Hunt, On $L(p,q)$ spaces. L'Einsegnement Math. **12**, 249–276 (1966)
31. A.V. Ivanov, P.Z. Mkrtychyan, On the existence of Hölder-continuous generalized solutions of the first boundary value problem for quasilinear doubly degenerate parabolic equations. J. Soviet Math. **62**, 2725–2740 (1992)
32. A.S. Kalashnikov, Some problems of the qualitative theory of second-order nonlinear degenerate parabolic equations. Rus. Math. Surv. **42**, 169–222 (1987)
33. T. Kilpeläinen, J. Malý, The Wiener test and potential estimates for quasilinear elliptic equations. Acta Math. **172**, 137–161 (1994)
34. T. Kuusi, G. Mingione, Potential estimates and gradient boundedness for nonlinear parabolic systems. Rev. Mat. Iberoam. **28**, 535–576 (2012)
35. T. Kuusi, G. Mingione, Gradient regularity for nonlinear parabolic equations. Ann. Scuola Norm. Sup. Pisa **12**, 755–822 (2013)
36. T. Kuusi, G. Mingione, The Wolff gradient bound for degenerate parabolic equations. J. Eur. Math. Soc. **16**, 835–892 (2014)
37. T. Kuusi, G. Mingione, Riesz potentials and nonlinear parabolic equations. Arch. Ration. Mech. Anal. 212, 727–780 (2014)
38. T. Kuusi, G. Mingione, Borderline gradient continuity for nonlinear parabolic systems. Math. Ann. **360**, 937–993 (2014)
39. V. Liskevich, I.I. Skrypnik, Harnack inequality and continuity of solutions to quasi-linear degenerate parabolic equations with coefficients from Kato-type classes. J. Differ. Equ. **247**, 2740–2777 (2009)
40. V. Liskevich, I.I. Skrypnik, Pointwise estimates for solutions of singular quasi-linear parabolic equations. Discret. Contin. Dyn. Syst. Ser. S **6**, 1029–1042 (2013)
41. V. Liskevich, I.I. Skrypnik, Pointwise estimates for solutions to the porous medium equation with measure as a forcing term. Israel J. Math. **194**, 259–275 (2013)
42. V. Liskevich, I.I. Skrypnik, Z. Sobol, Potential estimates for quasilinear parabolic equations. Adv. Nonlinear Stud. **11**, 905–915 (2011)
43. V. Liskevich, I.I. Skrypnik, Z. Sobol, Gradient estimates for degenerate quasi-linear parabolic equations. J. Lond. Math. Soc. **84**(2), 446–474 (2011)

44. V. Liskevich, I.I. Skrypnik, Z. Sobol, Estimates of solutions for the parabolic p-Laplacian equation with measure via parabolic nonlinear potentials. Commun. Pure Appl. Anal. **12**, 1731–1744 (2013)
45. T. Lukkari, The porous medium equation with measure data. J. Evol. Equ. **10**, 711–729 (2010)
46. T. Lukkari, The fast diffusion equation with measure data. NoDEA Nonlinear Differ. Equ. Appl. **19**, 329–343 (2012)
47. M. Pierre, Uniqueness of the solutions of $u_t - \Delta\Phi(u) = 0$ with initial datum a measure. Nonlinear Anal. **6**, 175–187 (1982)
48. E.M. Stein, G. Weiss, Introduction to Fourier Analysis on Euclidean Spaces. Princeton Math. Ser., 32, Princeton University Press, Princeton, NJ (1971)
49. J.L. Vázquez, Smoothing and decay estimates for nonlinear diffusion equations. Oxford Lecture Series in Mathematics and its Applications. Oxford University Press, Oxford (2006)
50. J.L. Vázquez, *The porous medium equation. Oxford Mathematical Monographs.* (The Clarendon Press/Oxford University Press, Oxford, 2007)
51. V. Vespri, On the local behaviour of solutions of a certain class of doubly nonlinear parabolic equations. Manuscr. Math. **75**, 65–80 (1992)

Aspects of the Mathematical Analysis of Nonlinear Stratified Water Waves

David Henry and Bogdan-Vasile Matioc

Abstract The mathematical analysis of stratified water waves, where the density distribution of the flow is free to fluctuate, is a highly intractable subject which is of great physical and geophysical importance. In this chapter, we present an overview of some recently derived rigorous analytical results for nonlinear steady periodic stratified water waves.

1 Introduction

The study of stratified water waves, or flows which exhibit a variable density distribution, is a subject which is of great physical interest, since fluid density may be caused to fluctuate by a plethora of factors – for example, salinity, temperature, pressure, topography, oxygenation [54, 55]. Furthermore, stratification plays a prominent role for geophysical fluid phenomena where the physical scales of the fluid motion are such that the effects of the Earth's rotation are significant [16]. Mathematically, allowing for heterogeneity in the fluid adds severe complications to the governing equations – for instance, stratified flows are inherently rotational [54, 55] – rendering them highly intractable to mathematical analysis. Accordingly, there has been a marked paucity of rigorous mathematical results for the full governing equations for stratified water waves, even in the inviscid regime and where the effects of the Earth's rotation are neglected.

This survey aims to give an overview of some recent analytical results for periodic, steady stratified water waves. The primary emphasis in this survey will be on existence results for the fully nonlinear governing equations in the inviscid regime. Due to space considerations, and due to the vastness of the subject, in the considerations of this survey chapter we must omit a number of important aspects of stratified water waves which are currently active areas of mathematical research, among these

D. Henry (✉)
School of Mathematical Sciences, University College Cork, Cork, Ireland
e-mail: d.henry@ucc.ie

B.-V. Matioc
Institute for Applied Mathematics, Leibniz University of Hannover, 30167 Hannover, Germany
e-mail: matioc@ifam.uni-hannover.de

J. Escher et al., (eds.), *Elliptic and Parabolic Equations*, Springer Proceedings in Mathematics & Statistics 119, DOI 10.1007/978-3-319-12547-3_7

being the study of internal waves. Also, aside from outlining some Gerstner-type exact solutions for geophysical water waves in Sect. 4, we neglect the effect of the Earth's rotation. We note that the survey [54] and book [55] are good expositions of the vast subject of stratified water waves, while [16] provides a nice introduction to stratification in geophysical flows.

The first results concerning the existence of small-amplitude waves for stratified flows were obtained by Dubreil-Jacotin [19] in 1937. In Sect. 2, we survey recent rigorous existence results [21, 27, 28] where the authors use local and global bifurcation theory to prove the existence of both small-, and large-, amplitude stratified water waves. A novel feature of this work is that the existence of critical points and layers is not precluded from the flows. Stagnation points have long been a source of great interest and fascination in hydrodynamical research, dating back to Kelvin's work concerning "cat's eyes" [29] and Stokes conjecture on the wave of greatest height, cf. [5, 43] for an overview of Stokes' waves.

In Sect. 3, we outline the existence results of Walsh [48–50] for stratified water waves which do not contain critical points. In the absence of critical or stagnation points, the framework of local and global bifurcation which was presented in the breakthrough paper [12] for homogeneous water waves with vorticity, is successfully adapted to prove the existence of large-amplitude stratified gravity water waves in [48]. In [49, 50], Walsh then proved the existence of large-amplitude stratified capillary-gravity water waves, where surface tension is a significant restoration force at the free boundary.

In Sect. 4, we outline some recently derived exact Gerstner-type solutions for geophysical water waves which incorporate stratification in the underlying flow, and in Sect. 5, we present some results concerning qualitative properties of stratified flows.

1.1 The Mathematical Model

For the analytic study of stratified water waves, the Long-Yih formulation [31, 44, 52] for the motion of a two-dimensional inviscid, incompressible fluid with variable density is essential. The governing equations are formulated in terms of the unknown free-surface of the wave $y = \eta(t,x)$, the velocity field (u, v) of the fluid, the pressure distribution P, the variable density ρ, and the gravity constant g. The equations are defined in the fluid domain

$$\Omega_\eta := \{(x, y) : x \in \mathbb{R} \quad \text{and} \quad -d < y < \eta(t,x)\},$$

where d is a positive constant and $y = -d$ models a flat impermeable bed. Taking $y = 0$ to represent the location of the undisturbed water surface, we assume for any fixed time t that $\eta(t, \cdot)$ has zero integral mean over a period.

This survey concerns traveling wave solutions of the governing equations, with all functions having an (x, y, t) dependence of the form $(x - ct, y)$, and so

$$\eta(t,x) = \eta(x - ct), \qquad (\rho, u, v, P)(t, x, y) = (\rho, u, v, P)(x - ct, y),$$

with the positive constant $c > 0$ the *wavespeed*. Under the assumption of constant temperature and zero viscosity, from a reference frame moving with the wave the flow is steady and is described by the steady two-dimensional Euler equations:

$$\begin{cases} \rho(u-c)u_x + \rho v u_y = -P_x & \text{in } \Omega_\eta, \\ \rho(u-c)v_x + \rho v v_y = -P_y - g\rho & \text{in } \Omega_\eta, \\ \rho_x(u-c) + \rho_y v = 0 & \text{in } \Omega_\eta, \\ u_x + v_y = 0 & \text{in } \Omega_\eta. \end{cases} \tag{1a}$$

The first two relations of (1a) are the conservation of momentum equations, the last relation expresses the fact that the flow is incompressible, and the second last relation is a reformulation of the steady continuity equation. Since some of the results we mention herein apply in the setting of waves with surface tension effects, the following boundary conditions have to be imposed

$$\begin{cases} v = (u-c)\eta' & \text{on } y = \eta(x), \\ P = P_0 - \sigma\eta''/(1+\eta'^2)^{3/2} & \text{on } y = \eta(x), \\ v = 0 & \text{on } y = -d. \end{cases} \tag{1b}$$

Here, P_0 is the constant atmospheric pressure and σ is the surface tension coefficient. In the cases where we neglect the effects of surface tension, we simply set $\sigma = 0$ in the above. The first and last equations of (1b) are kinematic conditions which express the fact that the wave surface moves along with the fluid and that the fluid bed is impermeable. The second relation of (1b) is the dynamic boundary condition which says that the pressure jump along the wave surface obeys the Laplace–Young law: it is proportional to the curvature of the surface and the constant of proportionality is the surface tension coefficient [30].

We reformulate system (1a, 1b) into a more useful, malleable form as follows. An important consequence of the last two equations of (1a) is that

$$\nabla \cdot (\sqrt{\rho}(u-c), \sqrt{\rho}v) = 0 \qquad \text{in } \Omega_\eta.$$

This relation enables us to associate to system (1a, 1b) the *pseudo-stream function* ψ defined by

$$\psi(x,y) := \lambda + \int_{-d}^{y} \sqrt{\rho}(x,s)(u(x,s)-c)\,\mathrm{d}s, \qquad (x,y) \in \overline{\Omega}_\eta.$$

The constant λ is defined by the property that ψ vanishes at the wave surface, with $-\lambda$ being the *relative mass flux* of the flow. Since $\nabla\psi = (-\sqrt{\rho}v, \sqrt{\rho}(u-c))$ is orthogonal to the steady velocity field $(u-c, v)$ throughout the fluid domain, it follows that the level sets of the ψ function are the streamlines of the steady flow.

An important assumption which is often invoked when studying traveling water waves is that

$$\partial_y\psi = \sqrt{\rho}(u-c) < 0 \qquad \text{in } \overline{\Omega}_\eta, \tag{2}$$

a condition which excludes the presence of stagnation points in the flow, that is, that points for which $\nabla\psi = 0$. This is a physically reasonable assumption for water waves, without underlying currents containing strong nonuniformities, and which are not near breaking [30]. Nevertheless, stagnation points are a very interesting phenomenon and ideally we would not wish to exclude them from our picture. For fluid motions where (2) holds we have $\lambda > 0$, the relative mass flux being negative. The condition (2) can be used to formulate the hydrodynamical problem in the rectangular domain $\Omega_\lambda := \mathbb{R} \times (-\lambda, 0)$, where the streamlines become straight horizontal lines in the Ω_λ. This property provides us with a convenient means of analyzing how certain quantities change either along fixed streamlines or as we vary the streamlines. To present this new formulation, we first recall that the Dubreil-Jacotin [18] semi-hodograph transformation:

$$\mathcal{H}(x, y) := (q, p)(x, y) := (x, -\psi(x, y)), \qquad (x, y) \in \overline{\Omega}_\eta \tag{3}$$

is a diffeomorphism $\mathcal{H} : \Omega_\eta \to \Omega_\lambda$. Using the second last equations of (1a), it is easy to see that the density is constant on each streamline, since

$$\partial_q \left(\rho \circ \mathcal{H}^{-1}\right) \circ \mathcal{H} = \rho_x + \rho_y \frac{v}{u - c} = \frac{(u - c)\rho_x + v\rho_y}{u - c} = 0.$$

This means that there exists a function $\overline{\rho} : [-\lambda, 0] \to (0, \infty)$, the so-called *streamline density function* such that $\rho \circ \mathcal{H}^{-1} = \overline{\rho}$, or equivalently $\rho(x, y) = \overline{\rho}(-\psi(x, y))$ for all $(x, y) \in \overline{\Omega}_\eta$. Furthermore, by Bernoulli's principle, the energy

$$E := P + \rho \frac{(u - c)^2 + v^2}{2} + g\rho y \tag{4}$$

is constant along streamlines, that is $\partial_q \left(E \circ \mathcal{H}^{-1}\right) = 0$. Particularly, when $p = 0$, we obtain

$$|\nabla\psi|^2 - 2\sigma \frac{\eta''}{(1 + \eta'^2)^{3/2}} + 2g\overline{\rho}(0)y = Q \qquad \text{on } y = \eta(x),$$

for some constant $Q \in \mathbb{R}$. Moreover, a direct calculation shows that

$$\Delta\psi - gy\overline{\rho}' \circ \mathcal{H} = -\partial_p(E \circ \mathcal{H}^{-1}) \circ \mathcal{H} \qquad \text{in } \Omega_\eta.$$

Since $\partial_q \left(E \circ \mathcal{H}^{-1}\right) = 0$, there exists a function $\beta = \beta(p)$, called *Bernoulli's function*, such that $-\partial_p(E \circ \mathcal{H}^{-1}) = \beta$ in Ω_λ. Piecing all this together leads to the Long-Yih formulation for steady stratified water waves [31, 44, 52] where ψ solves the following free boundary problem:

$$\begin{cases} \Delta\psi = gy\overline{\rho}'(-\psi) + \beta(-\psi) & \text{in } \Omega_\eta, \\ \psi = 0 & \text{on } y = \eta(x), \\ \psi = \lambda & \text{on } y = -d, \\ |\nabla\psi|^2 - 2\sigma \dfrac{\eta''}{(1 + \eta'^2)^{3/2}} + 2g\overline{\rho}(0)y = Q & \text{on } y = \eta(x). \end{cases} \tag{5}$$

A further equivalent formulation of the problem (1a, 1b) and (2) can be derived in terms of the height function $h : \overline{\Omega}_\lambda \to \mathbb{R}$ given by

$$h(q, p) = y + d \qquad \text{for } (q, p) \in \overline{\Omega}_\lambda.$$

In terms of h, the system (5) can be recast as a set of equations in a fixed rectangular domain:

$$\begin{cases} (1 + h_q^2)h_{pp} - 2h_q h_p h_{pq} + h_p^2 h_{qq} - (g\overline{\rho}'(h-d) + \beta)h_p^3 = 0 & \text{in } \Omega_\lambda, \\ h = 0 & \text{on } p = -\lambda, \\ 1 + h_q^2 + h_p^2\left[2g\overline{\rho}(0)(h-d) - 2\sigma \dfrac{h_{qq}}{(1+h_q^2)^{3/2}} - Q\right] = 0 & \text{on } p = 0, \end{cases} \tag{6}$$

the relation (2) taking the form

$$\inf_{\Omega_\lambda} h_p > 0. \tag{7}$$

The problem (6) consists of a quasilinear elliptic equation subjected to nonlinear boundary conditions. This formulation gives an insight into the flow because the streamlines in the moving frame are parametrized by the mappings $x \mapsto h(x, p) - d$. The formulations (1a, 1b), (2), and (5) are equivalent in the setting of classical solutions when (6) and (7) holds, cf. e.g., [48].

2 Existence Results for Flows Admitting Stagnation Points

When allowing for stagnation points to exist, that is, when dropping assumption (2), the three formulations – the Euler equations, the stream function formulation, and the height function formulation – are no longer equivalent. However, it is not difficult to see that if we are given a Bernoulli function $\beta : \mathbb{R} \to \mathbb{R}$ and a streamline density function $\overline{\rho} : \mathbb{R} \to (0, \infty)$, then any solution (η, ψ) of (5) defines a solution (η, u, v, P) of (1a, 1b). Furthermore, if $\nabla\psi(P) = 0$ at some point $P \in \overline{\Omega}_\eta$, then P is a stagnation point of the flow. This aspect has been exploited to construct small-amplitude gravity waves with a linear density distribution and constant Bernoulli function that contain stagnation points in [21]. In the case of more general streamline density and Bernoulli functions, small- as well as large-amplitude capillary-gravity waves that may posses stagnation points where shown to exist in [27, 28]. In the references [21, 27, 28], the fluid bed is taken to be located at $y = -1$ and the waves that are found are 2π-periodic and bifurcate from laminar flows with the wave surface being located at $y = 0$. Nevertheless, the methods can be applied in the context of flows with arbitrary finite depth and arbitrary wavelength.

2.1 Stratified Flows with Linear Density Distribution and Constant Bernoulli Function

The analysis in [21] was devoted to the classification of the gravity waves (that is $\sigma = 0$) with stagnation points that bifurcate from laminar flows when considering a linear streamline density and a constant Bernoulli function. A precise description of the flow pattern for the classes of flows that were found was also given. To recall the results of [21] we assume that

$$\overline{\rho}(p) = Ap + B \quad \text{and} \quad \beta(p) = C \qquad \text{for } p \in \mathbb{R}.$$

Of course, the constants $A, B, C \in \mathbb{R}$ have to be chosen such that the bifurcating solutions (η, ψ) of (5), that is in this particular case of

$$\begin{cases} \Delta\psi = Agy + C & \text{in } \Omega_\eta, \\ \psi = 0 & \text{on } y = \eta(x), \\ \psi = \lambda & \text{on } y = -1, \\ |\nabla\psi|^2 + 2Bgy = Q & \text{on } y = \eta(x), \end{cases} \tag{8}$$

have positive density $\rho(x, y) = \overline{\rho}(-\psi(x, y)) > 0$ for $(x, y) \in \Omega_\eta$. As a result of this and recalling that $\psi = 0$ at the wave surface, the constant B has to be taken positive. The laminar flow solutions of (8) with the free surface given by $\eta = 0$ can be parametrized as a family $\{(0, \psi_\mu) : \mu \in \mathbb{R}\}$, with the constants λ and Q being also function of μ, that is

$$\psi_\mu(y) := \frac{Agy^3}{6} + \frac{Cy^2}{2} + \mu y, \quad Q = \mu^2, \quad \lambda = -\frac{Ag}{6} + \frac{C}{2} - \mu.$$

The constant μ satisfies $\mu = \partial_y\psi|_{y=0}$, so that it represents the horizontal speed in the moving frame of the laminar flows at the wave surface. In fact, the stream function ψ can be uniquely determined when knowing μ and η, so that (μ, η) are the only unknowns in (8) when considering Q and λ as functions of μ as defined above.

The dispersion relation for such waves is determined as follows:

$$\mu_k^{\pm} := \frac{C}{2}\frac{\tanh(k)}{k} \pm \sqrt{\frac{C^2}{4}\frac{\tanh^2(k)}{k^2} + \frac{2Bg}{2}\frac{\tanh(k)}{k}},$$

the integer $|k|$ representing the wave number. Each of the points $(\mu_k^{\pm}, 0)$ is a critical bifurcation point.

Theorem 1 ([21, Theorem 2.1]) *Let $k \in \mathbb{N} \setminus \{0\}$, $\mu_* \in \{\mu_k^{\pm}\}$, $\alpha \in (0, 1)$, and $B > 0$ be given. Then, there exists $\varepsilon > 0$ and a real-analytic curve*

$$(\mu, \eta) : (-\varepsilon, \varepsilon) \to \mathbb{R} \times C^{2+\alpha}(\mathbb{R}),$$

consisting only of solutions $(\mu(s), \eta(s))$ of (8) of minimal period $2\pi/k$, having exactly one crest and trough per period, and with a real-analytic and symmetric free surface.

These are the only nonlaminar solutions of (8) close to $(\mu_*, 0)$, *and, for* $s \to 0$,

$$\mu(s) = \mu_* + O(s^2),$$
$$\eta(s) = -s\cos(kx) + O(s^2) \quad in\ C^{2+\alpha}(\mathbb{R}).$$

There are no bifurcation points $(\mu, 0)$ *other than* $\{(\mu_k^{\pm}, 0) : k \in \mathbb{N} \setminus \{0\}\}$.

Because of the asymptotic expansion derived for the bifurcation curves, it is not difficult to see that the flow determined by $(\mu(s), \eta(s))$ contains no stagnation points if $\partial_y \psi_{\mu_*}$ does not vanish in $[-1, 0]$ and ε is small. Moreover, it is shown in [21] that if

$$\partial_y \psi_{\mu_*}(y) = \frac{Agy^2}{2} + Cy + \mu = 0 \tag{9}$$

has solutions in $[-1, 0]$, then the bifurcating nonlaminar flows will also contain stagnation points. Hence, the classification of the flow pattern of the bifurcating small-amplitude waves with stagnation points has to take into account the location of the solutions y_1, y_2 of (9). Only the following classes of stratified waves bifurcating from laminar flows with stagnation points are possible:

(I) Stratified flows with exactly one critical layer. They bifurcate from the laminar flows $(\mu_*, 0)$ with $\mu_* \in \{(\mu_k^{\pm}, 0) : k \in \mathbb{N} \setminus \{0\}\}$ in one of the following cases:

- (a) The equation $\partial_y \psi_{\mu_*} = 0$ has a unique solution in $(-1, 0)$.
- (b) The equation $\partial_y \psi_{\mu_*} = 0$ has $y = -1$ as the unique solution within $[-1, 0)$ (or the solutions are both equal to -1).
- (c) The equation $\partial_y \psi_{\mu_*} = 0$ has two solutions in $(-1, 0)$, and they coincide.

(II) Stratified flows with two critical layers. They bifurcate from the laminar flows $(\mu_*, 0)$ with $\mu_* \in \{(\mu_k^{\pm}, 0) : k \in \mathbb{N} \setminus \{0\}\}$ in one of the following cases.

- (d) The equation $\partial_y \psi_{\mu_*} = 0$ has two solutions $-1 = y_2 < y_1 < 0$.
- (e) The equation $\partial_y \psi_{\mu_*} = 0$ has two solutions $-1 < y_2 < y_1 < 0$.

These five classes of stratified gravity waves with stagnation points are pictured in Fig. 1. We note that the stagnation points are located either beneath the wave crest or beneath the wave trough. Very interesting is the degenerate case (c) where bifurcation occurs when the solutions of $\partial_y \psi_{\mu_*} = 0$ satisfy $y_1 = y_2 \in (-1, 0)$. In this case, there is only a compact (in Ω_η) critical layer consisting of closed streamlines right beneath the wave crest. Such a flow pattern is not encountered in homogeneous flows with constant or linear vorticity, cf. [13, 20, 46], but the situation slightly resembles to that of background flows for tsunamis which contain isolated regions of vorticity surrounded by still water [7, 23]. For the solutions in all five classes, the density has an extremum at the stagnation point located in the center of the vortices.

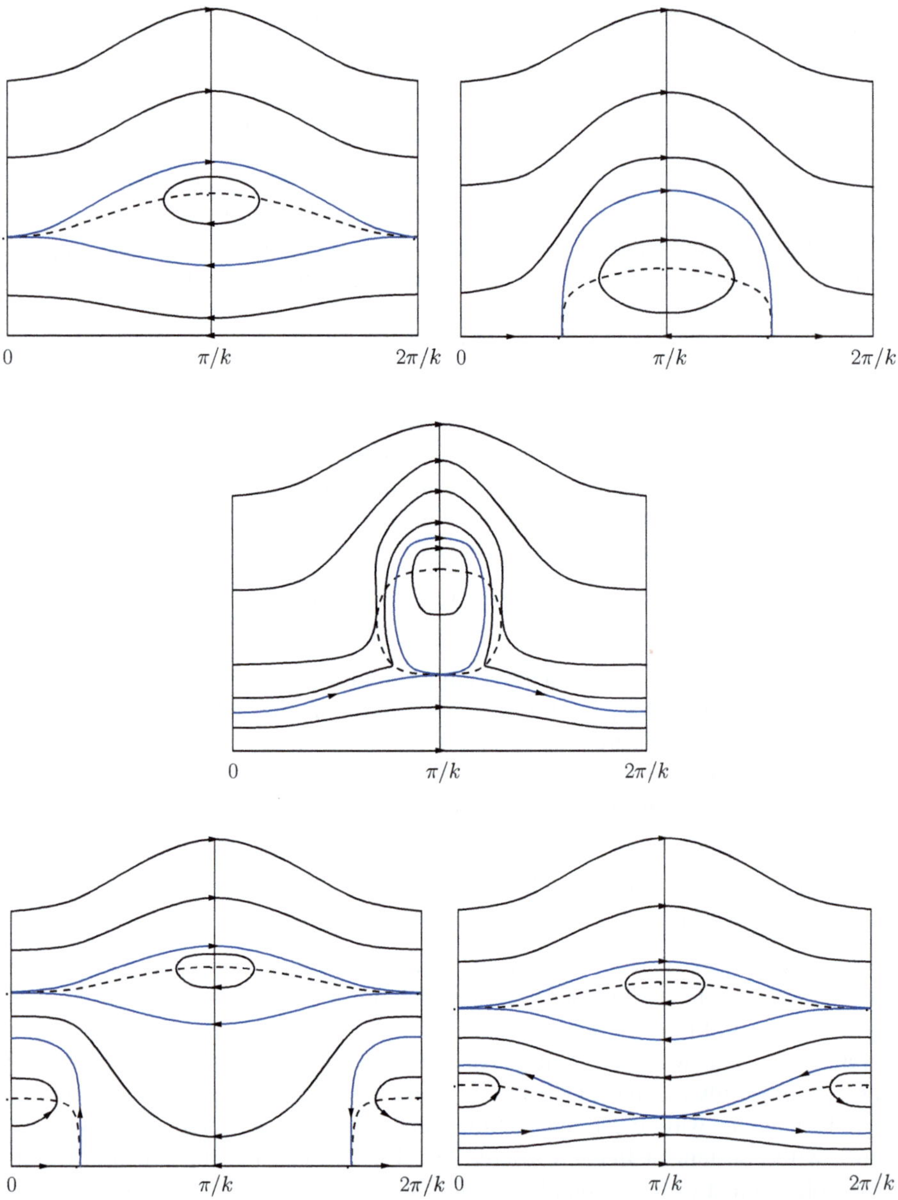

Fig. 1 The flow pattern for the stratified waves bifurcating from laminar flow solution with stagnation points for waves in the classes (*a*), (*b*), (*c*), (*d*), and (*e*) (from the *upper-left* corner to the *lower-right* corner). The dashed lines are curves where $\partial_y \psi$ vanishes, and the solid dots are the stagnation points. The blue curves connecting the stagnation points are separatrices that enclose (with the exception of the lower separatrix in the third figure) critical layers of closed streamlines

2.2 *Stratified Flows with More General Density Distribution and Bernoulli Function*

We address now the results established in [27, 28] in the context of more general streamline density and Bernoulli functions for waves with capillarity. To this end, we make the following notation:

$$f(y,\psi) := gy\overline{\rho}'(-\psi) + \beta(-\psi), \qquad (y,\psi) \in [-1,1] \times \mathbb{R}.$$

The solutions (η, ψ) of (5) are to be found such that $\max_{\mathbb{R}} |\eta| < 1$. In fact, relying on the divergence structure of the curvature operator in Bernoulli's condition, the constant Q can be eliminated from the equations, if the integral mean of η is zero, by integrating the Bernoulli condition over $[0, 2\pi]$,

$$Q = \frac{1}{2\pi} \int_0^{2\pi} |\nabla\psi|^2|_{y=\eta(x)}\, \mathrm{d}x.$$

One arrives at the following problem:

$$\begin{cases} \Delta\psi = f(y,\psi) & \text{in } \Omega_\eta, \\ \psi = 0 & \text{on } y = \eta(x), \\ \psi = \lambda & \text{on } y = -1, \\ |\nabla\psi|^2 - 2\sigma \dfrac{\eta''}{(1+\eta'^2)^{3/2}} + 2g\overline{\rho}(0)y = \dfrac{1}{2\pi}\displaystyle\int_0^{2\pi} |\nabla\psi|^2|_{y=\eta(x)}\, \mathrm{d}x & \text{on } y = \eta(x). \end{cases} \tag{10}$$

As a result of this choice for Q, for any solution (η, ψ) of (10), η has zero integral mean over $[0, 2\pi]$, and therefore the solutions obtained by bifurcation describe flows over the same volume of fluid (the mean depth of the fluid is constant along the bifurcation branch).

The following assumptions are made on $\overline{\rho}$, β, and f:

(A1) $\overline{\rho} > 0$.

(A2) $\overline{\rho} \in C^{4-}(\mathbb{R}), \beta \in C^{3-}(\mathbb{R})$, and $\beta, \overline{\rho}' \in BC^2(\mathbb{R})$.

(A3) $\partial_\psi f(y,\psi) \geq 0$ for all $(y,\psi) \in [-1,1] \times \mathbb{R}$.

(A4) $2f(y,\psi) + g(1+y)\overline{\rho}'(-\psi) \leq 0$ for all $(y,\psi) \in [-1,0] \times (-\infty, 0]$.

In the context of homogeneous flows, we have that $\overline{\rho}' \equiv 0$ and the Bernoulli function β is identified with the vorticity function γ of the rotational flow. In this context, the conditions $(A1) - (A4)$ are equivalent to

$$\gamma \in C^{3-}(\mathbb{R}) \cap BC^2(\mathbb{R}), \quad \gamma \leq 0 \text{ in } [0,\infty), \quad \text{and} \quad \gamma' \leq 0. \tag{11}$$

While the first assumption is physical, the assumptions $(A2)$ and $(A3)$ ensure that the semilinear elliptic problem for ψ that consists of the first three equations of (10) has a solution $\psi \in C^{2+\alpha}(\overline{\Omega}_\eta)$ ($\alpha \in (0,1)$ is fixed), that is uniquely determined by the pair $(\lambda, \eta) \in \mathbb{R} \times \text{Ad}$, whereby

$$\text{Ad} := \left\{\eta \in C^{2+\alpha}(\mathbb{R}) \,:\, \eta \text{ is } 2\pi-\text{periodic}, \int_0^{2\pi} \eta \, dx = 0, \text{ and } \max_{\mathbb{R}}|\eta| < 1\right\}.$$

Therefore, the pairs $(\lambda, \eta) \in \mathbb{R} \times \text{Ad}$ are the true unknowns of (10). This unique solvability property of the semilinear elliptic problem for ψ was used in [27], after transforming the problem on a fixed reference domain, to recast the hydrodynamical problem as a nonlinear and nonlocal equation with (λ, η) as unknowns. Because for (λ, η) with $\eta = 0$, the solution ψ of (10) depends only upon y, it is easy to see that $(\lambda, 0)$ is a solution of (10) for any $\lambda \in \mathbb{R}$. This trivial branch of solution contains countably many bifurcation points $(\lambda_*, 0)$ where branches consisting of nonlaminar solutions of (10) emerge.

Theorem 2 ([27, Theorem 4.6]). *Let $\sigma > 0$ be fixed and assume additionally to $(A1) - (A4)$ assume that:*

(B1) $\partial_{\psi\psi} f \geq 0$ *on* $[-1, 0] \times (-\infty, 0]$.

(B2) $2\partial_\psi f(y, \psi) - g(1+y)\overline{\rho}''(-\psi) \geq 0$ *for all* $(y, \psi) \in [-1, 0] \times (-\infty, 0]$.

(B3) $\displaystyle\int_{-1}^{0}\int_{t}^{0} \partial_\psi f(s, 0)\, ds\, dt \leq 1.$

Then there exists a positive integer $K \in \mathbb{N}$ and for all $k \geq K$ a sequence $(\lambda_m)_{m\geq 1} \subset \mathbb{R}$ with $\lambda_m \to -\infty$ such that:

(i) *Given $m \geq 1$, there exists a curve $(\lambda_m, \eta_m) : (-\varepsilon, \varepsilon) \to \mathbb{R} \times (\text{Ad} \cap C_{2\pi/k}(\mathbb{R}))$ which is continuously differentiable in $\mathbb{R} \times (C^{2+\alpha}(\mathbb{R}) \cap C_{2\pi/k}(\mathbb{R}))$ and consists only of solutions $(\lambda_m(s), \eta_m(s))$ of (10). The wave determined by $(\lambda_m(s), \eta_m(s))$ with $s \neq 0$ has minimal period $2\pi/(mk)$, exactly one crest and trough per period and is symmetric with respect to the crest line.*

(ii) *For $s \to 0$,*

$$\lambda_m(s) = \lambda_m + O(s), \quad \eta_m(s) = -s\cos(mkx) + O(s^2).$$

All the solutions of (10) in $\mathbb{R} \times (\text{Ad} \cap C_{2\pi/k}(\mathbb{R}))$ that are close to $(\lambda_m, 0)$ are either laminar flows or belong to the curve (λ_m, η_m). Moreover, there exists a constant $\Lambda_- \in \mathbb{R}$ with the property that if $(\lambda, 0)$ is a bifurcation point of (10) with $\lambda \in (-\infty, \Lambda_-)$, then $\lambda \in \{\lambda_m : m \geq 1\}$.

Hereby, $C_{2\pi/k}(\mathbb{R})$ stands for the Banach space of $2\pi/k$-periodic functions. That the wave number k should be sufficiently large is a condition that appears frequently when dealing with waves with surface tension effects, cf. e.g., [32, 45, 49]. Lastly, let us note that in the context of homogeneous capillary-gravity waves the conditions $(B1) - (B3)$ mean that, additionally to (11), we assume

$$\gamma'' \geq 0 \text{ in } [0, \infty) \qquad \text{and} \qquad \gamma'(0) \geq -2. \tag{12}$$

The local bifurcation branches obtained in Theorem 2 can be continued by using global bifurcation theory, cf. [28]. The authors of [28] observed that the quasilinear curvature term in Bernoulli's condition can be inverted (see also [36]), the problem (10) being equivalent to an operator equation for a compact (nonlocal and nonlinear) perturbation of the identity. This facilitates the use of the Rabinowitz global bifurcation theorem, cf. e.g., Theorem II.3.3 in [23].

Theorem 3 ([28, Theorem 4.1]) *Let the assumptions of Theorem 2 be satisfied, let $k \geq K$, and fix $m \in \mathbb{N}$ with $m \geq 1$. Moreover, let $\mathcal{C}_M$ be the maximal connected component of the closure of the set:*

$$\left\{(\lambda, \eta) \;:\; (\lambda, \eta) \in \mathbb{R} \times (\mathrm{Ad} \cap C_{2\pi/k}(\mathbb{R})) \text{ is a solution of (10) with } \eta \not\equiv 0\right\}$$

in $\mathbb{R} \times (C^{2+\alpha}(\mathbb{R}) \cap C_{2\pi/k}(\mathbb{R}))$ that contains $(\lambda_m, 0)$. Then, we have:

(i) $\mathcal{C}_M$ is unbounded in $(-\infty, \Lambda_-) \times (C^{2+\alpha}(\mathbb{R}) \cap C_{2\pi/k}(\mathbb{R}))$, or
(ii) $\sup_{(\lambda,\eta)\in\mathcal{C}_M} \max_{\mathbb{R}} |\eta| = 1.$

Finally, let us emphasize that the Theorems 2 and 3 were stated for a fixed surface tension coefficient. There is a second possibility when dealing with the problem (10): To consider the mass flux fixed and use the surface tension coefficient as a bifurcation parameter. From mathematical point of view this choice is more useful: one can show [27, Theorem 4.3] that there exists a sequence of bifurcation points $(\sigma_m, 0)$ with $m \in \mathbb{N}$, $m \geq 1$, for (10) with $\sigma_m \searrow 0$. Each of the points $(\sigma_m, 0)$ belongs to a continuously differentiable local curve of nonlaminar solutions in $\mathbb{R} \times \mathrm{Ad}$. There is no restriction on the wave number any longer, and there is no need of the assumptions $(B1) - (B3)$. These local branches can be continued to global continua as in [28, Theorem 4.1].

3 Existence Results for Flows Without Critical Points

For flows without stagnation or critical points, the assumption (2) is valid throughout the fluid domain. Accordingly, the Dubreil-Jacotin semi-hodograph transformation (3) represents a change of variables and the stratified capillary-gravity water wave problem may be recast as system (6)—the inequality (7) ensures that system (6) is uniformly elliptic. In this framework, Walsh has recently proved the existence of finite amplitude steady periodic stratified water waves, in [48] for the setting of pure gravity waves, and in [49, 50] for capillary-gravity waves.

3.1 Gravity Stratified Waves

For the setting of pure gravity stratified water waves, which we may obtain by setting $\sigma = 0$ in (6), Walsh proved the existence of large-amplitude steady periodic stratified water waves [48]. The approach used in [48] successfully implemented a local and global bifurcation analysis along the lines of the seminal work [12] which proved existence for homogeneous, gravity water waves with vorticity. However, the application of this method for stratified water waves is highly nontrivial, and one must overcome a number of significant technical hurdles induced by the fluid stratification. A benefit of this approach is the derivation of an explicit size condition (13) on the physical variables which is sufficient for bifurcation to occur. The local bifurcation analysis employs the Crandall–Rabinowitz bifurcation theorem [15], whereby a necessary and sufficient condition, denoted (L-B), for local bifurcation to occur is derived depending on $\lambda, \beta, \overline{\rho}$. An explicit version of this condition, which is sufficient for local bifurcation to occur, may be stated as follows. First, we define

$$B(p) = -\int_0^p \beta(s)\mathrm{d}s, \qquad -\lambda \le p \le 0,$$

and let

$$\varepsilon_0 = \max\left\{\left(2g\|\overline{\rho}'\|_\infty \lambda^2 e^\lambda\right)^{2/3}, (2g\|\overline{\rho}'\|_\infty)^2, (4\|\overline{\rho}'\|_\infty)^2, (8g\lambda\overline{\rho}(0))^{2/3}\right\}.$$

Then the condition which is sufficient for local bifurcation to occur is given by

$$\begin{aligned} g\overline{\rho}(0)\lambda^2 > \int_{-\lambda}^0 \Bigg\{ & \frac{4\pi^2}{L^2}(2B(p) - 2B_{\min} + 2\varepsilon_0)^{3/2} \\ & + (p+\lambda)^2\left((2B(p) - 2B_{\min} + 2\varepsilon_0)^{1/2} + g\overline{\rho}'(p)\right)\Bigg\}\mathrm{d}p. \end{aligned} \tag{13}$$

Following the local existence result, the bifurcation curve may then be extended to a global continuum using the global bifurcation approach of Healey-Simpson and Kielhofer, cf. [6] for an overview of this approach. Using the notation of this survey chapter, the main result for stratified gravity water waves in [48] may be stated as follows.

Theorem 4 ([48]) *Fix a wavespeed $c > 0$, wavelength $L > 0$, and $\lambda > 0$. Fix any $\alpha \in (0, 1)$, and let the functions $\beta \in C^{1+\alpha}([-\lambda, 0])$ and $\overline{\rho} \in C^{1+\alpha}([-\lambda, 0])$ be given such that the (L-B) condition holds. Also, we assume the streamline density function $\overline{\rho}$ is nonincreasing. Consider traveling solutions to the stratified water wave problem (1a, 1b) of speed c, relative mass flux $-\lambda$, Bernoulli function β, and streamline density function $\overline{\rho}$ such that $u < c$ throughout the fluid. There exists a connected set $\mathcal{C}$ of solutions (u, v, ρ, η) in the space $C^{2+\alpha}_{per}(\overline{\Omega}_\eta) \times C^{2+\alpha}_{per}(\overline{\Omega}_\eta) \times C^{2+\alpha}_{per}(\overline{\Omega}_\eta) \times C^{3+\alpha}_{per}(\mathbb{R})$ with the following properties:*

1. *$\mathcal{C}$ contains a laminar flow (with a flat surface $\eta \equiv 0$ and all streamlines parallel to the bed).*
2. *Along some sequence $(u_n, v_n, \rho_n, \eta_n) \in \mathcal{C}$, either $\max_{\overline{\Omega}_{\eta_n}} u_n \uparrow c$, $\min_{\overline{\Omega}_{\eta_n}} u_n \downarrow -\infty$, or $\mathcal{C}$ contains more than one distinct laminar solution.*

Furthermore, each nonlaminar flow $(u, v, \rho, \eta) \in \mathcal{C}$ *is regular in the sense that:*

1. u, v, ρ, *and* η *each have period* L *in* x.
2. *Within each period the wave profile* η *has a single crest and trough; say, the crest occurs at* $x = 0$.
3. u, ρ, *and* η *are symmetric,* v *antisymmetric across the line* $x = 0$.
4. *A water particle located at* (x, y), *with* $0 < x < L$ *and* $y > d$ *has positive vertical velocity* $v > 0$.
5. $\eta'(x) < 0$ *on* $(0, L)$.

The symmetry of solutions ensured by Theorem 4 is interesting, and it will be considered further in Sect. 5.

3.2 Capillary-Gravity Stratified Waves

When surface tension is taken into account in the governing equations (6), Walsh has also proven the existence of large-amplitude waves [50]. The global bifurcation analysis in [50] extends the local bifurcation curves derived in [49], adapting the methods first applied for the local existence of homogeneous capillary-gravity waves with vorticity in [45]. In [50], the author also uses Dancer's analytic global bifurcation theory [2, 6] to extend the local bifurcation curves. In [49], the author proves local bifurcation from both simple eigenvalues, and from double eigenvalues where additional non-degeneracy conditions are satisfied. For the simple eigenvalue setting, we let ε_0 be as above, then a sufficient condition for local bifurcation to occur is given by

$$(g\overline{\rho}(0) + \sigma)\,\lambda^2 > \int_{-\lambda}^{0} \left\{ \frac{4\pi^2}{L^2}\,(2B(p) - 2B_{\min} + 2\varepsilon_0)^{3/2} \right.$$
$$\left. + (p+\lambda)^2\big(\,(2B(p) - 2B_{\min} + 2\varepsilon_0)^{1/2} + g\overline{\rho}'(p)\big) \right\} dp.$$

The main results for bifurcation from a simple eigenvalue in [49, 50] may then be stated (combined) as follows:

Theorem 5 *Fix any* $\alpha \in (0, 1)$, *and let the wavespeed* $c > 0$, *wavelength* $L > 0$, *and* $\lambda > 0$ *be given, along with* $\beta \in C^{1+\alpha}([-\lambda, 0])$ *and* $\overline{\rho} \in C^{2+\alpha}([-\lambda, 0])$, *such that the (L-B) condition holds. Also, let the given coefficient of surface tension* $\sigma \in \Sigma_1$, *where* $(\sigma_c, \infty) \subset \Sigma_1 \subset (0, \infty)$, *and we assume the streamline density function* $\overline{\rho}$ *is nonincreasing. Consider traveling solutions to the stratified water wave problem (1a, 1b) of speed* c, *relative mass flux* $-\lambda$, *Bernoulli function* β, *and streamline density function* $\overline{\rho}$ *such that* $u < c$ *throughout the fluid. There exists a connected set* $\mathcal{C}$ *of solutions* (u, v, ρ, η) *in the space* $C^{2+\alpha}_{per}(\overline{\Omega}_\eta) \times C^{2+\alpha}_{per}(\overline{\Omega}_\eta) \times C^{2+\alpha}_{per}(\overline{\Omega}_\eta) \times C^{3+\alpha}_{per}(\mathbb{R})$ *with the following properties:*

1. $\mathcal{C}$ *contains a laminar flow (with a flat surface* $\eta \equiv 0$ *and all streamlines parallel to the bed).*

2. *Along some sequence* $(u_n, v_n, \rho_n, \eta_n) \in \mathcal{C}$, *either* $\max_{\overline{\Omega}_{\eta_n}} u_n \uparrow c$, $\min_{\overline{\Omega}_{\eta_n}} u_n \downarrow -\infty$, *or* $\mathcal{C}$ *contains more than one distinct laminar solution.*

Additionally, there exists a path-connected subset $\mathcal{K} \subset \mathcal{C}$ *such that:*

1. $\mathcal{K}$ *admits a global, locally injective continuous parameterization with a locally injective* C_1 *reparametrization.*
2. *Either* $\mathcal{K}$ *is a closed loop, or it is unbounded in the sense that along some sequence* $(u_n, v_n, \rho_n, \eta_n) \in \mathcal{K}$, *either* $\max_{\overline{\Omega}_{\eta_n}} u_n \uparrow c$, $\min_{\overline{\Omega}_{\eta_n}} u_n \downarrow -\infty$.

Furthermore, each nonlaminar flow $(u, v, \rho, \eta) \in \mathcal{C}$ *is regular in the sense of Theorem 4.*

A mathematical idiosyncrasy of the effects of surface tension, rather than stratification, is the possibility of bifurcation from double eigenvalues, and we therefore refer the reader to [49, 50] for an exposition of this phenomenon in stratified capillary-gravity waves.

4 Stratification for Gerstner-Type Flows

Interestingly, prior to her work on the existence of small-amplitude stratified water waves [18, 19], Dubreil-Jacotin showed that the celebrated Gerstner's water wave solution [3, 24] could be adapted to allow for continuous stratification in the underlying fluid [17]. Gerstner's wave is one of the few examples of an explicit solution for the fully nonlinear governing equations, since the formulation of the solution is explicit in the Lagrangian formulation [1]. The form of Gerstner's solution can be modified in order to obtain an explicit exact solution for edge waves – this was first achieved in [4]. In [37, 40, 53], it was shown that stratification is also possible in the prescribed flow for edge waves.

Recently, a number of exact solutions for geophysical water waves have been derived, beginning with [8] and extending to [9, 25, 33, 34], which are explicit in the Lagrangian representation and which admit continuous stratification. We present the solution derived in [8] for three-dimensional trapped equatorial water waves, since it can be reduced to the setting of the stratified Gerstner's solution when the geophysical parameter is set to zero. The β-plane approximation of the geophysical governing equations, which applies in regions close to of the equator, is given by [16]:

$$(u-c)u_x + vu_y + wu_z + 2\Omega w - \beta y v = -\frac{1}{\rho}P_x, \tag{14a}$$

$$(u-c)v_x + vv_y + wv_z + \beta y u = -\frac{1}{\rho}P_y, \tag{14b}$$

$$(u-c)w_x + vw_y + ww_z - 2\Omega u = -\frac{1}{\rho}P_z - g. \tag{14c}$$

Here, we take the Earth to be a perfect sphere of radius $R = 6378$ km, which has a constant rotational speed of $\Omega = 73 \times 10^{-6}$ *rad* s^{-1} . Then

$g = 9.8\ ms^{-2}$ is the standard gravitational acceleration at the Earth's surface, and $\beta = 2\Omega/R = 2.28 \times 10^{-11}\ m^{-1}s^{-1}$ is a geophysical parameter. The boundary conditions for the fluid are given by

$$w = (u - c)\eta_x + v\eta_y \qquad \text{on } y = \eta(x, y, t), \tag{15}$$

$$P = P_0 \qquad \text{on } y = \eta(x, y, t). \tag{16}$$

Then, the Eulerian coordinates of fluid particles (x, y, z) are expressed as functions of the Lagrangian labeling variables $(q, r, s) \in \mathbb{R}\times(-\infty, r_0)\times\mathbb{R}$, and time t, as follows:

$$x = q - \frac{1}{k}e^{k[r-f(s)]}\sin[k(q - ct)], \tag{17a}$$

$$y = s, \tag{17b}$$

$$z = r + \frac{1}{k}e^{k[r-f(s)]}\cos[k(q - ct)], \tag{17c}$$

where $r_0 < 0$ and k is the wave number. The function $f(s)$ is given by

$$f(s) = \frac{c\beta}{2g}s^2, \tag{18}$$

and it determines the decay of the particle oscillation as it moves in the latitudinal direction away from the equator. Let us prescribe the density function by

$$\rho(r, s) = F\left(\frac{e^{2k[r-f(s)]}}{2k} - r\right),$$

where $F : (0, \infty) \to (0, \infty)$ is continuously differentiable and nondecreasing. Then the formula (17) prescribes a solution of the governing equations (14) if we define the pressure function

$$P = g\mathcal{F}\left(\frac{e^{2k[r-f(s)]}}{2k} - r\right) + P_0 - g\mathcal{F}\left(\frac{e^{2kr_0}}{2k} - r_0\right), \tag{19}$$

where $\mathcal{F}' = F$ and $\mathcal{F}(0) = 0$, cf. [8]. We observe that setting $\beta = 0$ in the previous considerations reduces us to the case of Gerstner's water wave, and taking F constant we get the homogeneous non-stratified setting. We finally mention that Gerstner-type solutions have been derived which model discontinuous stratification, [10, 39, 41, 42], with these solutions giving an explicit formulation for internal waves in a fluid.

5 Qualitative Properties for General Stratified Flows

In this section, we recall recent results concerning the regularity of the streamlines and the symmetry of the wave profile for stratified waves without stagnation points. In the first paper [26], it was shown that the streamlines and the wave profile of stratified flows for which $\beta, \overline{\rho}' \in C^\alpha([-\lambda, 0])$, for some $\alpha \in (0,1)$, are smooth curves. The streamlines are even real-analytic curves provided that the density decreases with depth. These results of [26] were improved by Wang [51, Theorems 2.2 and 5.1] as follows:

Theorem 6 (Regularity properties). *Let $\alpha \in (0,1)$ and assume that $\sigma \geq 0$. If $\beta, \overline{\rho}' \in C^\alpha([-\lambda, 0])$ and $h \in C^{2+\alpha}(\overline{\Omega}_\lambda)$ is a classical solution of (6) and (7) then, the wave profile together with all the other streamlines are real-analytic curves.*

The symmetry of steady water waves is an old problem, investigated first by Garabedian [22]. It was recently shown that, when excluding stagnation, periodic gravity water waves that possess a single crest per period have to be symmetric [11, 14, 38]. Surprisingly, the symmetry property of such waves can be characterized intrinsically in terms of the flow beneath the wave [35, Theorem 2.1]: the wave profile is symmetric and has only one crest and trough per period if and only if there exists a vertical line within the fluid domain such that all the fluid particles located on that line minimize there their distance to the fluid bed.

Because maximum principles do not apply directly in this context, the symmetry problem for stratified flows is more delicate. An intrinsic characterization of the symmetric waves in terms of the underlying flow as in the case of gravity waves is not yet known. There exist though criteria which ensure the symmetry of stratified waves with a single crest per period [47]. To present them we introduce first some notation:

$$A_0 := \max_{\overline{\Omega}_\eta} \sqrt{\overline{\rho}}(c-u), \quad \eta_M := \max_{\mathbb{R}} \eta + d, \quad \eta_m := \min_{\mathbb{R}} \eta + d,$$

$$a_0 := \min_{\overline{\Omega}_\eta} \frac{\sqrt{\overline{\rho}}(c-u)}{2}\Big(1 + |\nabla\psi|^2 - \sqrt{(1+|\nabla\psi|^2)^2 - 4(\sqrt{\overline{\rho}}(c-u))^2}\Big),$$

$$M := \max\Big\{\max_{\overline{\Omega}_\eta}\Big|\frac{v}{u-c}\Big|, \max_{\overline{\Omega}_\eta}\Big|\Big(\partial_x + \frac{v}{u-c}\partial_y\Big)\frac{v}{u-c}\Big|, \max_{\overline{\Omega}_\eta}\Big|\frac{1}{\sqrt{\overline{\rho}}(u-c)}\partial_y\frac{v}{u-c}\Big|\Big\},$$

$$\varepsilon := \frac{A_0^3 M(1+2M^2)}{a_0} + \frac{(1+3a_0)A_0^3}{4a_0^2}(\max|\beta| + g\eta_M \max|\overline{\rho}'|)$$
$$+ 2A_0^2 M(1+a_0 M) + \frac{A_0^3 M^2}{2a_0^2}.$$

The result in Theorem 7 below was obtained by Walsh [47] and presents two criteria for symmetry.

Theorem 7 (Symmetry result). *Consider a stably stratified steady train propagating at fixed speed c over a flat bed at $y = -d$ with relative pseudo-mass flux $-\lambda < 0$. Let $\overline{\rho} \in C^2([-\lambda, 0])$ and $\beta \in C^1([-\lambda, 0])$ be the streamline density function and Bernoulli function associated with the flow, and let $(u, v) \in (C^2(\overline{\Omega}_\eta))^2$ be the vector field. Assume that the wave profile $y = \eta(x)$ is monotonic between crests and troughs with period L, has integral mean zero over a period, and that* $\max_{\overline{\Omega}_\eta} u < c$. *Each of* *the following is a sufficient condition for the wave to be symmetric:*

$$(i)\quad \max\beta' + g\eta_M \max|\overline{\rho}''| < \max\left\{\frac{\pi^2}{\eta_M^2}, \exp\left(-\min\{L, \eta_m\}\right)\right\}, \qquad or$$

$$(ii)\quad \begin{cases} \varepsilon\sqrt{\frac{\varepsilon}{a_0}} + g\max|\overline{\rho}'| < \exp(-a_0^{-1/2}\min\{L, \lambda\}), \\ \varepsilon < a_0. \end{cases}$$

Acknowledgments The authors wish to thank the organizers of the "Elliptic and Parabolic Equations" workshop for the very stimulating and pleasant research environment which they created.

References

1. A. Bennett, *Lagrangian Fluid Dynamics* (Cambridge University Press, Cambridge, 2006)
2. B. Buffoni, J.F. Toland, *Analytic Theory of Global Bifurcation* (Princeton University Press, Princeton and Oxford, 2003)
3. A. Constantin, On the deep water wave motion. J. Phys. A **34**, 1405–1417 (2001)
4. A. Constantin, Edge waves along a sloping beach. J. Phys. A **34**, 9723–9731 (2001)
5. A. Constantin, The trajectories of particles in Stokes waves. Invent. Math. **166**, 523–535 (2006)
6. A. Constantin, in *Nonlinear Water Waves with Applications to Wave-Current Interactions and Tsunamis*. CBMS-NSF Conference Series in Applied Mathematics, vol. 81 (SIAM, Philadelphia, 2011)
7. A. Constantin, A dynamical systems approach towards isolated vorticity regions for tsunami background states. Arch. Ration. Mech. Anal. **200**, 239–253 (2011)
8. A. Constantin, An exact solution for equatorially trapped waves. J. Geophys. Res. **117**, C05029 (2012)
9. A. Constantin, Some three-dimensional nonlinear equatorial flows. J. Phys. Oceanogr. **43**, 165–175 (2013)
10. A. Constantin, Some nonlinear, equatorially trapped, nonhydrostatic internal geophysical waves. J. Phys. Oceanogr. **44**, 781–789 (2014)
11. A. Constantin, J. Escher, Symmetry of steady periodic surface water waves with vorticity. J. Fluid Mech. **498**, 171–181 (2004)
12. A. Constantin, W. Strauss, Exact steady periodic water waves with vorticity. Comm. Pure Appl. Math. **57**(4), 481–527 (2004)
13. A. Constantin, E. Varvaruca, Steady periodic water waves with constant vorticity: regularity and local bifurcation. Arch. Rational Mech. Anal. **199**, 33–67 (2011)
14. A. Constantin, M. Ehrnström, E. Wahlén, Symmetry of steady periodic gravity water waves with vorticity. Duke Math. J. **140**, 591–603 (2007)
15. M.G. Crandall, P.H. Rabinowitz, Bifurcation from simple eigenvalues. J. Funct. Anal. **8**, 321–340 (1971)
16. B. Cushman-Roisin, J.-M. Beckers, *Introduction to Geophysical Fluid Dynamics: Physical and Numerical Aspects* (Academic, Waltham, 2011)
17. M.L. Dubreil-Jacotin, Sur les ondes de type permanent dans les liquides hétérogènes. Atti Accad. Naz. Lincei Rend. **6**, 814–819 (1932)
18. M.L. Dubreil-Jacotin, Sur la détermination rigoureuse des ondes permanentes périodiques d'ampleur finie. J. Math. Pures et Appl. **13**, 217–291 (1934)
19. M.L. Dubreil-Jacotin, Sur les théorèmes d'existence relatifs aux ondes permanentes périodiques a deux dimensions dans les liquides hétérogènes. J. Math. Pures et Appl. **9**, 43–67 (1937)

20. M. Ehrnström, J. Escher, E. Wahlén, Steady water waves with multiple critical layers. SIAM J. Math. Anal. **43**(3), 1436–1456 (2011)
21. J. Escher, A.-V. Matioc, B.-V. Matioc, On stratified steady periodic water waves with linear density distribution and stagnation points. J. Differential Equations. **251**, 2932–2949 (2011)
22. P.R. Garabedian, Surface waves of finite depth. J. Anal. Math. **14**, 161–169 (1965)
23. A. Geyer, A note on uniqueness and compact support of solutions in a recent model for tsunami background flows. Comm. Pure Appl. Anal. **11**(4), 1431–1438 (2012)
24. D. Henry, On Gerstner's water wave. J. Nonlinear Math. Phys. **15**, 87–95 (2008)
25. D. Henry, An exact solution for equatorial geophysical water waves with an underlying current. Eur. J. Mech. B Fluids. **38**, 18–21 (2013)
26. D. Henry, B.-V. Matioc, On the regularity of steady periodic stratified water waves. Comm. Pure Appl. Anal. **11** (4), 1453–1464 (2012)
27. D. Henry, B.-V. Matioc, On the existence of steady periodic capillary-gravity stratified water waves. Ann. Scuola Norm. Sup. Pisa **XII** (4), 955–974 (2013)
28. D. Henry, A.-V. Matioc, Global bifurcation of capillary-gravity stratified water waves. Proc. Roy. Soc. Edinburgh Sect. A **114**, 775–786 (2014)
29. L. Kelvin, Vibrations of a columnar vortex. Phil. Mag. **10**, 155–168 (1880)
30. J. Lighthill, *Waves in Fluids* (Cambridge University Press, Cambridge, 1978)
31. R.R. Long, Some aspects of the flow of stratified fluids. Part I : a theoretical investigation. Tellus **5**, 42–57 (1953)
32. C.I. Martin, B.-V. Matioc, Steady periodic water waves with unbounded vorticity: equivalent formulations and existence results. J. Nonlinear Sci. **24**, 633–659 (2014)
33. A.-V. Matioc, An exact solution for geophysical equatorial edge waves over a sloping beach. J. Phys. A **45**, 365501 (2012)
34. A.-V. Matioc, Exact geophysical waves in stratified fluids. Appl. Anal. **92** (11), 2254–2261 (2014)
35. B.-V. Matioc, A characterization of the symmetric steady water waves in terms of the underlying flow. Discrete Contin. Dyn. Syst. Ser. A. **34** (8), 3125–3133 (2014)
36. B.-V. Matioc, Global bifurcation for water waves with capillary and constant vorticity. Monatsh. Math. **174** (3), 459–475 (2014)
37. E. Mollo-Christensen, Edge waves in a rotating stratified fluid, an exact solution. J. Phys. Oceanogr. **9**, 226–229 (1979)
38. H. Okamoto, M. Shōji, in *The Mathematical Theory of Permanent Progressive Water-Waves*. Adv. Ser. Nonlinear Dynam., vol. 20 (World Scientific, River Edge,(2001)
39. M. Stiassnie, R. Stuhlmeier, Progressive waves on a blunt interface. Discrete Contin. Dyn. Syst. Ser. A. **34**, 3171–3182 (2014)
40. R. Stuhlmeier, On edge waves in stratified water along a sloping beach. J. Nonlinear Math. Phys. **18**, 127–137 (2011)
41. R. Stuhlmeier, Internal Gerstner waves on a sloping bed. Discrete Contin. Dyn. Syst. Ser. A. **34**, 3183–3192 (2014)
42. R. Stuhlmeier, Internal Gerstner waves: applications to dead water. Appl. Anal. **34**, 3183–3192 (2014)
43. J.F. Toland, Stokes waves. Topol. Meth. Nonl. Anal. **7**, 1–48 (1996)
44. R.E. Turner, Internal waves in fluids with rapidly varying density, Ann. Scuola Norm. Sup. Pisa Cl. Sci. **8**, 513–573 (1981)
45. E. Wahlén, Steady periodic capillary-gravity waves with vorticity. SIAM J. Math. Anal. **38** (3), 921–943 (2006)
46. E. Wahlén, Steady water waves with a critical layer. J. Differential Equations. **246**, 2468–2483 (2009)
47. S. Walsh, Some criteria for the symmetry of stratified water waves. Wave Motion. **46**, 350–362 (2009)
48. S. Walsh, Stratified steady periodic water waves. SIAM J. Math. Anal. **41**(3), 1054–1105 (2009)

49. S. Walsh, Steady stratified periodic gravity waves with surface tension I: local bifurcation. Discrete Cont. Dyn. Syst. Ser. A **8**, 3287–3315 (2014)
50. S. Walsh, Steady stratified periodic gravity waves with surface tension II: global bifurcation. Discrete Cont. Dyn. Syst. Ser. A **8**, 3241–3285 (2014)
51. L.-J. Wang, Regularity of traveling periodic stratified water waves with vorticity. Nonlinear Anal. **81**, 247–263 (2013)
52. C.-S. Yih, Exact solutions for steady two-dimensional flow of a stratified fluid. J. Fluid Mech. **9**, 161–174 (1960)
53. C.-S. Yih, Note on edge waves in a stratified fluid. J. Fluid Mech. **24**, 765–767 (1966)
54. C.-S. Yih, Stratified flows. Ann. Rev. Fluid Mech. **1**, 73–110 (1969)
55. C.-S. Yih, *Stratified Flows* (Academic, New York, 1980)

A Calculus of Abstract Edge Pseudodifferential Operators of Type ρ, δ

Thomas Krainer

Abstract In this paper, we expand on B.-W. Schulze's abstract edge pseudodifferential calculus and introduce a larger class of operators that is modeled on Hörmander's ρ, δ calculus, where $0 \le \delta < \rho \le 1$. This expansion is motivated by recent work on boundary value problems for elliptic wedge operators with variable indicial roots by G. Mendoza and the author, where operators of type $1, \delta$ for $0 < \delta < 1$ appear naturally. Some of the results of this chapter also represent improvements over the existing literature on the standard abstract edge calculus of operators of type $1, 0$, such as trace class mapping properties of operators in abstract wedge Sobolev spaces. The presentation in this paper is largely self-contained to allow for an independent reading.

1 Introduction

In [17], Schulze introduced a calculus of pseudodifferential operators with operator-valued symbols that satisfy symbol estimates that are twisted by strongly continuous group actions. This calculus, which is generally referred to as the abstract edge pseudodifferential calculus, turned out to be very useful to describe and analyze the local structure of elliptic partial differential equations and their parametrices on manifolds with boundary near the boundary, and, more generally, on manifolds with incomplete edge singularities near the edges, thus motivating the name for the calculus. The abstract edge calculus has been further developed over the years by many contributors, see for example [2, 3, 5, 16, 18–21]. A feature of the symbol estimates that is common to all these works is that differentiation with respect to the variables does not worsen the growth estimates (i.e., the order) in the covariables. In other words, the symbols exhibit estimates resembling Hörmander's type $\rho, 0$ symbols, twisted by strongly continuous group actions, where generally $\rho = 1$ (except in J. Seiler's boundedness theorem from [21] where type $0, 0$ symbols are considered).

In this chapter, we expand on the abstract edge pseudodifferential calculus and introduce a larger class of operators that is modeled on Hörmander's ρ, δ calculus,

T. Krainer (✉)
Department of Mathematics and Statistics, Pennsylvania State University, 3000 Ivyside Park, Altoona Campus, Altoona, PA 16601, USA
e-mail: krainer@psu.edu

J. Escher et al., (eds.), *Elliptic and Parabolic Equations*, Springer Proceedings in Mathematics & Statistics 119, DOI 10.1007/978-3-319-12547-3_8

where $0 \leq \delta < \rho \leq 1$. This expansion is motivated by recent work on boundary value problems for elliptic wedge operators with variable indicial roots [7–9], where operators of type $1, \delta$ for $0 < \delta < 1$ appear naturally. Some of the results obtained here for the general ρ, δ class also represent improvements over the existing literature on the standard abstract edge calculus of operators of type $1, 0$. These improvements pertain, in particular, to trace class mapping properties of operators in abstract wedge Sobolev spaces, and independence of the operator class with respect to any τ-quantization (see [22]) of the symbols, in particular with respect to the Kohn–Nirenberg ($\tau = 0$) and Weyl ($\tau = 1/2$) quantizations. The presentation in this chapter is largely self-contained. Where appropriate, some key arguments of the existing literature are reproduced to allow for an independent reading.

The structure of the chapter is as follows:

In Sect. 2, we review the definition and some of the properties of the abstract wedge Sobolev space of E-valued distributions, where E is a Hilbert space. Unless stated otherwise, all Hilbert spaces in this work are assumed to be complex and separable. We also give an alternative characterization of the wedge Sobolev spaces that is based on Littlewood–Paley theory. This characterization proves to be very useful to study mapping properties of pseudodifferential operators in such spaces. Effectively, this characterization and the composition theorem in the abstract edge calculus will allow us to reduce most arguments to ordinary pseudodifferential operators (i.e., those based on symbols that are not twisted by group actions) acting in Hilbert space valued L^2-spaces.

Section 3 contains the elements of the abstract edge calculus of operators of type ρ, δ. We allow the symbols to exhibit polynomial growth in the variables and to take values in general Banach operator ideals. The latter is useful for applications in index and spectral theory, especially symbols taking values in Schatten–von Neumann classes. We consider general τ-quantizations, see [22], and prove independence of the pseudodifferential operator class from the specific τ-quantization. We stick to the range $0 \leq \delta < \rho \leq 1$ because we want to retain asymptotic expansion formulas for the operations of the calculus.

Sections 4 and 5 are concerned with mapping properties of the pseudodifferential operators from Sect. 3 in the scale of abstract wedge Sobolev spaces. Section 4 addresses boundedness and compactness, while Sect. 5 is concerned with trace class mapping properties (i.e., membership in the Schatten–von Neumann class $\mathcal{C}_1$).

Finally, we include an appendix on Banach operator ideals in the Hilbert space category. In this appendix, we also review some of the properties of the Schatten–von Neumann classes.

2 Abstract Wedge Sobolev Spaces

Abstract wedge Sobolev spaces were introduced by Schulze in [17], see also [5, 18, 19]. In this section, we review the standard definition of these function spaces, and give a different elementary characterization for them that is based on Littlewood–Paley theory. This characterization is useful since it allows to reduce the analysis of operators acting in abstract wedge Sobolev spaces to standard analysis of operators in L^2-spaces. We will take advantage of this in subsequent sections.

Let E be a Hilbert space, and let $\{\kappa_\rho\}_{\rho>0}$ be a strongly continuous group action on E, i.e., $\mathbb{R}_+ \ni \rho \mapsto \kappa_\rho \in \mathcal{L}(E)$ is a representation of the multiplicative group $(\mathbb{R}_+, \cdot)$ in $\mathcal{L}(E)$ that is continuous with respect to the strong operator topology. By the uniform boundedness principle, there exist constants $c, M \geq 0$ such that

$$\|\kappa_\rho\|_{\mathcal{L}(E)} \leq c\max\{\rho, \rho^{-1}\}^M \tag{1}$$

for all $\rho > 0$.

Definition 1 ([18, Sect. 3.1.2]). For $s \in \mathbb{R}$, the abstract wedge Sobolev space is defined as

$$\begin{aligned}\mathcal{W}^s(\mathbb{R}^q, (E, \kappa)) = \{u \in \mathcal{S}'(\mathbb{R}^q, E) : \mathcal{F}u \text{ is regular, and}\\ [\eta \mapsto \langle\eta\rangle^s \kappa_{\langle\eta\rangle}^{-1} \mathcal{F}u(\eta)] \in L^2(\mathbb{R}^q, E)\}.\end{aligned}$$

Here, $\mathcal{F}$ denotes the Fourier transform on $\mathcal{S}'(\mathbb{R}^q, E)$, and $\langle\eta\rangle = (1 + |\eta|^2)^{1/2}$, where $|\eta|$ is the Euclidean norm of $\eta \in \mathbb{R}^q$. It is custom to simply write $\mathcal{W}^s(\mathbb{R}^q, E)$ if the group action κ_ρ on E is clear from the context.

$\mathcal{W}^s(\mathbb{R}^q, E)$ is a Hilbert space with inner product

$$\langle u, v\rangle_{\mathcal{W}^s} = \int_{\mathbb{R}^q} \langle\eta\rangle^{2s} \langle\kappa_{\langle\eta\rangle}^{-1} \mathcal{F}u(\eta), \kappa_{\langle\eta\rangle}^{-1} \mathcal{F}v(\eta)\rangle_E \, d\eta.$$

The space $\mathcal{S}(\mathbb{R}^q, E)$ of rapidly decreasing E-valued functions is dense in the space $\mathcal{W}^s(\mathbb{R}^q, E)$, and we have

$$H^{s+M}(\mathbb{R}^q, E) \subset \mathcal{W}^s(\mathbb{R}^q, E) \subset H^{s-M}(\mathbb{R}^q, E)$$

with continuous embeddings with the growth constant $M \geq 0$ for the group action $\{\kappa_\rho\}_{\rho>0}$ from (1). Note that multiplication by the strongly continuous operator functions $\eta \mapsto \kappa_{\langle\eta\rangle}$ and $\eta \mapsto \kappa_{\langle\eta\rangle}^{-1}$ preserves strong measurability of E-valued functions, and that both these operator functions are of tempered growth by (1). The density of $\mathcal{S}(\mathbb{R}^q, E)$ in $\mathcal{W}^s(\mathbb{R}^q, E)$ follows most easily from the characterization of the $\mathcal{W}^s$-spaces based on Littlewood–Paley theory that is given below. Alternatively, it follows from the denseness of C^∞-elements for the group action κ_ρ in E and a tensor product argument.

The $\mathcal{W}^s$-spaces form a function space scale based on $\mathcal{W}^0(\mathbb{R}^q, E)$ in the sense that

$$\langle D_y\rangle^\mu = \mathcal{F}_{\eta\to y}^{-1} \langle\eta\rangle^\mu \mathcal{F}_{y'\to\eta} : \mathcal{W}^s(\mathbb{R}^q, E) \to \mathcal{W}^{s-\mu}(\mathbb{R}^q, E) \tag{2}$$

is an isomorphism for all $s, \mu \in \mathbb{R}$. Moreover, setting $\kappa_\rho^{(s)} = \rho^{-s}\kappa_\rho$, we also see that

$$\mathcal{W}^s(\mathbb{R}^q, (E, \kappa)) = \mathcal{W}^0(\mathbb{R}^q, (E, \kappa^{(s)})). \tag{3}$$

This allows to reduce many considerations regarding the $\mathcal{W}^s$-scale to the case when $s = 0$.

The following class of operator-valued symbols is well adapted for Fourier multipliers in the $\mathcal{W}^s$-space scale.

Definition 2 ([18, Sect. 3.2.1]) Let (E, κ_ρ) and $(\tilde{E}, \tilde{\kappa}_\rho)$ be Hilbert spaces that come equipped with strongly continuous group actions. For $\mu \in \mathbb{R}$ define

$$S^{\mu}_{1,0}(\mathbb{R}^q; (E,\kappa), (\tilde{E},\tilde{\kappa}))$$

as the space of all $a(\eta) \in C^\infty(\mathbb{R}^q, \mathcal{L}(E,\tilde{E}))$ such that for all $\alpha \in \mathbb{N}_0^q$ there exists a constant $C_\alpha > 0$ such that

$$\|\tilde{\kappa}^{-1}_{\langle\eta\rangle}[\partial^\alpha_\eta a(\eta)]\kappa_{\langle\eta\rangle}\|_{\mathcal{L}(E,\tilde{E})} \leq C_\alpha \langle\eta\rangle^{\mu-|\alpha|}$$

for all $\eta \in \mathbb{R}^q$.

If any of the Hilbert spaces E or $\tilde{E}$ carries the trivial group action $\kappa_\rho \equiv \mathrm{Id}_E$ or $\tilde{\kappa}_\rho \equiv \mathrm{Id}_{\tilde{E}}$, then that action is dropped from the notation of the symbol class, i.e., one simply writes $S^{\mu}_{1,0}(\mathbb{R}^q; E, (\tilde{E},\tilde{\kappa}))$ or $S^{\mu}_{1,0}(\mathbb{R}^q; (E,\kappa), \tilde{E})$ (or both).

For any $a \in S^{\mu}_{1,0}(\mathbb{R}^q; (E,\kappa), (\tilde{E},\tilde{\kappa}))$, the Fourier multiplier

$$\mathcal{F}^{-1}_{\eta\to y} a(\eta)\mathcal{F}_{y'\to\eta} : \mathcal{W}^s(\mathbb{R}^q, E) \to \mathcal{W}^{s-\mu}(\mathbb{R}^q, \tilde{E}) \tag{4}$$

is evidently continuous (the estimates on the derivatives on $a(\eta)$ from Definition 2 are not required for this). The operator (2) is a special case of (4) with $a(\eta) = \langle\eta\rangle^\mu \mathrm{Id}_E \in S^{\mu}_{1,0}(\mathbb{R}^q; (E,\kappa), (E,\kappa))$.

We now proceed to give an elementary characterization of the $\mathcal{W}^0$-space that is based on Littlewood–Paley theory. As already mentioned, this characterization is useful since it allows to reduce Fourier analytic considerations of operators acting in the $\mathcal{W}^s$-function space scale to standard Fourier analysis of operators in L^2-spaces.

Let $\phi_0 \in C^\infty(\mathbb{R}^q)$ such that $\phi_0 \equiv 1$ in a neighborhood of $|\eta| \leq 1$, $\mathrm{supp}(\phi_0) \subset \{\eta : |\eta| < 2\}$, and $0 \leq \phi_0 \leq 1$. Define

$$\phi_j(\eta) = \phi_0(2^{-j}\eta) - \phi_0(2^{-j+1}\eta) \quad \text{for} j \in \mathbb{N}.$$

Then the $\{\phi_j\}_{j=0}^\infty$ form a dyadic resolution of the identity on $\mathbb{R}^q$ with the following properties:

1. $\phi_j \in C^\infty(\mathbb{R}^q)$ with $\mathrm{supp}(\phi_j) \subset \{\eta : 2^{j-1} < |\eta| < 2^{j+1}\}$ for $j \in \mathbb{N}$.
2. $0 \leq \phi_j \leq 1$.
3. $\sum_{j=0}^\infty \phi_j(\eta) = 1$ for all $\eta \in \mathbb{R}^q$. This is a locally finite sum, and for each $\eta \in \mathbb{R}^q$ at most two consecutive summands are nonzero.
4. For every multi-index $\alpha \in \mathbb{N}_0^q$, there exists a constant $C_\alpha > 0$ that is independent of $j \in \mathbb{N}_0$ such that $|\partial^\alpha_\eta \phi_j(\eta)| \leq C_\alpha 2^{-j|\alpha|}$ for all $\eta \in \mathbb{R}^q$.
   Because $|\eta|/2^j \leq 2$ on $\mathrm{supp}(\phi_j)$, this implies the estimate

$$|\partial^\alpha_\eta \phi_j(\eta)| \leq 2^{|\alpha|} C_\alpha |\eta|^{-|\alpha|} \quad \text{for} |\eta| \geq 1.$$

Let $\psi_0, \psi_1 \in C^\infty(\mathbb{R}^q)$ be such that $\mathrm{supp}(\psi_0) \subset \{\eta : |\eta| < 2\}$ and $\psi_0 \equiv 1$ in a neighborhood of $\mathrm{supp}(\phi_0)$, and correspondingly $\mathrm{supp}(\psi_1) \subset \{\eta : 1 < |\eta| < 4\}$ with

$\psi_1 \equiv 1$ in a neighborhood of $\mathrm{supp}(\phi_1)$. Define $\psi_j(\eta) = \psi_1(2^{-j+1}\eta)$ for $j = 2, 3, \ldots$ Then $\mathrm{supp}(\psi_j) \subset \{\eta : 2^{j-1} < |\eta| < 2^{j+1}\}$, and $\psi_j \equiv 1$ in a neighborhood of $\mathrm{supp}(\phi_j)$ for all $j \in \mathbb{N}$. Moreover, for every multi-index $\alpha \in \mathbb{N}_0^q$ there exists a constant $D_\alpha > 0$ that is independent of $j \in \mathbb{N}_0$ such that $|\partial_\eta^\alpha \psi_j(\eta)| \leq D_\alpha 2^{-j|\alpha|}$ for all $\eta \in \mathbb{R}^q$, and consequently $|\partial_\eta^\alpha \psi_j(\eta)| \leq 2^{|\alpha|} D_\alpha |\eta|^{-|\alpha|}$ for $|\eta| \geq 1$.

Let $\ell^2(\mathbb{N}_0, E)$ be the Hilbert space of square summable sequences with entries in E with inner product

$$\langle (e_k)_{k=0}^\infty, (f_k)_{k=0}^\infty \rangle_{\ell^2} = \sum_{k=0}^\infty \langle e_k, f_k \rangle_E.$$

For $j \in \mathbb{N}_0$, define

$$\iota_j : E \to \ell^2(\mathbb{N}_0, E),\ \iota_j(e) = (0, \ldots, 0, e, 0, \ldots),$$

where the element $e \in E$ is injected as the jth component of the sequence, and

$$\pi_j : \ell^2(\mathbb{N}_0, E) \to E,\ \pi_j[(e_k)_{k=0}^\infty] = e_j.$$

Then $\iota_j \in \mathcal{L}(E, \ell^2(\mathbb{N}_0, E))$ and $\pi_j \in \mathcal{L}(\ell^2(\mathbb{N}_0, E), E)$ with operator norm $\|\iota_j\| = \|\pi_j\| = 1$ for all j, and $\pi_k \iota_j = \delta_{jk} \mathrm{Id}_E$ with the Kronecker delta δ_{jk}.

Lemma 1 *For $\eta \in \mathbb{R}^q$, let*

$$s(\eta) = \sum_{j=0}^\infty \phi_j(\eta) \iota_j \kappa_{2^{-j}} : E \to \ell^2(\mathbb{N}_0, E),$$

$$t(\eta) = \sum_{k=0}^\infty \psi_k(\eta) \kappa_{2^k} \pi_k : \ell^2(\mathbb{N}_0, E) \to E.$$

Then

$$s \in S_{1,0}^0(\mathbb{R}^q, (E, \kappa), \ell^2(\mathbb{N}_0, E)),$$
$$t \in S_{1,0}^0(\mathbb{R}^q, \ell^2(\mathbb{N}_0, E), (E, \kappa)),$$

and $t(\eta)s(\eta) = \mathrm{Id}_E$ for all $\eta \in \mathbb{R}^q$.

Proof Since both sums are locally finite, $s(\eta)$ and $t(\eta)$ are C^∞ operator functions. We have

$$t(\eta)s(\eta) = \sum_{k,j=0}^\infty \psi_k(\eta)\phi_j(\eta) \underbrace{\kappa_{2^k} \pi_k \iota_j \kappa_{2^{-j}}}_{=\delta_{jk}\mathrm{Id}_E} = \sum_{j=0}^\infty \underbrace{\psi_j(\eta)\phi_j(\eta)}_{=\phi_j(\eta)} \mathrm{Id}_E = \mathrm{Id}_E.$$

It remains to show the symbol estimates for $s(\eta)$ and $t(\eta)$. For $|\eta| \geq 2$, we have

$$\|[\partial_\eta^\alpha s(\eta)] \kappa_{|\eta|}\| = \left\| \sum_{j=0}^\infty \partial_\eta^\alpha \phi_j(\eta) \iota_j \kappa_{|2^{-j}\eta|} \right\|$$

$$\leq \sum_{j=0}^{\infty} |\partial_\eta^\alpha \phi_j(\eta)| \|\kappa_{|2^{-j}\eta|}\| \leq 2^{|\alpha|+1} C_\alpha K |\eta|^{-|\alpha|},$$

$$\|\kappa_{|\eta|}^{-1}[\partial_\eta^\alpha t(\eta)]\| = \Big\| \sum_{k=0}^{\infty} \partial_\eta^\alpha \psi_k(\eta) \kappa_{|2^{-k}\eta|^{-1}} \pi_k \Big\|$$

$$\leq \sum_{k=0}^{\infty} |\partial_\eta^\alpha \psi_k(\eta)| \|\kappa_{|2^{-k}\eta|^{-1}}\| \leq 2^{|\alpha|+1} D_\alpha K |\eta|^{-|\alpha|},$$

where $K = \sup\{\|\kappa_\rho\| : 2^{-1} \leq \rho \leq 2\} < \infty$ (recall that this is finite by the uniform boundedness principle in view of the strong continuity of κ_ρ). Note that $|2^{-j}\eta|$ and $|2^{-k}\eta|^{-1}$ are in the interval $[2^{-1}, 2]$ for all η in the support of $\partial_\eta^\alpha \phi_j$ and $\partial_\eta^\alpha \psi_k$, respectively, and that at most two summands in the infinite sums are nonzero for each η by construction of the ϕ_j and ψ_k. This completes the proof. □

By Lemma 1 and (4),

$$\begin{aligned} S &= \mathrm{Op}(s) = \mathcal{F}_{\eta \to y}^{-1} s(\eta) \mathcal{F}_{y' \to \eta} : \mathcal{W}^0(\mathbb{R}^q, E) \to L^2(\mathbb{R}^q, \ell^2(\mathbb{N}_0, E)) \\ T &= \mathrm{Op}(t) = \mathcal{F}_{\eta \to y}^{-1} t(\eta) \mathcal{F}_{y' \to \eta} : L^2(\mathbb{R}^q, \ell^2(\mathbb{N}_0, E)) \to \mathcal{W}^0(\mathbb{R}^q, E) \end{aligned} \tag{5}$$

are continuous, and $T \circ S = \mathrm{Id}_{\mathcal{W}^0}$. This shows that S maps $\mathcal{W}^0(\mathbb{R}^q, E)$ isomorphically to a closed subspace of $L^2(\mathbb{R}^q, \ell^2(\mathbb{N}_0, E))$, and thus $\|Su\|_{L^2(\mathbb{R}^q, \ell^2(\mathbb{N}_0, E))}$ for $u \in \mathcal{W}^0(\mathbb{R}^q, E)$ is an equivalent norm on $\mathcal{W}^0(\mathbb{R}^q, E)$. Consequently, we have proved that $\mathcal{W}^0(\mathbb{R}^q, E)$ is isomorphic to a generalized E-valued Triebel–Lizorkin space (or E-valued Besov space which is equivalent here) with operator-valued weight given by the inverse group action κ_ρ^{-1}. More generally, it would make sense to define spaces $B_{p,q}^s(\mathbb{R}^q, (E, \kappa))$ and $F_{p,q}^s(\mathbb{R}^q, (E, \kappa))$ with that operator-valued weight, generalizing the standard E-valued Besov and Triebel–Lizorkin spaces (see [1, 14, 15, 24] for the latter). In this case, we would recover

$$\mathcal{W}^s(\mathbb{R}^q, (E, \kappa)) = B_{2,2}^s(\mathbb{R}^q, (E, \kappa)) = F_{2,2}^s(\mathbb{R}^q, (E, \kappa))$$

from the above. However, we will not pursue this here, but merely summarize what we have proved as follows.

Proposition 1 *A distribution $u \in \mathcal{S}'(\mathbb{R}^q, E)$ belongs to $\mathcal{W}^0(\mathbb{R}^q, E)$ if and only if $Su \in L^2(\mathbb{R}^q, \ell^2(\mathbb{N}_0, E))$ for the operator S from (5). More generally, $u \in \mathcal{W}^s(\mathbb{R}^q, E)$ if and only if*

$$\sum_{j=0}^{\infty} \|2^{js} \kappa_{2^j}^{-1} \phi_j(D_y) u\|_{L^2(\mathbb{R}^q, E)}^2 < \infty.$$

In particular, as mentioned after Definition 1, it is easy now to see the denseness of $\mathcal{S}(\mathbb{R}^q, E)$ in $\mathcal{W}^s(\mathbb{R}^q, E)$: In view of (3), we just have to prove this for $s = 0$. Note that the Fourier multipliers S and T from (5) map rapidly decreasing functions to rapidly decreasing functions. Thus, for $u \in \mathcal{W}^0(\mathbb{R}^q, E)$, choose a sequence $(v_k)_k$

in $\mathcal{S}(\mathbb{R}^q, \ell^2(\mathbb{N}_0, E))$ such that $v_k \to Su$ in $L^2(\mathbb{R}^q, \ell^2(\mathbb{N}_0, E))$. Then $u_k = Tv_k \in \mathcal{S}(\mathbb{R}^q, E)$, and $u_k \to TSu = u$ in $\mathcal{W}^0(\mathbb{R}^q, E)$.

The operators S and T from (5) are particularly useful to study functional analytic properties of operators in the $\mathcal{W}^s$-scale. To this end, recall that $\mathcal{I} \subset \mathcal{L}$ is an operator ideal in the bounded operators acting between Hilbert spaces if for all Hilbert spaces E, $\tilde{E}$, and $\hat{E}$ we have that $\mathcal{I}(E, \tilde{E}) \subset \mathcal{L}(E, \tilde{E})$ is a subspace that contains the finite-rank operators, and for any $A \in \mathcal{I}(E, \tilde{E})$ and $G \in \mathcal{L}(\tilde{E}, \hat{E})$ and $G' \in \mathcal{L}(\hat{E}, E)$ we have $GA \in \mathcal{I}(E, \hat{E})$ and $AG' \in \mathcal{I}(\hat{E}, \tilde{E})$. Examples of interest include $\mathcal{I} = \mathcal{L}$ (all bounded operators), $\mathcal{I} = \mathcal{K}$ (compact operators), and $\mathcal{I} = \mathcal{C}_p$ for $1 \leq p < \infty$ (the Banach operator ideal of Schatten–von Neumann operators with p-summable approximation numbers). See the appendix for additional information.

Proposition 2 *Let (E, κ) and $(\tilde{E}, \tilde{\kappa})$ be Hilbert spaces with strongly continuous group actions $\{\kappa_\rho\}_{\rho>0}$ and $\{\tilde{\kappa}_\rho\}_{\rho>0}$, and let S, T be the operators associated with (E, κ) and $\tilde{S}, \tilde{T}$ be the operators associated with $(\tilde{E}, \tilde{\kappa})$ according to (5), respectively.*

Then an operator $A : \mathcal{S}(\mathbb{R}^q, E) \to \mathcal{S}'(\mathbb{R}^q, \tilde{E})$ is continuous in the spaces

$$A : \mathcal{W}^0(\mathbb{R}^q, E) \to \mathcal{W}^0(\mathbb{R}^q, \tilde{E})$$

if and only if

$$\tilde{S}AT : L^2(\mathbb{R}^q, \ell^2(\mathbb{N}_0, E)) \to L^2(\mathbb{R}^q, \ell^2(\mathbb{N}_0, \tilde{E}))$$

is continuous.

More generally, for any operator ideal in $\mathcal{I} \subset \mathcal{L}$, we have the equivalence

$$A \in \mathcal{I}(\mathcal{W}^0(\mathbb{R}^q, E), \mathcal{W}^0(\mathbb{R}^q, \tilde{E}))$$

$$\Longleftrightarrow$$

$$\tilde{S}AT \in \mathcal{I}(L^2(\mathbb{R}^q, \ell^2(\mathbb{N}_0, E)), L^2(\mathbb{R}^q, \ell^2(\mathbb{N}_0, \tilde{E}))).$$

Proof If $A : \mathcal{W}^0(\mathbb{R}^q, E) \to \mathcal{W}^0(\mathbb{R}^q, \tilde{E})$ is continuous (or belongs to $\mathcal{I}$), then

$$\tilde{S}AT : L^2(\mathbb{R}^q, \ell^2(\mathbb{N}_0, E)) \to L^2(\mathbb{R}^q, \ell^2(\mathbb{N}_0, \tilde{E})) \tag{6}$$

is continuous as a composition of continuous operators (or belongs to $\mathcal{I}$ because of the ideal property). Conversely, if (6) is continuous (or belongs to $\mathcal{I}$), then

$$A = \tilde{T}[\tilde{S}AT]S : \mathcal{W}^0(\mathbb{R}^q, E) \to \mathcal{W}^0(\mathbb{R}^q, \tilde{E})$$

is continuous as a composition of continuous operators (or belongs to $\mathcal{I}$ because of the ideal property). □

We conclude this section with the definition of weighted abstract wedge Sobolev spaces, see [3, 20].

Definition 3 For $\vec{s} = (s_1, s_2) \in \mathbb{R}^2$, define

$$\mathcal{W}^{\vec{s}}(\mathbb{R}^q, E) = \langle y \rangle^{-s_2} \mathcal{W}^{s_1}(\mathbb{R}^q, E).$$

3 Abstract Edge Pseudodifferential Operators

Operator-valued symbols twisted by strongly continuous group actions and associated pseudodifferential operators of type $1,0$ were introduced by Schulze [17]. The calculus has been further developed and applied in [2, 3, 5, 16, 18–21]. Recent work [7–9] on boundary value problems for elliptic wedge differential operators with variable indicial roots provides a motivation for considering symbols and operators of more general types, in particular those of type $1,\delta$ with $\delta > 0$. In this section, we therefore introduce a class of global operator-valued symbols and associated operators of type ρ,δ, $0 \leq \delta < \rho \leq 1$, and discuss the elements of the calculus for this class. We allow the symbols to take values in general Banach operator ideals $\mathcal{I}$ in the Hilbert space category, and prove independence of the pseudodifferential operator class with respect to all τ-quantizations [22], thus acquiring more flexibility for applications of the calculus.

We will utilize Kumano-go's technique [10], similar to the works [3, 16, 20, 21] that discuss various classes of twisted pseudodifferential operators of type $1,0$.

Definition 4 (Twisted operator-valued symbols). Fix $0 \leq \delta < \rho \leq 1$.

For $\vec{\mu} = (\mu_1, \mu_2) \in \mathbb{R}^2$, let $S^{\vec{\mu}}_{\rho,\delta}(\mathbb{R}^q \times \mathbb{R}^q; (E,\kappa), (\tilde{E},\tilde{\kappa}))_{\mathcal{I}}$ be the space of all $a(y,\eta) \in C^\infty(\mathbb{R}^q \times \mathbb{R}^q, \mathcal{I}(E,\tilde{E}))$ such that for each $\alpha, \beta \in \mathbb{N}_0^q$ there exists a constant $C_{\alpha,\beta} \geq 0$ with

$$\|\tilde{\kappa}^{-1}_{\langle\eta\rangle}[D^\beta_y \partial^\alpha_\eta a(y,\eta)]\kappa_{\langle\eta\rangle}\|_{\mathcal{I}(E,\tilde{E})} \leq C_{\alpha,\beta}\langle y\rangle^{\mu_2}\langle\eta\rangle^{\mu_1-\rho|\alpha|+\delta|\beta|}$$

for all $(y,\eta) \in \mathbb{R}^q \times \mathbb{R}^q$.

Moreover, for $\vec{\mu} = (\mu_1, \mu_2, \mu_3) \in \mathbb{R}^3$, let $S^{\vec{\mu}}_{\rho,\delta}(\mathbb{R}^q \times \mathbb{R}^q \times \mathbb{R}^q; (E,\kappa), (\tilde{E},\tilde{\kappa}))_{\mathcal{I}}$ be the space of all $a(y,y',\eta) \in C^\infty(\mathbb{R}^q \times \mathbb{R}^q \times \mathbb{R}^q, \mathcal{I}(E,\tilde{E}))$ such that for each $\alpha, \beta, \gamma \in \mathbb{N}_0^q$ there exists a constant $C_{\alpha,\beta,\gamma} \geq 0$ with

$$\|\tilde{\kappa}^{-1}_{\langle\eta\rangle}[D^\beta_y D^\gamma_{y'} \partial^\alpha_\eta a(y,y',\eta)]\kappa_{\langle\eta\rangle}\|_{\mathcal{I}(E,\tilde{E})} \leq C_{\alpha,\beta,\gamma}\langle y\rangle^{\mu_2}\langle y'\rangle^{\mu_3}\langle\eta\rangle^{\mu_1-\rho|\alpha|+\delta(|\beta|+|\gamma|)}$$

for all $(y, y,' \eta) \in \mathbb{R}^q \times \mathbb{R}^q \times \mathbb{R}^q$.

The symbol spaces carry natural Fréchet topologies induced by the seminorms given by the best constants in the estimates.

In the case $\mathcal{I} = \mathcal{L}$, the operator ideal subscript is dropped from the symbol space notation. Likewise, if E or $\tilde{E}$ carry the trivial group action $\kappa_\varrho = \mathrm{Id}_E$ or $\tilde{\kappa}_\varrho = \mathrm{Id}_{\tilde{E}}$, respectively, we will just write E instead of the pair (E, Id_E) or $\tilde{E}$ instead of $(\tilde{E}, \mathrm{Id}_{\tilde{E}})$. If $\mu_2 = 0$ (for simple symbols) or $\mu_2 = \mu_3 = 0$ (for double symbols), it is customary to revert to the single superscript μ_1 instead of the vector $(\mu_1, 0)$ or $(\mu_1, 0, 0)$ to indicate the order of the symbol.

A technically convenient observation for the twisted calculus is that there are trivial inclusions

$$S^{\vec{\mu}-(M+\tilde{M},0,0)}_{\rho,\delta}(\mathbb{R}^q \times \mathbb{R}^q \times \mathbb{R}^q; E, \tilde{E})_{\mathcal{I}} \subset S^{\vec{\mu}}_{\rho,\delta}(\mathbb{R}^q \times \mathbb{R}^q \times \mathbb{R}^q; (E,\kappa), (\tilde{E},\tilde{\kappa}))_{\mathcal{I}}$$
$$\subset S^{\vec{\mu}+(M+\tilde{M},0,0)}_{\rho,\delta}(\mathbb{R}^q \times \mathbb{R}^q \times \mathbb{R}^q; E, \tilde{E})_{\mathcal{I}},$$

as well as

$$S^{\vec{\mu}}_{\rho,\delta}(\mathbb{R}^q \times \mathbb{R}^q \times \mathbb{R}^q; E, \tilde{E})_{\mathcal{I}} \subset S^{\vec{\mu}}_{\rho,\delta}(\mathbb{R}^q \times \mathbb{R}^q \times \mathbb{R}^q; E, \tilde{E}),$$

where M and $\tilde{M}$ are the growth constants for $\{\kappa_\rho\}_{\rho>0}$ and $\{\tilde{\kappa}_\rho\}_{\rho>0}$ in (1), respectively. Hence, the impact of the group actions is in essence limited to determining a filtration by order that is different from the standard filtration of the Hörmander calculus of type ρ, δ [6, 10].

With any double symbol $a(y, y', \eta) \in S^{\vec{\mu}}_{\rho,\delta}(\mathbb{R}^q \times \mathbb{R}^q \times \mathbb{R}^q; (E, \kappa), (\tilde{E}, \tilde{\kappa}))$, we associate a pseudodifferential operator $\mathrm{Op}(a) : \mathcal{S}(\mathbb{R}^q, E) \to \mathcal{S}(\mathbb{R}^q, \tilde{E})$ as usual via

$$[\mathrm{Op}(a)u](y) = (2\pi)^{-q} \int_{\mathbb{R}^q} \int_{\mathbb{R}^q} e^{i(y-y')\eta} a(y, y', \eta) u(y') \mathrm{d}y' \mathrm{d}\eta.$$

Moreover, for each $\tau \in \mathbb{R}$ and $a \in S^{\vec{\mu}}_{\rho,\delta}(\mathbb{R}^q \times \mathbb{R}^q; (E, \kappa), (\tilde{E}, \tilde{\kappa}))$, we associate a τ-quantized pseudodifferential operator $\mathrm{Op}_\tau(a) : \mathcal{S}(\mathbb{R}^q, E) \to \mathcal{S}(\mathbb{R}^q, \tilde{E})$ via

$$\mathrm{Op}_\tau(a)u(y) = (2\pi)^{-q} \int_{\mathbb{R}^q} \int_{\mathbb{R}^q} e^{i(y-y')\eta} a(\tau y' + (1-\tau)y, \eta) u(y') \mathrm{d}y' \mathrm{d}\eta.$$

Note that $\mathrm{Op}_0(a) = a(Y, D)$ is the Kohn–Nirenberg-quantized pseudodifferential operator with (left) symbol $a(y, \eta)$; for $\tau = 1/2$ we obtain a Weyl-quantized operator $\mathrm{Op}_{1/2}(a) = a^w(Y, D)$ with Weyl symbol a, and for $\tau = 1$ we obtain the quantization of the right symbol $a(y', \eta)$.

For $\vec{\mu} = (\mu_1, \mu_2)$, define

$$\Psi^{\vec{\mu}}_{\rho,\delta}(\mathbb{R}^q; (E, \kappa), (\tilde{E}, \tilde{\kappa}))_{\mathcal{I}} = \{\mathrm{Op}_\tau(a) : \mathcal{S}(\mathbb{R}^q, E) \to \mathcal{S}(\mathbb{R}^q, \tilde{E}) : \\ a \in S^{\vec{\mu}}_{\rho,\delta}(\mathbb{R}^q \times \mathbb{R}^q; (E, \kappa), (\tilde{E}, \tilde{\kappa}))_{\mathcal{I}}\}.$$

We will see in Theorem 1 below that $\Psi^{\vec{\mu}}_{\rho,\delta}(\mathbb{R}^q; (E, \kappa), (\tilde{E}, \tilde{\kappa}))_{\mathcal{I}}$ is independent of $\tau \in \mathbb{R}$, and that the τ-quantization map

$$S^{\vec{\mu}}_{\rho,\delta}(\mathbb{R}^q \times \mathbb{R}^q; (E, \kappa), (\tilde{E}, \tilde{\kappa}))_{\mathcal{I}} \ni a \mapsto \mathrm{Op}_\tau(a) \in \Psi^{\vec{\mu}}_{\rho,\delta}(\mathbb{R}^q; (E, \kappa), (\tilde{E}, \tilde{\kappa}))_{\mathcal{I}}$$

is a bijection. We equip the operator space $\Psi^{\vec{\mu}}_{\rho,\delta}(\mathbb{R}^q; (E, \kappa), (\tilde{E}, \tilde{\kappa}))_{\mathcal{I}}$ with the symbol topology transferred by any τ-quantization map, making it a Fréchet space. This topology turns out to be independent of τ.

Lemma 2 *Fix $\tau \in \mathbb{R}$. Let $a(y, y', \eta) \in S^{(\mu_1,\mu_2,\mu_3)}_{\rho,\delta}(\mathbb{R}^q \times \mathbb{R}^q \times \mathbb{R}^q; (E, \kappa), (\tilde{E}, \tilde{\kappa}))_{\mathcal{I}}$, and define*

$$a^\tau_\theta(y, \eta) = \frac{1}{(2\pi)^q} \iint e^{-ix\xi} a(y - \tau x, y + (1-\tau)x, \eta + \theta\xi)\, dx\, d\xi,\ 0 \le \theta \le 1,$$

where the integral is an $\mathcal{I}(E, \tilde{E})$-valued oscillatory integral.

Then $a_\theta^\tau \in C([0,1]_\theta, S_{\rho,\delta}^{(\mu_1,\mu_2+\mu_3)}(\mathbb{R}^q \times \mathbb{R}^q;(E,\kappa),(\tilde{E},\tilde{\kappa}))_{\mathcal{I}})$*, and the map* $a \mapsto a_\theta^\tau$ *is continuous. For each* $N \in \mathbb{N}$*, we have*

$$a_\theta^\tau(y,\eta) = \sum_{|\alpha+\beta|<N} \frac{\theta^{|\alpha+\beta|}}{\alpha!\beta!}\tau^{|\alpha|}(1-\tau)^{|\beta|}\partial_\eta^{\alpha+\beta}(-D_y)^\alpha D_{y'}^\beta a(y,y',\eta)|_{y'=y} + r_{N,\theta}^\tau(y,\eta), \tag{7}$$

where $r_{N,\theta}^\tau(y,\eta) \in C([0,1]_\theta, S_{\rho,\delta}^{(\mu_1-(\rho-\delta)N,\mu_2+\mu_3)}(\mathbb{R}^q \times \mathbb{R}^q;(E,\kappa),(\tilde{E},\tilde{\kappa}))_{\mathcal{I}})$ *is given by*

$$r_{N,\theta}^\tau(y,\eta) = N\theta^N \sum_{|\alpha+\beta|=N} \int_0^1 \frac{(1-s)^{N-1}}{\alpha!\beta!}\tau^{|\alpha|}(1-\tau)^{|\beta|} \cdot$$

$$\cdot \Big[\frac{1}{(2\pi)^q}\iint e^{-ix\xi}(\partial_\eta^{\alpha+\beta}(-D_y)^\alpha D_{y'}^\beta a)(y-\tau x, y+(1-\tau)x, \eta+s\theta\xi)\,dx\,d\xi\Big]\,ds.$$

Proof We first argue that it is enough to prove the lemma in the case of trivial group actions $\kappa_\varrho \equiv \mathrm{Id}_E$ and $\tilde{\kappa}_\varrho \equiv \mathrm{Id}_{\tilde{E}}$. Thus, suppose that the lemma is proved in that case. Let M and $\tilde{M}$ be the growth constants for $\{\kappa_\varrho\}_{\varrho>0}$ and $\{\tilde{\kappa}_\varrho\}_{\varrho>0}$ in (1), respectively. Because

$$S_{\rho,\delta}^{(\mu_1,\mu_2,\mu_3)}(\mathbb{R}^q\times\mathbb{R}^q\times\mathbb{R}^q;(E,\kappa),(\tilde{E},\tilde{\kappa}))_{\mathcal{I}} \subset S_{\rho,\delta}^{(\mu_1+M+\tilde{M},\mu_2,\mu_3)}(\mathbb{R}^q\times\mathbb{R}^q\times\mathbb{R}^q;E,\tilde{E})_{\mathcal{I}}$$

we can apply the lemma to $a(y,y',\eta)$ as an element of the latter class. Choose $N \in \mathbb{N}$ so large that $(\rho-\delta)N \geq 2(M+\tilde{M})$ and apply (7). Then

$$S_{\rho,\delta}^{(\mu_1+M+\tilde{M}-(\rho-\delta)N,\mu_2+\mu_3)}(\mathbb{R}^q\times\mathbb{R}^q;E,\tilde{E})_{\mathcal{I}} \hookrightarrow S_{\rho,\delta}^{(\mu_1,\mu_2+\mu_3)}(\mathbb{R}^q\times\mathbb{R}^q;(E,\kappa),(\tilde{E},\tilde{\kappa}))_{\mathcal{I}},$$

and consequently

$$r_{N,\theta}^\tau \in C([0,1]_\theta, S_{\rho,\delta}^{(\mu_1,\mu_2+\mu_3)}(\mathbb{R}^q\times\mathbb{R}^q;(E,\kappa),(\tilde{E},\tilde{\kappa}))_{\mathcal{I}}).$$

But since

$$\sum_{|\alpha+\beta|<N} \frac{\theta^{|\alpha+\beta|}}{\alpha!\beta!}\tau^{|\alpha|}\big(1-\tau\big)^{|\beta|}\partial_\eta^{\alpha+\beta}\big(-D_y\big)^\alpha D_{y'}^\beta a\big(y,y',\eta\big)\Big|_{y'=y}$$
$$\in C([0,1]_\theta, S_{\rho,\delta}^{(\mu_1,\mu_2+\mu_3)}(\mathbb{R}^q\times\mathbb{R}^q;(E,\kappa),(\tilde{E},\tilde{\kappa}))_{\mathcal{I}}),$$

we also get that $a_\theta^\tau \in C([0,1]_\theta, S_{\rho,\delta}^{(\mu_1,\mu_2+\mu_3)}(\mathbb{R}^q\times\mathbb{R}^q;(E,\kappa),(\tilde{E},\tilde{\kappa}))_{\mathcal{I}})$ as desired. Continuity of $a \mapsto a_\theta^\tau$ in case of general group actions follows from the case of trivial group actions via an application of the closed graph theorem. By what we have already shown we now obtain that, for all $\alpha,\beta \in \mathbb{N}_0^q$,

$$(s,\theta) \mapsto \frac{1}{(2\pi)^q}\iint e^{-ix\xi}(\partial_\eta^{\alpha+\beta}(-D_y)^\alpha D_{y'}^\beta a)(y-\tau x, y+(1-\tau)x, \eta+s\theta\xi)\mathrm{d}x\,\mathrm{d}\xi$$

is continuous on $[0,1] \times [0,1]$ taking values in

$$S^{(\mu_1-(\rho-\delta)|\alpha+\beta|,\mu_2+\mu_3)}_{\rho,\delta}(\mathbb{R}^q \times \mathbb{R}^q; (E,\kappa), (\tilde{E},\tilde{\kappa}))_{\mathcal{I}}.$$

Thus, the formula for the remainders $r^\tau_{N,\theta}(y,\eta)$ implies that

$$r^\tau_{N,\theta}(y,\eta) \in C([0,1]_\theta, S^{(\mu_1-(\rho-\delta)N,\mu_2+\mu_3)}_{\rho,\delta}(\mathbb{R}^q \times \mathbb{R}^q; (E,\kappa), (\tilde{E},\tilde{\kappa}))_{\mathcal{I}})$$

for each $N \in \mathbb{N}$, which completes the argument.

It remains to consider the case of trivial group actions. The proof in this case follows along the lines of Kumano-go [10, Chap. 2, Lemma 2.4]. Set

$$b^\tau_\theta(y,\eta,x,\xi) = a(y - \tau x, y + (1-\tau)x, \eta + \theta\xi).$$

For any $k \in \mathbb{N}_0$, there exists a continuous seminorm $|\cdot|_k$ on $S^{(\mu_1,\mu_2,\mu_3)}_{\rho,\delta}(\mathbb{R}^q \times \mathbb{R}^q \times \mathbb{R}^q; E, \tilde{E})_{\mathcal{I}}$ such that, for all $|\alpha'|, |\alpha| \le k$ and $|\beta'|, |\beta| \le k$,

$$\begin{aligned} &\left\| D^{\beta'}_x D^\beta_y \partial^{\alpha'}_\xi \partial^\alpha_\eta b^\tau_\theta(y,\eta,x,\xi) \right\|_{\mathcal{I}(E,\tilde{E})} \\ &\qquad \lesssim |a|_k \langle y - \tau x\rangle^{\mu_2} \langle y + (1-\tau)x\rangle^{\mu_3} \langle \eta + \theta\xi\rangle^{\mu_1 - \rho(|\alpha|+|\alpha'|) + \delta(|\beta|+|\beta'|)} \\ &\qquad \lesssim |a|_k \langle y\rangle^{\mu_2+\mu_3} \langle x\rangle^{|\mu_2|+|\mu_3|} \langle \eta + \theta\xi\rangle^{\mu_1 - \rho(|\alpha|+|\alpha'|) + \delta(|\beta|+|\beta'|)} \\ &\qquad \lesssim |a|_k \langle y\rangle^{\mu_2+\mu_3} \langle \langle\eta\rangle^\delta x\rangle^{|\mu_2|+|\mu_3|} \langle \eta + \theta\xi\rangle^{\mu_1 - \rho(|\alpha|+|\alpha'|) + \delta(|\beta|+|\beta'|)}. \end{aligned} \tag{8}$$

We are omitting constants in these estimates, therefore writing $\lesssim$ instead of $\le$. Here and in the sequel it is important to note that, whenever we write $\lesssim$, the implicit constants may depend on $k \in \mathbb{N}_0$, $\vec{\mu}$, the dimension q, as well as on $\tau \in \mathbb{R}$, but they are independent of all the variables (y,η,x,ξ), the parameter $0 \le \theta \le 1$, and the double symbol a.

These estimates show, in particular, that $b^\tau_\theta(y,\eta,x,\xi)$ is an $\mathcal{I}(E,\tilde{E})$-valued amplitude function in (x,ξ). Thus, $a^\tau_\theta(y,\eta)$ is well defined, and in fact is a C^∞-function of (y,η) taking values in $\mathcal{I}(E,\tilde{E})$; we can differentiate under the integral sign to find the derivatives of a^τ_θ. Taylor expansion shows that

$$\begin{aligned} a(y - \tau x, y + (1-\tau)x, \eta + \theta\xi) &= \sum_{|\gamma|<N} \frac{\theta^{|\gamma|}}{\gamma!} (\partial^\gamma_\eta a)(y - \tau x, y + (1-\tau)x, \eta)\xi^\gamma \\ &\quad + N \sum_{|\gamma|=N} \frac{\theta^N}{\gamma!} \xi^\gamma \int_0^1 (1-s)^{N-1} (\partial^\gamma_\eta a)(y - \tau x, y + (1-\tau)x, \eta + s\theta\xi) ds. \end{aligned}$$

We insert this in the oscillatory integral formula for $a^\tau_\theta(y,\eta)$. Using $\xi^\gamma e^{-ix\xi} = (-D_x)^\gamma e^{-ix\xi}$ and integrating by parts then gives (7).

It remains to show the asserted estimates for $a^\tau_\theta(y,\eta)$ and for $r^\tau_{N,\theta}(y,\eta)$. Because we already have (7), we see that both reduce to proving that $\{a^\tau_\theta : 0 \le \theta \le 1\}$ is a family of symbols, where the constants in the symbol estimates can be chosen to be continuous seminorms of a that are independent of $0 \le \theta \le 1$. For this proof, we may further assume without loss of generality by (7) that $\mu_1 < -q$, i.e., the amplitude function b_θ^τ is already integrable with respect to ξ (this uses again that $\delta < \rho$). Once these uniform estimates are established the continuity of $\theta \mapsto a_\theta^\tau$ follows from the mean value theorem, using the formula

$$\partial_\theta a_\theta^\tau(y,\eta) = \sum_{|\alpha+\beta|=1} \tau^{|\alpha|}(1-\tau)^{|\beta|}$$

$$\cdot \frac{1}{(2\pi)^q} \iint e^{-ix\xi}((-D_y)^\alpha D_{y'}^\beta \partial_\eta^{\alpha+\beta} a)(y-\tau x, y+(1-\tau)x, \eta+\theta\xi)\mathrm{d}x\,\mathrm{d}\xi,$$

which shows that $\{\partial_\theta a_\theta^\tau : 0 \le \theta \le 1\}$ is bounded in the symbol topology in view of the uniform symbol estimates.

As we can differentiate under the integral sign, it is enough to prove that $\|a_\theta^\tau(y,\eta)\|_{\mathcal{I}(E,\tilde{E})} \lesssim |a|_k \langle y\rangle^{\mu_2+\mu_3}\langle\eta\rangle^{\mu_1}$ for some $k \in \mathbb{N}_0$ as in (8).

We regularize the oscillatory integral for a_θ^τ and obtain

$$a_\theta^\tau(y,\eta) = \frac{1}{(2\pi)^q}\iint e^{-ix\xi} c_{\ell,\theta}^\tau(y,\eta,x,\xi)\mathrm{d}x\,\mathrm{d}\xi,$$

$$c_{\ell,\theta}^\tau(y,\eta,x,\xi) = \langle\langle\eta\rangle^\delta x\rangle^{-2\ell}(1-\langle\eta\rangle^{2\delta}\Delta_\xi)^\ell b_\theta^\tau(y,\eta,x,\xi)$$

for each $\ell \in \mathbb{N}$. This is integrable in (x,ξ) for $2\ell > |\mu_2| + |\mu_3| + q$ by (8), which we assume henceforth (recall that $\mu_1 < -q$). Let

$$\Omega_1 = \{\xi : |\xi| \le \langle\eta\rangle^\delta/2\},$$

$$\Omega_2 = \{\xi : \langle\eta\rangle^\delta/2 \le |\xi| \le \langle\eta\rangle/2\},$$

$$\Omega_3 = \{\xi : |\xi| \ge \langle\eta\rangle/2\},$$

and write

$$a_\theta^\tau(y,\eta) = \frac{1}{(2\pi)^q}\iint_{\mathbb{R}^{2q}} e^{-ix\xi} c_{\ell,\theta}^\tau(y,\eta,x,\xi)\,\mathrm{d}x\,\mathrm{d}\xi = I_1(y,\eta) + I_2(y,\eta) + I_3(y,\eta),$$

where

$$I_j(y,\eta) = \frac{1}{(2\pi)^q}\int_{\Omega_j}\int_{\mathbb{R}^q} e^{-ix\xi} c_{\ell,\theta}^\tau(y,\eta,x,\xi)\mathrm{d}x\,\mathrm{d}\xi, \quad j = 1,2,3.$$

We proceed to estimate the three integrals separately. On $\Omega_1 \cup \Omega_2$, we have

$$\langle\eta\rangle/2 \le \langle\eta+\theta\xi\rangle \le (3/2)\langle\eta\rangle,$$

thus we get from (8)

$$\left\|c_{\ell,\theta}^\tau(y,\eta,x,\xi)\right\|_{\mathcal{I}(E,\tilde{E})} \lesssim |a|_{2\ell}\langle y\rangle^{\mu_2+\mu_3}\langle\langle\eta\rangle^\delta x\rangle^{|\mu_2|+|\mu_3|-2\ell}\langle\eta\rangle^{\mu_1}.$$

Changing variables $x' = \langle\eta\rangle^\delta x$ shows that

$$\int_{\mathbb{R}^q}\langle\langle\eta\rangle^\delta x\rangle^{|\mu_2|+|\mu_3|-2\ell}\,\mathrm{d}x \lesssim \langle\eta\rangle^{-\delta q},$$

and we have

$$\int_{\Omega_1} \mathrm{d}\xi \le \int_{\{\xi : |\xi_j| \le \langle\eta\rangle^\delta/2\}} \mathrm{d}\xi = \langle\eta\rangle^{\delta q}.$$

Consequently,

$$\big\| I_1(y,\eta) \big\|_{\mathcal{I}(E,\tilde{E})} \lesssim |a|_{2\ell} \langle y\rangle^{\mu_2+\mu_3} \langle\eta\rangle^{\mu_1}.$$

In order to estimate $I_2(y,\eta)$, we first note that

$$\big\| D_x^{\beta'} c^\tau_{\ell,\theta}(y,\eta,x,\xi) \big\|_{\mathcal{I}(E,\tilde{E})} \lesssim |a|_{2\ell} \langle y\rangle^{\mu_2+\mu_3} \langle\langle\eta\rangle^\delta x\rangle^{|\mu_2|+|\mu_3|-2\ell} \langle\eta\rangle^{\mu_1+\delta|\beta'|}$$

for $|\beta'| \le 2\ell$ on $\Omega_1 \cup \Omega_2$. This follows from the Leibniz rule, (8), and the estimate

$$\big| D_x^\beta \langle\langle\eta\rangle^\delta x\rangle^{-2\ell} \big| \lesssim \langle\eta\rangle^{\delta|\beta|} \langle\langle\eta\rangle^\delta x\rangle^{-2\ell}.$$

On Ω_2, we write

$$\int_{\mathbb{R}^q} e^{-ix\xi} c^\tau_{\ell,\theta}(y,\eta,x,\xi)\,\mathrm{d}x = |\xi|^{-2\ell} \int_{\mathbb{R}^q} e^{-ix\xi} (-\Delta_x)^\ell c^\tau_{\ell,\theta}(y,\eta,x,\xi)\mathrm{d}x.$$

Thus,

$$\big\| I_2(y,\eta) \big\|_{\mathcal{I}(E,\tilde{E})} \lesssim |a|_{2\ell} \langle y\rangle^{\mu_2+\mu_3} \langle\eta\rangle^{\mu_1+2\ell\delta} \underbrace{\int_{\mathbb{R}^q} \langle\langle\eta\rangle^\delta x\rangle^{|\mu_2|+|\mu_3|-2\ell}\,\mathrm{d}x}_{\lesssim \langle\eta\rangle^{-\delta q}} \int_{\Omega_2} |\xi|^{-2\ell}\,\mathrm{d}\xi.$$

To estimate the ξ-integral, we pass to polar coordinates and obtain

$$\int_{\Omega_2} |\xi|^{-2\ell}\,\mathrm{d}\xi \lesssim \int_{\langle\eta\rangle^\delta/2}^\infty r^{q-1-2\ell}\,\mathrm{d}r \lesssim \langle\eta\rangle^{(q-2\ell)\delta},$$

and consequently

$$\big\| I_2(y,\eta) \big\|_{\mathcal{I}(E,\tilde{E})} \lesssim |a|_{2\ell} \langle y\rangle^{\mu_2+\mu_3} \langle\eta\rangle^{\mu_1}.$$

It remains to estimate $I_3(y,\eta)$. On Ω_3, we have $\langle\eta+\theta\xi\rangle \le \langle\eta\rangle + |\xi| \le 3|\xi|$, and consequently

$$\big\| D_x^{\beta'} c^\tau_{\ell,\theta}(y,\eta,x,\xi) \big\|_{\mathcal{I}(E,\tilde{E})} \lesssim |a|_k \langle y\rangle^{\mu_2+\mu_3} \langle\langle\eta\rangle^\delta x\rangle^{|\mu_2|+|\mu_3|-2\ell} |\xi|^{\delta(|\beta'|+2\ell)}$$

for $|\beta'| \le k$, where $k \ge 2\ell$. As above, we write

$$\int_{\mathbb{R}^q} e^{-ix\xi} c^\tau_{\ell,\theta}(y,\eta,x,\xi)\,\mathrm{d}x = |\xi|^{-2L} \int_{\mathbb{R}^q} e^{-ix\xi} (-\Delta_x)^L c^\tau_{\ell,\theta}(y,\eta,x,\xi)\mathrm{d}x$$

on Ω_3, where $L \ge \ell$ is chosen large enough, to be determined momentarily. Thus,

$$\big\| I_3(y,\eta) \big\|_{\mathcal{I}(E,\tilde{E})} \lesssim |a|_{2L} \langle y\rangle^{\mu_2+\mu_3} \underbrace{\int_{\mathbb{R}^q} \langle\langle\eta\rangle^\delta x\rangle^{|\mu_2|+|\mu_3|-2\ell}\,\mathrm{d}x}_{\lesssim \langle\eta\rangle^{-\delta q} \le 1} \int_{\Omega_3} |\xi|^{-2L(1-\delta)+2\ell\delta}\,\mathrm{d}\xi.$$

We have

$$\int_{\Omega_3} |\xi|^{-2L(1-\delta)+2\ell\delta}\, d\xi \lesssim \int_{\langle\eta\rangle/2}^{\infty} r^{-2L(1-\delta)+2\ell\delta+q-1}\, dr \lesssim \langle\eta\rangle^{-2L(1-\delta)+2\ell\delta+q},$$

and consequently

$$\big\| I_3(y,\eta)\big\|_{\mathcal{I}(E,\tilde{E})} \lesssim |a|_{2L}\langle y\rangle^{\mu_2+\mu_3}\langle\eta\rangle^{-2L(1-\delta)+2\ell\delta+q} \lesssim |a|_{2L}\langle y\rangle^{\mu_2+\mu_3}\langle\eta\rangle^{\mu_1}$$

provided that $L \geq \ell$ is chosen so large that $-2L(1-\delta)+2\ell\delta+q \leq \mu_1$ (recall that $\mu_1 < -q$). This completes the proof of the lemma. □

Definition 5 For $\vec{\mu} \in \mathbb{R}^3$ let

$$S^{\vec{\mu}}_{\rho,\delta}(\mathbb{R}^q \times \mathbb{R}^q \times \mathbb{R}^q; (E,\kappa),(\tilde{E},\tilde{\kappa}))_{\mathcal{I},(0)}$$

be the closure of $C_c^\infty(\mathbb{R}^q \times \mathbb{R}^q \times \mathbb{R}^q, \mathcal{I}(E,\tilde{E}))$ in the symbol space $S^{\vec{\mu}}_{\rho,\delta}(\mathbb{R}^q \times \mathbb{R}^q \times \mathbb{R}^q; (E,\kappa),(\tilde{E},\tilde{\kappa}))_{\mathcal{I}}$. We use the analogous notation for the classes of simple symbols.

Lemma 3 *We have*

$$S^{\vec{\mu}}_{\rho,\delta}(\mathbb{R}^q \times \mathbb{R}^q \times \mathbb{R}^q; (E,\kappa),(\tilde{E},\tilde{\kappa}))_{\mathcal{I}} \subset S^{\vec{\mu}'}_{\rho,\delta}(\mathbb{R}^q \times \mathbb{R}^q \times \mathbb{R}^q; (E,\kappa),(\tilde{E},\tilde{\kappa}))_{\mathcal{I},(0)}$$

if $\mu_j < \mu_j'$ *for all* $j = 1,2,3$. *The analogous result holds for the classes of simple symbols.*

Proof This follows by the standard argument: Let $\chi \in C^\infty(\mathbb{R}^{3q})$ be an excision function of the origin, i.e., $\chi \equiv 0$ near $(0,0,0) \in \mathbb{R}^{3q}$, and $\chi \equiv 1$ for large $|y,y',\eta|$. More specifically, assume that $\chi(y,y',\eta) \equiv 0$ for $|y,y',\eta| \leq 1$ and $\chi(y,y',\eta) \equiv 1$ for $|y,y',\eta| \geq 2$. For $a \in S^{\vec{\mu}}_{\rho,\delta}(\mathbb{R}^q \times \mathbb{R}^q \times \mathbb{R}^q; (E,\kappa),(\tilde{E},\tilde{\kappa}))_{\mathcal{I}}$ and $j \in \mathbb{N}$ define

$$a_j(y,y',\eta) = (1-\chi(y/j,y'/j,\eta/j))a(y,y',\eta) \in C_c^\infty(\mathbb{R}^q \times \mathbb{R}^q \times \mathbb{R}^q, \mathcal{I}(E,\tilde{E})).$$

Then

$$\begin{aligned}(a-a_j)(y,y',\eta) &= \chi(y/j,y'/j,\eta/j)a(y,y',\eta)\\ &\in S^{\vec{\mu}'}_{\rho,\delta}(\mathbb{R}^q \times \mathbb{R}^q \times \mathbb{R}^q; (E,\kappa),(\tilde{E},\tilde{\kappa}))_{\mathcal{I}},\end{aligned}$$

and $a - a_j \to 0$ as $j \to \infty$ in $S^{\vec{\mu}'}_{\rho,\delta}(\mathbb{R}^q \times \mathbb{R}^q \times \mathbb{R}^q; (E,\kappa),(\tilde{E},\tilde{\kappa}))_{\mathcal{I}}$ which proves the claim.

To see this convergence, set $\varepsilon = \min\{\mu_j' - \mu_j : j = 1,2,3\} > 0$ and use the Leibniz rule and basic inequalities to estimate

$$\begin{aligned}&\langle y\rangle^{-\mu_2'}\langle y'\rangle^{-\mu_3'}\langle\eta\rangle^{-\mu_1'+\rho|\alpha_3|-\delta(|\alpha_1|+|\alpha_2|)}\,\cdot\\ &\qquad\cdot\big\|\tilde{\kappa}^{-1}_{\langle\eta\rangle}\big[D_y^{\alpha_1}D_{y'}^{\alpha_2}\partial_\eta^{\alpha_3}[\chi(y/j,y'/j,\eta/j)a(y,y',\eta)]\big]\kappa_{\langle\eta\rangle}\big\|_{\mathcal{I}}\\ &\leq \langle y,y',\eta\rangle^{-\varepsilon}\cdot \sum_{\substack{\gamma_1\leq\alpha_1\\ \gamma_2\leq\alpha_2\\ \gamma_3\leq\alpha_3}} \binom{\alpha_1}{\gamma_1}\binom{\alpha_2}{\gamma_2}\binom{\alpha_3}{\gamma_3}\cdot\end{aligned}$$

$$\cdot \langle y\rangle^{-\mu_2} \langle y'\rangle^{-\mu_3} \langle \eta\rangle^{-\mu_1+\rho|\gamma_3|-\delta(|\gamma_1|+|\gamma_2|)} \left\| \tilde{\kappa}_{\langle\eta\rangle}^{-1} \left[D_y^{\gamma_1} D_{y'}^{\gamma_2} \partial_\eta^{\gamma_3} a(y,y',\eta) \right] \kappa_{\langle\eta\rangle} \right\|_{\mathcal{I}} \cdot$$

$$\cdot \left| (D_y^{\alpha_1-\gamma_1} D_{y'}^{\alpha_2-\gamma_2} \partial_\eta^{\alpha_3-\gamma_3} \chi)(y/j, y'/j, \eta/j) \right| \langle y/j, y'/j, \eta/j \rangle^{|\alpha_1-\gamma_1|+|\alpha_2-\gamma_2|+|\alpha_3-\gamma_3|}.$$

On the support of $(D_y^{\alpha_1-\gamma_1} D_{y'}^{\alpha_2-\gamma_2} \partial_\eta^{\alpha_3-\gamma_3} \chi)(y/j, y'/j, \eta/j)$, we have

$$\langle y/j, y'/j, \eta/j \rangle^{|\alpha_1-\gamma_1|+|\alpha_2-\gamma_2|+|\alpha_3-\gamma_3|} \leq 3^{|\alpha_1|+|\alpha_2|+|\alpha_3|}$$

and $\langle y, y', \eta\rangle^{-\varepsilon} \leq j^{-\varepsilon}$ by the choice of χ. Summing up, this shows that for every $\alpha_1, \alpha_2, \alpha_3 \in \mathbb{N}_0^q$ there exists a continuous seminorm $|\cdot|$ on $S_{\rho,\delta}^{\vec{\mu}}(\mathbb{R}^q \times \mathbb{R}^q \times \mathbb{R}^q; (E,\kappa), (\tilde{E},\tilde{\kappa}))_{\mathcal{I}}$ such that

$$\sup \left\{ \langle y\rangle^{-\mu_2'} \langle y'\rangle^{-\mu_3'} \langle \eta\rangle^{-\mu_1'+\rho|\alpha_3|-\delta(|\alpha_1|+|\alpha_2|)} \cdot \right.$$

$$\left. \cdot \left\| \tilde{\kappa}_{\langle\eta\rangle}^{-1} \left[D_y^{\alpha_1} D_{y'}^{\alpha_2} \partial_\eta^{\alpha_3} [\chi(y/j, y'/j, \eta/j) a(y,y',\eta)] \right] \kappa_{\langle\eta\rangle} \right\|_{\mathcal{I}} : y, y', \eta \in \mathbb{R}^q \right\} \leq \frac{1}{j^\varepsilon} |a|,$$

thus proving the claimed convergence. □

Lemma 4 *For every $\tau \in \mathbb{R}$ the map*

$$T_\tau : S_{\rho,\delta}^{\vec{\mu}}(\mathbb{R}^q \times \mathbb{R}^q; (E,\kappa), (\tilde{E},\tilde{\kappa}))_{\mathcal{I}} \to S_{\rho,\delta}^{\vec{\mu}}(\mathbb{R}^q \times \mathbb{R}^q; (E,\kappa), (\tilde{E},\tilde{\kappa}))_{\mathcal{I}}$$

given by

$$T_\tau(a)(y,\eta) = \frac{1}{(2\pi)^q} \iint e^{-ix\xi} a(y-\tau x, \eta+\xi)\, dx\, d\xi$$

is a topological isomorphism with inverse $T_{-\tau}$.

We have

$$T_\tau(a)(y,\eta) \sim \sum_{\alpha\in\mathbb{N}_0^q} \frac{(-\tau)^{|\alpha|}}{\alpha!} D_y^\alpha \partial_\eta^\alpha a(y,\eta).$$

Proof The continuity of T_τ and the asymptotic expansion follow from Lemma 2. It remains to show that T_τ is invertible with inverse $T_{-\tau}$. It is easy to see that

$$T_\tau : \mathcal{S}(\mathbb{R}^q \times \mathbb{R}^q, \mathcal{I}(E,\tilde{E})) \to \mathcal{S}(\mathbb{R}^q \times \mathbb{R}^q, \mathcal{I}(E,\tilde{E}))$$

is invertible with inverse $T_{-\tau}$. By continuity and Lemma 3, we then get that

$$T_\tau : S_{\rho,\delta}^{\vec{\mu}+\vec{\varepsilon}}(\mathbb{R}^q \times \mathbb{R}^q; (E,\kappa), (\tilde{E},\tilde{\kappa}))_{\mathcal{I},(0)} \to S_{\rho,\delta}^{\vec{\mu}+\vec{\varepsilon}}(\mathbb{R}^q \times \mathbb{R}^q; (E,\kappa), (\tilde{E},\tilde{\kappa}))_{\mathcal{I},(0)}$$

is invertible with inverse $T_{-\tau}$, where $\vec{\varepsilon} = (\varepsilon,\varepsilon)$ for some $\varepsilon > 0$. Using again Lemma 3, this implies that T_τ is invertible with inverse $T_{-\tau}$ on the space $S_{\rho,\delta}^{\vec{\mu}}(\mathbb{R}^q \times \mathbb{R}^q; (E,\kappa), (\tilde{E},\tilde{\kappa}))_{\mathcal{I}}$ as desired. □

Theorem 1

1. *Let* $a(y, y', \eta) \in S^{(\mu_1,\mu_2,\mu_3)}_{\rho,\delta}(\mathbb{R}^q \times \mathbb{R}^q \times \mathbb{R}^q; (E, \kappa), (\tilde{E}, \tilde{\kappa}))_{\mathcal{I}}$. *Then*

$$\operatorname{Op}(a) = \operatorname{Op}_\tau(b) : \mathcal{S}(\mathbb{R}^q, E) \to \mathcal{S}(\mathbb{R}^q, \tilde{E}),$$

where $b \in S^{(\mu_1,\mu_2+\mu_3)}_{\rho,\delta}(\mathbb{R}^q \times \mathbb{R}^q, (E, \kappa), (\tilde{E}, \tilde{\kappa}))_{\mathcal{I}}$ *is given by*

$$b(y, \eta) = \frac{1}{(2\pi)^q} \iint e^{-ix\xi} a(y - \tau x, y + (1 - \tau)x, \eta + \xi)\, dx\, d\xi.$$

We have the asymptotic expansion

$$b(y, \eta) \sim \sum_{\alpha,\beta \in \mathbb{N}_0^q} \frac{1}{\alpha!\beta!} \tau^{|\alpha|}(1 - \tau)^{|\beta|} \partial_\eta^{\alpha+\beta}(-D_y)^\alpha D_{y'}^\beta a(y, y', \eta)\big|_{y'=y}.$$

2. *The class*

$$\Psi^{\vec{\mu}}_{\rho,\delta}(\mathbb{R}^q; (E, \kappa), (\tilde{E}, \kappa))_{\mathcal{I}} = \{\operatorname{Op}_\tau(a) : \mathcal{S}(\mathbb{R}^q, E) \to \mathcal{S}(\mathbb{R}^q, \tilde{E}) : \\ a \in S^{\vec{\mu}}_{\rho,\delta}(\mathbb{R}^q \times \mathbb{R}^q; (E, \kappa), (\tilde{E}, \kappa))_{\mathcal{I}}\}$$

is independent of $\tau \in \mathbb{R}$, *and the map*

$$S^{\vec{\mu}}_{\rho,\delta}(\mathbb{R}^q \times \mathbb{R}^q; (E, \kappa), (\tilde{E}, \tilde{\kappa}))_{\mathcal{I}} \ni a \mapsto \operatorname{Op}_\tau(a) \in \Psi^{\vec{\mu}}_{\rho,\delta}(\mathbb{R}^q; (E, \kappa), (\tilde{E}, \tilde{\kappa}))_{\mathcal{I}}$$

is a bijection. The operator space $\Psi^{\vec{\mu}}_{\rho,\delta}(\mathbb{R}^q; (E, \kappa), (\tilde{E}, \tilde{\kappa}))_{\mathcal{I}}$ *carries a Fréchet topology transferred from the symbol topology by any such* τ*-quantization map. This topology is independent of* τ.

Proof The symbol $b(y, \eta)$ given by the oscillatory integral in the statement of part 1 belongs to $S^{(\mu_1,\mu_2+\mu_3)}_{\rho,\delta}(\mathbb{R}^q \times \mathbb{R}^q, (E, \kappa), (\tilde{E}, \tilde{\kappa}))_{\mathcal{I}}$ and has the asserted asymptotic expansion by Lemma 2. Thus, it remains to show that

$$\operatorname{Op}(a) = \operatorname{Op}_\tau(b) : \mathcal{S}(\mathbb{R}^q, E) \to \mathcal{S}(\mathbb{R}^q, \tilde{E}).$$

Note that this is certainly the case whenever $a(y, y', \eta) \in \mathcal{S}(\mathbb{R}^q \times \mathbb{R}^q \times \mathbb{R}^q, \mathcal{I}(E, \tilde{E}))$ by an elementary argument. Now, let $a(y, y', \eta)$ be arbitrary as in the statement of the theorem. Choose $\varepsilon > 0$ and a sequence $a_j \in \mathcal{S}(\mathbb{R}^q \times \mathbb{R}^q \times \mathbb{R}^q, \mathcal{I}(E, \tilde{E}))$ such that $a_j \to a$ in $S^{(\mu_1+\varepsilon,\mu_2+\varepsilon,\mu_3+\varepsilon)}_{\rho,\delta}(\mathbb{R}^q \times \mathbb{R}^q \times \mathbb{R}^q; (E, \kappa), (\tilde{E}, \tilde{\kappa}))_{\mathcal{I}}$. This is possible by Lemma 3. Let $b_j \in \mathcal{S}(\mathbb{R}^q \times \mathbb{R}^q, \mathcal{I}(E, \tilde{E}))$ be associated to a_j by the oscillatory integral formula stated in the theorem. Then

$$b_j \to b \in S^{(\mu_1+\varepsilon,\mu_2+\mu_3+2\varepsilon)}(\mathbb{R}^q \times \mathbb{R}^q; (E, \kappa), (\tilde{E}, \tilde{\kappa}))_{\mathcal{I}}$$

by Lemma 2. For any $u \in \mathcal{S}(\mathbb{R}^q, E)$, we then obtain, with convergence in $\mathcal{S}(\mathbb{R}^q, \tilde{E})$,

$$\operatorname{Op}(a)u \longleftarrow \operatorname{Op}(a_j)u = \operatorname{Op}_\tau(b_j)u \longrightarrow \operatorname{Op}_\tau(b)u,$$

proving part 1.

The injectivity of the Kohn–Nirenberg quantization map $a \mapsto \mathrm{Op}_0(a)$ is standard, see [10]. It remains to note that, by part 1 of the theorem, $\mathrm{Op}_0(a) = \mathrm{Op}_\tau(b)$, where $b = T_\tau(a)$ and $a = T_{-\tau}b$ with the isomorphisms T_τ and $T_{-\tau}$ from Lemma 4. This proves all claims in part 2. □

As expected, composition of operators is well behaved for the twisted calculus. Typical examples for the three Banach operator ideals that appear in the statement of Theorem 2 below are $\mathcal{I}_1 = \mathcal{I}_3 = \mathcal{I}$ and $\mathcal{I}_2 = \mathcal{L}$, or $\mathcal{I}_2 = \mathcal{I}_3 = \mathcal{I}$ and $\mathcal{I}_1 = \mathcal{L}$, where $\mathcal{I}$ is some fixed Banach operator ideal, and also $\mathcal{I}_1 = \mathcal{C}_p$, $\mathcal{I}_2 = \mathcal{C}_{p'}$, and $\mathcal{I}_3 = \mathcal{C}_1$, where $1 < p, p' < \infty$ with $\frac{1}{p} + \frac{1}{p'} = 1$.

Theorem 2 *(Composition). Let $\mathcal{I}_j$, $j = 1, 2, 3$, be Banach operator ideals such that composition of operators is continuous in $\mathcal{I}_1 \times \mathcal{I}_2 \to \mathcal{I}_3$. Let*

$$A \in \Psi^{\vec{\mu}_1}_{\rho,\delta}(\mathbb{R}^q; (\tilde{E}, \tilde{\kappa}), (\hat{E}, \hat{\kappa}))_{\mathcal{I}_1},$$

$$B \in \Psi^{\vec{\mu}_2}_{\rho,\delta}(\mathbb{R}^q; (E, \kappa), (\tilde{E}, \tilde{\kappa}))_{\mathcal{I}_2}.$$

Then $A \circ B \in \Psi^{\vec{\mu}_1+\vec{\mu}_2}_{\rho,\delta}(\mathbb{R}^q; (E, \kappa), (\hat{E}, \hat{\kappa}))_{\mathcal{I}_3}$, and the map $(A, B) \mapsto A \circ B$ is bilinear and continuous in the indicated operator space topologies.

If $A = \mathrm{Op}_0(a)$ and $B = \mathrm{Op}_0(b)$, then $A \circ B = \mathrm{Op}_0(a\#b)$, where the Leibniz product:

$$(a\#b)(y, \eta) \in S^{\vec{\mu}_1+\vec{\mu}_2}_{\rho,\delta}(\mathbb{R}^q \times \mathbb{R}^q; (E, \kappa), (\hat{E}, \hat{\kappa}))_{\mathcal{I}_3}$$

has the asymptotic expansion

$$a\#b \sim \sum_{\alpha \in \mathbb{N}_0^q} \frac{1}{\alpha!} (\partial_\eta^\alpha a)(D_y^\alpha b).$$

Proof By Theorem 1 write $A = \mathrm{Op}_0(a)$ and $B = \mathrm{Op}_0(b) = \mathrm{Op}_1(T_1(b))$ with T_1 from Lemma 4. Then

$$A \circ B = \mathrm{Op}(c), \quad c(y, y', \eta) = a(y, \eta) T_1(b)(y', \eta).$$

By Theorem 1 again, $A \circ B = \mathrm{Op}_0(a\#b)$, and all assertions of the theorem follow. □

Let E_0 and E_1 be separable complex Hilbert spaces, and let

$$[\cdot, \cdot] : E_0 \times E_1 \to \mathbb{C}$$

be a sesquilinear map that satisfies the following properties:

1. (Continuity) There exists a constant $C > 0$ such that
$$|[e_0, e_1]| \leq C \|e_0\|_{E_0} \|e_1\|_{E_1}$$
for all $e_0 \in E_0$ and all $e_1 \in E_1$.
2. (Nondegeneracy) There exists a constant $c > 0$ such that
$$\sup \left\{ \left|[e_0, e_1]\right| : \|e_0\|_{E_0} \leq 1 \right\} \geq c \|e_1\|_{E_1}$$

for all $e_1 \in E_1$, and such that

$$\sup\left\{\left|[e_0, e_1]\right| : \|e_1\|_{E_1} \le 1\right\} \ge c\|e_0\|_{E_0}$$

for all $e_0 \in E_0$.

As can be readily seen, these conditions on $[\cdot,\cdot]$ are equivalent to the existence of a topological isomorphism $J : E_1 \to E_0$ such that

$$[e_0, e_1] = \langle e_0, Je_1\rangle_{E_0}$$

for all $e_0 \in E_0$ and $e_1 \in E_1$.

Now let E_0 and E_1 be Hilbert spaces equipped with such a nondegenerate continuous sesquilinear pairing $[\cdot,\cdot] : E_0 \times E_1 \to \mathbb{C}$, and let $\tilde{E}_0$ and $\tilde{E}_1$ be Hilbert spaces equipped with the nondegenerate continuous sesquilinear pairing $[\cdot,\cdot]_\sim : \tilde{E}_0 \times \tilde{E}_1 \to \mathbb{C}$. Any bounded operator $G : E_0 \to \tilde{E}_0$ then has an adjoint with respect to these pairings, the bounded operator $G^\# : \tilde{E}_1 \to E_1$ that is defined by the relation

$$[Ge_0, \tilde{e}_1]_\sim = [e_0, G^\#\tilde{e}_1] \quad \text{for } e_0 \in E_0 \text{ and } \tilde{e}_1 \in \tilde{E}_1.$$

If $[e_0, e_1] = \langle e_0, Je_1\rangle_{E_0}$ and $[\tilde{e}_0, \tilde{e}_1]_\sim = \langle \tilde{e}_0, \tilde{J}\tilde{e}_1\rangle_{\tilde{E}_0}$ with $J : E_1 \to E_0$ and $\tilde{J} : \tilde{E}_1 \to \tilde{E}_0$ as above, then $G^\# = J^{-1}G^*\tilde{J}$, where $G^* \in \mathcal{L}(\tilde{E}_0, E_0)$ is the Hilbert space adjoint to G. In particular, if $G \in \mathcal{I}(E_0, \tilde{E}_0)$, and the operator ideal $\mathcal{I}$ is closed under taking Hilbert space adjoints, then $G^\# \in \mathcal{I}(\tilde{E}_1, E_1)$.

Finally, if $\kappa_\rho : E_0 \to E_0$, $\rho > 0$, is a strongly continuous group action, then $[\kappa^\#]^{-1}$, which as indicated by the notation is defined as $[\kappa^\#_\rho]^{-1} : E_1 \to E_1$, $\rho > 0$, is a strongly continuous group action on E_1 (this follows from the strong continuity of the group of Hilbert space adjoints κ^*_ρ, see [4, Chap. I.5.b]).

Theorem 3 *(Adjoints). Let $\mathcal{I}$ be a Banach operator ideal that is closed under taking Hilbert space adjoints, and let $A \in \Psi^{\vec{\mu}}_{\rho,\delta}(\mathbb{R}^q; (E_0,\kappa), (\tilde{E}_0,\tilde{\kappa}))_{\mathcal{I}}$.*

Let $[\cdot,\cdot] : E_0 \times E_1 \to \mathbb{C}$ and $[\cdot,\cdot]_\sim : \tilde{E}_0 \times \tilde{E}_1 \to \mathbb{C}$ be continuous nondegenerate sesquilinear pairings.

Then the formal adjoint operator $A^\# : \mathcal{S}(\mathbb{R}^q, \tilde{E}_1) \to \mathcal{S}(\mathbb{R}^q, E_1)$, defined by the relation

$$\int_{\mathbb{R}^q} [Au(y), v(y)]_\sim\, dy = \int_{\mathbb{R}^q} [u(y), A^\# v(y)]dy$$

for $u \in \mathcal{S}(\mathbb{R}^q, E_0)$ and $v \in \mathcal{S}(\mathbb{R}^q, \tilde{E}_1)$, is well defined, and we have

$$A^\# \in \Psi^{\vec{\mu}}_{\rho,\delta}(\mathbb{R}^q; (\tilde{E}_1, [\tilde{\kappa}^\#]^{-1}), (E_1, [\kappa^\#]^{-1}))_{\mathcal{I}}.$$

If $A = \mathrm{Op}_0(a)$, then $A^\# = \mathrm{Op}_0(b)$, where $b(y,\eta)$ has the asymptotic expansion

$$b(y,\eta) \sim \sum_{\alpha\in\mathbb{N}_0^q} \frac{1}{\alpha!} D_y^\alpha \partial_\eta^\alpha a(y,\eta)^\#.$$

Proof Consider first the case that $E_1 = E_0$ and $\tilde{E}_1 = \tilde{E}_0$, and that both pairings are merely the inner products. In this case, $A^\#$ is the standard formal adjoint A^*, and it is evident that $A^* = \mathrm{Op}_1(c)$, where $c(y,\eta) = a(y,\eta)^* : \tilde{E}_0 \to E_0$ is the Hilbert space adjoint for each $(y,\eta) \in \mathbb{R}^q \times \mathbb{R}^q$. Note that

$$c(y,\eta) \in S^{\vec{\mu}}_{\rho,\delta}(\mathbb{R}^q \times \mathbb{R}^q; (\tilde{E}_0, [\tilde{\kappa}^*]^{-1}), (E_0, [\kappa^*]^{-1}))_{\mathcal{I}}.$$

This follows from the symbol estimates for $a(y,\eta)$, the properties of the Hilbert space adjoint, and the continuity of $* : \mathcal{I}(E_0, \tilde{E}_0) \to \mathcal{I}(\tilde{E}_0, E_0)$ in the $\mathcal{I}$-norm, where the latter is a consequence of the closed graph theorem. Consequently, all assertions follow from Theorem 1 in this case.

The general case follows from $A^\# = \mathrm{Op}_0(J^{-1}) \circ A^* \circ \mathrm{Op}_0(\tilde{J})$ with the isomorphisms $J : E_1 \to E_0$ and $\tilde{J} : \tilde{E}_1 \to \tilde{E}_0$ associated with the sesquilinear pairings discussed above. Note that

$$\tilde{J} \in S^{(0,0)}_{\rho,\delta}(\mathbb{R}^q \times \mathbb{R}^q; (\tilde{E}_1, [\tilde{\kappa}^\#]^{-1}), (\tilde{E}_0, [\tilde{\kappa}^*]^{-1})),$$

$$J^{-1} \in S^{(0,0)}_{\rho,\delta}(\mathbb{R}^q \times \mathbb{R}^q; (E_0, [\kappa^*]^{-1}), (E_1, [\kappa^\#]^{-1})),$$

and both are independent of (y,η). □

4 Boundedness and Compactness

We remind the reader about our standing assumption that $0 \le \delta < \rho \le 1$.

Theorem 4 *Every $A \in \Psi^{\vec{\mu}}_{\rho,\delta}(\mathbb{R}^q; (E,\kappa), (\tilde{E},\tilde{\kappa}))$ extends by continuity to a bounded operator*

$$A : \mathcal{W}^{\vec{s}}(\mathbb{R}^q, E) \to \mathcal{W}^{\vec{s}-\vec{\mu}}(\mathbb{R}^q, \tilde{E})$$

for all $\vec{s} \in \mathbb{R}^2$, and the induced map

$$\Psi^{\vec{\mu}}_{\rho,\delta}(\mathbb{R}^q; (E,\kappa), (\tilde{E},\tilde{\kappa})) \to \mathcal{L}(\mathcal{W}^{\vec{s}}(\mathbb{R}^q, E), \mathcal{W}^{\vec{s}-\vec{\mu}}(\mathbb{R}^q, \tilde{E}))$$

is continuous.

Proof We first observe that by Theorem 2, it suffices to consider only the case that $\vec{s} = \vec{\mu} = (0,0)$. Otherwise, we replace A by

$$B = \langle D_y \rangle^{s_1-\mu_1} \circ \langle y \rangle^{s_2-\mu_2} \circ A \circ \langle y \rangle^{-s_2} \circ \langle D_y \rangle^{-s_1} \in \Psi^{(0,0)}_{\rho,\delta}(\mathbb{R}^q; (E,\kappa), (\tilde{E},\tilde{\kappa}));$$

note that the map $A \mapsto B$ is continuous in the operator space topologies. Moreover, the proof of the theorem reduces further to the case of trivial group actions $\kappa_\varrho \equiv \mathrm{Id}_E$ and $\tilde{\kappa}_\varrho \equiv \mathrm{Id}_{\tilde{E}}$. To see this, let S and T be the operators associated with (E,κ) from (5), and let $\tilde{S}$ and $\tilde{T}$ be the ones associated with $(\tilde{E},\tilde{\kappa})$, respectively. By Proposition 2, we need to prove that

$$\tilde{S}AT : L^2(\mathbb{R}^q, \ell^2(\mathbb{N}_0, E)) \to L^2(\mathbb{R}^q, \ell^2(\mathbb{N}_0, \tilde{E}))$$

is continuous. By Theorem 2, we have

$$\tilde{S}AT \in \Psi_{\rho,\delta}^{(0,0)}(\mathbb{R}^q, \ell^2(\mathbb{N}_0, E), \ell^2(\mathbb{N}_0, \tilde{E})),$$

and the map $A \mapsto \tilde{S}AT$ is continuous. This effectively eliminates the group actions, and we may therefore assume from the beginning that the group actions on both Hilbert spaces E and $\tilde{E}$ are trivial.

The remaining case of $\vec{\mu} = \vec{s} = (0,0)$ and trivial group actions, however, is standard. In view of $\delta < \rho$, it follows from Hörmander's elegant argument of the L^2-boundedness of basic pseudodifferential operators, see [6, Theorem 18.1.11], which we proceed to outline for the sake of completeness.

Let first $A = \mathrm{Op}_0(a)$ with $a(y,\eta) \in S_{\rho,\delta}^{(-q-1,0)}(\mathbb{R}^q \times \mathbb{R}^q; E, \tilde{E})$. Then

$$Au(y) = \int_{\mathbb{R}^q} k(y,y')u(y')\,\mathrm{d}y', \ u \in C_c^\infty(\mathbb{R}^q, E),$$

where

$$k(y,y') = (2\pi)^{-q} \int_{\mathbb{R}^q} e^{i(y-y')\eta} a(y,\eta)\,\mathrm{d}\eta \in \mathcal{L}(E,\tilde{E}).$$

The function $k(y,y')$ is continuous on $\mathbb{R}^q \times \mathbb{R}^q$, and

$$\|k(y,y')\|_{\mathcal{L}(E,\tilde{E})} \leq (2\pi)^{-q} \int_{\mathbb{R}^q} \|a(y,\eta)\|_{\mathcal{L}(E,\tilde{E})}\,\mathrm{d}\eta \leq C_q \cdot |a|_{S_{\rho,\delta}^{-q-1}}$$

for a continuous seminorm $|\cdot|_{S_{\rho,\delta}^{-q-1}}$ on $S_{\rho,\delta}^{(-q-1,0)}(\mathbb{R}^q \times \mathbb{R}^q; E, \tilde{E})$. In view of

$$(y-y')^\alpha k(y,y') = \frac{(-1)^{|\alpha|}}{(2\pi)^q} \int_{\mathbb{R}^q} e^{i(y-y')\eta} D_\eta^\alpha a(y,\eta)\,\mathrm{d}\eta$$

for all $\alpha \in \mathbb{N}_0^q$, we see that there exists a continuous seminorm $|\cdot|$ on $S_{\rho,\delta}^{(-q-1,0)}(\mathbb{R}^q \times \mathbb{R}^q; E, \tilde{E})$ such that

$$\left(1 + |y-y'|\right)^{q+1} \|k(y,y')\|_{\mathcal{L}(E,\tilde{E})} \leq |a|.$$

In particular,

$$\sup_{y \in \mathbb{R}^q} \int_{\mathbb{R}^q} \|k(y,y')\|_{\mathcal{L}(E,\tilde{E})}\,\mathrm{d}y' \leq C_q|a|, \quad \sup_{y' \in \mathbb{R}^q} \int_{\mathbb{R}^q} \|k(y,y')\|_{\mathcal{L}(E,\tilde{E})}\,\mathrm{d}y \leq C_q|a|,$$

and Schur's Lemma therefore implies the L^2-continuity of A with operator norm bounded by $C_q|a|$.

We next prove the L^2-continuity of all operators of class $\Psi_{\rho,\delta}^{(\mu,0)}(\mathbb{R}^q; E, \tilde{E})$ for any $\mu < 0$. This follows inductively by considering $\mu_j = -(q+1)/2^j$ for $j \in \mathbb{N}_0$. The case $j = 0$ was just discussed. Generally, if $A \in \Psi_{\rho,\delta}^{(\mu_{j+1},0)}(\mathbb{R}^q; E, \tilde{E})$, then $A^*A \in \Psi_{\rho,\delta}^{(\mu_j,0)}(\mathbb{R}^q; E, E)$ by Theorems 2 and 3, and thus

$$\|Au\|_{L^2(\mathbb{R}^q,\tilde{E})}^2 = \langle Au, Au\rangle_{L^2(\mathbb{R}^q,\tilde{E})} = \langle A^*Au, u\rangle_{L^2(\mathbb{R}^q,E)} \leq \|A^*A\|\,\|u\|_{L^2(\mathbb{R}^q,E)}^2,$$

where we used the induction hypothesis according to which A^*A is L^2-continuous. This completes the inductive argument.

Finally, let $A = \mathrm{Op}_0(a)$ with $a(y,\eta) \in S^{(0,0)}_{\rho,\delta}(\mathbb{R}^q \times \mathbb{R}^q; E, \tilde{E})$, and choose

$$M > 2 \sup_{(y,\eta)\in\mathbb{R}^q\times\mathbb{R}^q} \|a(y,\eta)^* a(y,\eta)\|_{\mathcal{L}(E)}.$$

Then

$$c(y,\eta) = (M - a(y,\eta)^* a(y,\eta))^{1/2} \in S^{(0,0)}_{\rho,\delta}(\mathbb{R}^q \times \mathbb{R}^q; E, E).$$

To see this observe that $M - a(y,\eta)^* a(y,\eta) \in \mathcal{L}(E)$ is self-adjoint with spectrum contained in $[M/2, M]$, so we can write

$$c(y,\eta) = \frac{1}{2\pi i} \int_\Gamma \lambda^{1/2} (\lambda - M + a(y,\eta)^* a(y,\eta))^{-1} \, d\lambda$$

with a fixed contour Γ that is contained in the right half-plane $\mathfrak{R}(\lambda) > 0$, encloses $[M/2, M]$, and is independent of $(y,\eta) \in \mathbb{R}^q \times \mathbb{R}^q$. We have

$$M - a(y,\eta)^* a(y,\eta) \in S^{(0,0)}_{\rho,\delta}(\mathbb{R}^q \times \mathbb{R}^q; E, E),$$

and by the spectral theorem we have

$$\sup\left\{ \left\| (\lambda - M + a(y,\eta)^* a(y,\eta))^{-1} \right\|_{\mathcal{L}(E)} : (y,\eta) \in \mathbb{R}^q \times \mathbb{R}^q,\ \lambda \in \Gamma \right\} < \infty.$$

The Dunford integral representation for $c(y,\eta)$ and differentiation under the integral sign therefore show that $c(y,\eta) \in S^{(0,0)}_{\rho,\delta}(\mathbb{R}^q \times \mathbb{R}^q; E, E)$ as was claimed. From the symbol and operator calculus (Theorems 2 and 3), we now obtain

$$\mathrm{Op}_0(c)^* \circ \mathrm{Op}_0(c) = M - \mathrm{Op}_0(a)^* \circ \mathrm{Op}_0(a) + R$$

for some $R \in \Psi^{(\mu,0)}_{\rho,\delta}(\mathbb{R}^q; E, E)$ with $\mu < 0$. In particular,

$$\left\| \mathrm{Op}_0(a) u \right\|^2_{L^2(\mathbb{R}^q,\tilde{E})} \leq M \left\| u \right\|^2_{L^2(\mathbb{R}^q,E)} + \langle Ru, u \rangle_{L^2(\mathbb{R}^q,E)},$$

and by the L^2-boundedness of R proved earlier, we conclude that $A = \mathrm{Op}_0(a)$ is L^2-bounded. Lastly, the continuity of the map

$$\Psi^{(0,0)}_{\rho,\delta}(\mathbb{R}^q; E, \tilde{E}) \to \mathcal{L}(L^2(\mathbb{R}^q, E), L^2(\mathbb{R}^q, \tilde{E}))$$

follows readily from the closed graph theorem, which finishes the proof. □

To address compactness, we need two lemmas. Let $\mathcal{K}$ denote the operator ideal of compact operators in the sequel.

Lemma 5 *Let $k(y,y') \in C_c(\mathbb{R}^q \times \mathbb{R}^q, \mathcal{L}(E,\tilde{E}))$, and suppose that $k(y,y') \in \mathcal{K}(E,\tilde{E})$ for all $y, y' \in \mathbb{R}^q \times \mathbb{R}^q$. Then the integral operator*

$$Au(y) = \int_{\mathbb{R}^q} k(y,y') u(y') \, dy'$$

belongs to $\mathcal{K}(L^2(\mathbb{R}^q, E), L^2(\mathbb{R}^q, \tilde{E}))$.

Proof We recall the standard proof (see also [11, Proposition 2.1]).

Choose $R > 0$ such that $\operatorname{supp}(k) \subset (-R, R)^q \times (-R, R)^q$, and consider the operator

$$A\big|_{[-R,R]^q} : L^2([-R, R]^q, E) \to L^2([-R, R]^q, \tilde{E}).$$

Clearly, $A = \operatorname{Ext} \circ A|_{[-R,R]^q} \circ r$, where $r : L^2(\mathbb{R}^q, E) \to L^2([-R, R]^q, E)$ is the (continuous) restriction operator, and $\operatorname{Ext} : L^2([-R, R]^q, \tilde{E}) \to L^2(\mathbb{R}^q, \tilde{E})$ is the (continuous) trivial extension operator. Hence, the claim reduces to proving that any integral operator B with continuous kernel $k_B(y, y') \in C([-R, R]^q \times [-R, R]^q, \mathcal{K}(E, \tilde{E}))$ is compact in $\mathcal{L}(L^2([-R, R]^q, E), L^2([-R, R]^q, \tilde{E}))$.

In view of

$$\|B\|_{\mathcal{L}(L^2,L^2)} \leq (2R)^q \sup\{\|k_B(y, y')\| : y, y' \in [-R, R]^q\},$$

the map

$$C([-R, R]^q \times [-R, R]^q, \mathcal{K}(E, \tilde{E})) \ni k_B \mapsto B \in \mathcal{L}(L^2, L^2)$$

is continuous. Using compactness, uniform continuity of the kernel, and partitions of unity shows that every compact kernel k_B is the uniform limit of kernels in the algebraic tensor product $C([-R, R]^q) \otimes \mathcal{K}(E, \tilde{E}) \otimes C([-R, R]^q)$. If $k_B(y, y') = \phi(y)K\psi(y')$ is a pure tensor, then

$$B = M_\phi \circ K \circ Q_\psi : L^2([-R, R]^q, E) \to L^2([-R, R]^q, \tilde{E}), \tag{9}$$

where

$$Q_\psi : L^2([-R, R]^q, E) \to E,\ Q_\psi u = \int \psi(y')u(y')\,dy',$$

$$M_\phi : \tilde{E} \to L^2([-R, R]^q, \tilde{E}),\ [M_\phi \tilde{e}](y) = \phi(y)\tilde{e},$$

are continuous, and $K : E \to \tilde{E}$ is compact. Thus, (9) is compact, which shows that $k_B \mapsto B$ maps

$$C([-R, R]^q) \otimes \mathcal{K}(E, \tilde{E}) \otimes C([-R, R]^q) \to \mathcal{K}(L^2, L^2).$$

Consequently, by density and continuity, all integral operators with continuous compact kernels are compact. □

Lemma 6 *Every* $a(y, \eta) \in S^{(0,0)}_{\rho,\delta}(\mathbb{R}^q \times \mathbb{R}^q; E, \tilde{E})_{\mathcal{K},(0)}$ *induces a compact operator*

$$A = \operatorname{Op}_0(a) : L^2(\mathbb{R}^q, E) \to L^2(\mathbb{R}^q, \tilde{E}).$$

Proof (See [11, Proposition 2.1]). Because

$$S^{(0,0)}_{\rho,\delta}(\mathbb{R}^q \times \mathbb{R}^q; E, \tilde{E}) \ni a \mapsto \operatorname{Op}_0(a) \in \mathcal{L}(L^2(\mathbb{R}^q, E), L^2(\mathbb{R}^q, \tilde{E}))$$

is continuous it suffices to show that $\operatorname{Op}_0(a) \in \mathcal{K}(L^2(\mathbb{R}^q, E), L^2(\mathbb{R}^q, \tilde{E}))$ for any $a \in C_c^\infty(\mathbb{R}^q \times \mathbb{R}^q, \mathcal{K}(E, \tilde{E}))$. Because of the ideal property of compact operators and

Plancherel this is equivalent to the compactness of the operator

$$B = \mathrm{Op}_0(a) \circ \mathcal{F}^{-1} : L^2(\mathbb{R}^q, E) \to L^2(\mathbb{R}^q, \tilde{E}).$$

We have

$$Bu(y) = \int_{\mathbb{R}^q} k_B(y, y')u(y')\,\mathrm{d}y',$$

with $k_B(y, y') = (2\pi)^{-q} e^{iyy'} a(y, y') \in C_c^\infty(\mathbb{R}^q \times \mathbb{R}^q, \mathcal{K}(E, \tilde{E}))$. Thus, B is compact by Lemma 5.

Theorem 5 *Every $A \in \Psi^{\vec{\mu}}_{\rho,\delta}(\mathbb{R}^q; (E, \kappa), (\tilde{E}, \tilde{\kappa}))_{\mathcal{K}}$ induces a compact operator*

$$A : \mathcal{W}^{\vec{s}}(\mathbb{R}^q, E) \to \mathcal{W}^{\vec{s}'}(\mathbb{R}^q, \tilde{E})$$

for all $\vec{s}, \vec{s}' \in \mathbb{R}^2$ such that $\vec{\mu} < \vec{s} - \vec{s}'$, where this inequality is to hold componentwise.

Proof Arguing as in the proof of Theorem 4 reduces the claim to the case that $\vec{s} = \vec{s}' = (0, 0)$, $\vec{\mu} < (0, 0)$, and trivial group actions $\kappa_\rho \equiv \mathrm{Id}_E$ and $\tilde{\kappa}_\rho \equiv \mathrm{Id}_{\tilde{E}}$ (eliminating the group actions is based on Proposition 2). But this case is treated in Lemma 6, taking into account Lemma 3. □

5 Trace Class Operators

We remind the reader that $\mathcal{C}_1$ denotes the Banach operator ideal of Schatten–von Neumann operators with ℓ^1-summable approximation numbers (i.e., the trace class operators in case of operators acting on a single Hilbert space). Moreover, we always assume that $0 \le \delta < \rho \le 1$.

The proof of trace class properties presented here in Lemmas 7–9 and Proposition 3 follows Widom [25].

Lemma 7 *Let H and $\tilde{H}$ be separable complex Hilbert spaces, and let $\sigma : \mathbb{R}^q \to \mathcal{L}(H, \tilde{H})$ be a weakly measurable operator function. Suppose that $\sigma(\eta) \in \mathcal{C}_1(H, \tilde{H})$ for almost all $\eta \in \mathbb{R}^q$, and suppose further that there exists a function $g \in L^1(\mathbb{R}^q)$ such that $\|\sigma(\eta)\|_{\mathcal{C}_1} \le g(\eta)$ for almost all $\eta \in \mathbb{R}^q$. Define the operator $A_\sigma : H \to \tilde{H}$ via*

$$\langle A_\sigma h, \tilde{h}\rangle_{\tilde{H}} = \int_{\mathbb{R}^q} \langle \sigma(\eta)h, \tilde{h}\rangle_{\tilde{H}}\, d\eta$$

for $h \in H$ and $\tilde{h} \in \tilde{H}$. Then $A_\sigma \in \mathcal{C}_1(H, \tilde{H})$ with $\|A\|_{\mathcal{C}_1} \le \|g\|_{L^1}$.

Proof Because $\|\sigma(\eta)\|_{\mathcal{L}(H,\tilde{H})} \le \|\sigma(\eta)\|_{\mathcal{C}_1(H,\tilde{H})}$, we first obtain that

$$\left|\int_{\mathbb{R}^q} \langle \sigma(\eta)h, \tilde{h}\rangle_{\tilde{H}}\, \mathrm{d}\eta\right| \le \|g\|_{L^1} \|h\|_H \|\tilde{h}\|_{\tilde{H}},$$

which shows that $A_\sigma \in \mathcal{L}(H, \tilde{H})$ is well defined.

Now, let $\{h_j\} \subset H$ and $\{\tilde{h}_j\} \subset \tilde{H}$ be finite sets of orthonormal vectors. Then

$$\sum \left|\langle A_\sigma h_j, \tilde{h}_j\rangle_{\tilde{H}}\right| \leq \int_{\mathbb{R}^q} \underbrace{\sum \left|\langle \sigma(\eta) h_j, \tilde{h}_j\rangle_{\tilde{H}}\right|}_{\leq \|\sigma(\eta)\|_{\mathcal{C}_1}} \, \mathrm{d}\eta \leq \|g\|_{L^1}.$$

Passing to the supremum over all such sets of orthonormal vectors we obtain that $A_\sigma \in \mathcal{C}_1(H, \tilde{H})$ with $\|A\|_{\mathcal{C}_1} \leq \|g\|_{L^1}$ as asserted. □

Lemma 8 *Let $\varphi, \psi \in L^2(\mathbb{R}^q)$, and let both $a, \mathcal{F}a \in L^1(\mathbb{R}^q, \mathcal{C}_1(E, \tilde{E}))$. Consider the map $G : L^2(\mathbb{R}^q, E) \to L^2(\mathbb{R}^q, \tilde{E})$, $Gu(y) = \varphi(y)a(y)\int_{\mathbb{R}^q} \psi(y')u(y')dy'$. Then $G \in \mathcal{C}_1(L^2(\mathbb{R}^q, E), L^2(\mathbb{R}^q, \tilde{E}))$, and*

$$\left\|G\right\|_{\mathcal{C}_1(L^2,L^2)} \leq (2\pi)^{-q}\|\varphi\|_{L^2}\|\psi\|_{L^2}\left\|\mathcal{F}a\right\|_{L^1(\mathbb{R}^q,\mathcal{C}_1(E,\tilde{E}))}.$$

Proof By assumption on a, both $a(y)$ and $\hat{a}(\eta) = \mathcal{F}a(\eta)$ are continuous and bounded functions on $\mathbb{R}^q$, and we have $a(y) = (2\pi)^{-q}\int_{\mathbb{R}^q} e^{iy\eta}\hat{a}(\eta)\,d\eta$. Write

$$Gu(y) = (2\pi)^{-q}\int_{\mathbb{R}^q} e^{iy\eta}\varphi(y)\hat{a}(\eta)\,d\eta\int_{\mathbb{R}^q}\psi(y')u(y')\mathrm{d}y'.$$

For $\eta \in \mathbb{R}^q$, consider the operator

$$\sigma(\eta) : E \to L^2(\mathbb{R}^q, \tilde{E}),$$
$$\sigma(\eta)e = [y \mapsto (2\pi)^{-q}e^{iy\eta}\varphi(y)\hat{a}(\eta)e].$$

By Lebesgue's dominated convergence theorem, the function $\eta \mapsto \sigma(\eta)e$ is continuous with values in $L^2(\mathbb{R}^q, \tilde{E})$ for each $e \in E$. In particular, the operator function $\sigma : \mathbb{R}^q \to \mathcal{L}(E, L^2(\mathbb{R}^q, \tilde{E}))$ is weakly measurable. Note, furthermore, that $\sigma(\eta)$ is itself the composition of the operators $\tilde{E} \ni \tilde{e} \mapsto (2\pi)^{-q}e^{iy\eta}\varphi(y)\tilde{e} \in L^2(\mathbb{R}^q, \tilde{E})$ and $\hat{a}(\eta) : E \to \tilde{E}$. The latter belongs to $\mathcal{C}_1(E, \tilde{E})$ by assumption, and the former is bounded with operator norm at most $(2\pi)^{-q}\|\varphi\|_{L^2}$. Thus, $\sigma(\eta) \in \mathcal{C}_1(E, L^2(\mathbb{R}^q, \tilde{E}))$ for all $\eta \in \mathbb{R}^q$ with $\|\sigma(\eta)\|_{\mathcal{C}_1} \leq (2\pi)^{-q}\|\hat{a}(\eta)\|_{\mathcal{C}_1}\|\varphi\|_{L^2}$. By Lemma 7, the operator

$$E \ni e \mapsto [y \mapsto (2\pi)^{-q}\int_{\mathbb{R}^q} e^{iy\eta}\varphi(y)\hat{a}(\eta)e\,\mathrm{d}\eta] \in L^2(\mathbb{R}^q, \tilde{E})$$

belongs to $\mathcal{C}_1(E, L^2(\mathbb{R}^q, \tilde{E}))$ with $\mathcal{C}_1$-norm at most $(2\pi)^{-q}\|\varphi\|_{L^2}\|\hat{a}\|_{L^1(\mathbb{R}^q,\mathcal{C}_1(E,\tilde{E}))}$. Finally, the operator G is the composition of this map and the operator $L^2(\mathbb{R}^q, E) \ni u \mapsto \int_{\mathbb{R}^q}\psi(y')u(y')\,\mathrm{d}y' \in E$, which is continuous with operator norm at most $\|\psi\|_{L^2}$. The lemma is proved. □

Lemma 9 *Let $a(y, \eta) \in \mathcal{S}(\mathbb{R}^q \times \mathbb{R}^q, \mathcal{C}_1(E, \tilde{E}))$, $p = \lceil (q+1)/4 \rceil$, and $\mu' < -q$. Consider the operator*

$$Au(y) = (2\pi)^{-q}\int_{\mathbb{R}^q} e^{iy\eta}\langle y\rangle^{-2p}a(y, \eta)\int_{\mathbb{R}^q} e^{-iy'\eta}\langle y'\rangle^{-2p}u(y')\,dy'\,d\eta$$

for $u \in \mathcal{S}(\mathbb{R}^q, E)$. Then $A \in \mathcal{C}_1(L^2(\mathbb{R}^q, E), L^2(\mathbb{R}^q, \tilde{E}))$ with

$$\|A\|_{\mathcal{C}_1(L^2,L^2)} \leq (2\pi)^{-2q}\Big(\int_{\mathbb{R}^q}\langle y'\rangle^{-4p}\,dy'\Big)\left\|\mathcal{F}_{y\to\eta'}a(\eta', \eta)\right\|_{L^1(\mathbb{R}^q_{\eta'}\times\mathbb{R}^q_\eta,\mathcal{C}_1(E,\tilde{E}))}$$

$$\leq C(\mu', q) \sup_{(y,\eta)\in\mathbb{R}^{2q}} \langle y\rangle^{4p}\langle\eta\rangle^{-\mu'} \left\|(1-\Delta_y)^{2p} a(y,\eta)\right\|_{\mathcal{C}_1(E,\tilde{E})},$$

where

$$C(\mu', q) = (2\pi)^{-2q}\Big(\int_{\mathbb{R}^q} \langle y'\rangle^{-4p}\, dy'\Big)^3\Big(\int_{\mathbb{R}^q} \langle\eta\rangle^{\mu'}\, d\eta\Big).$$

Proof Fix $\eta \in \mathbb{R}^q$ and consider the map

$$\sigma(\eta) : L^2(\mathbb{R}^q, E) \to L^2(\mathbb{R}^q, \tilde{E}),$$
$$\sigma(\eta)u(y) = (2\pi)^{-q} e^{iy\eta}\langle y\rangle^{-2p} a(y,\eta) \int_{\mathbb{R}^q} e^{-iy'\eta}\langle y'\rangle^{-2p} u(y')dy'.$$

Lemma 8 is applicable to $\sigma(\eta)$. We obtain that $\sigma(\eta) \in \mathcal{C}_1(L^2(\mathbb{R}^q, E), L^2(\mathbb{R}^q, \tilde{E}))$ with

$$\|\sigma(\eta)\|_{\mathcal{C}_1(L^2,L^2)} \leq (2\pi)^{-2q}\Big(\int_{\mathbb{R}^q} \langle y'\rangle^{-4p}\, dy'\Big) \left\|\mathcal{F}_{y\to\eta'} a(\eta', \eta)\right\|_{L^1(\mathbb{R}^q_{\eta'}, \mathcal{C}_1(E,\tilde{E}))}.$$

An application of Lebesgue's dominated convergence theorem shows that the function $\eta \mapsto \sigma(\eta)u$ depends continuously on $\eta \in \mathbb{R}^q$ taking values in $L^2(\mathbb{R}^q, \tilde{E})$ for each $u \in L^2(\mathbb{R}^q, E)$. In particular, $\sigma : \mathbb{R}^q \to \mathcal{L}(L^2(\mathbb{R}^q, E), L^2(\mathbb{R}^q, \tilde{E}))$ is a weakly measurable operator function. By Lemma 7, the operator $A = A_\sigma$ defined by σ belongs to $\mathcal{C}_1(L^2(\mathbb{R}^q, E), L^2(\mathbb{R}^q, \tilde{E}))$ with

$$\|A\|_{\mathcal{C}_1(L^2,L^2)} \leq (2\pi)^{-2q}\Big(\int_{\mathbb{R}^q} \langle y'\rangle^{-4p}\, \mathrm{d}y'\Big) \left\|\mathcal{F}_{y\to\eta'} a(\eta', \eta)\right\|_{L^1(\mathbb{R}^q_{\eta'}\times\mathbb{R}^q_\eta, \mathcal{C}_1(E,\tilde{E}))},$$

proving the claim. □

Proposition 3 *Let $p = \lceil (q+1)/4 \rceil$, and let $\mu_1 < -q - 4p\delta$ and $\mu_2 < -8p$. For every $a(y,\eta) \in S^{(\mu_1,\mu_2)}_{\rho,\delta}(\mathbb{R}^q \times \mathbb{R}^q; E, \tilde{E})_{\mathcal{C}_1}$, the operator*

$$Au(y) = \mathrm{Op}_0(a)u(y) = (2\pi)^{-q}\int_{\mathbb{R}^q} e^{iy\eta} a(y,\eta)\hat{u}(\eta)d\eta, \quad u \in \mathcal{S}(\mathbb{R}^q, E),$$

belongs to $\mathcal{C}_1(L^2(\mathbb{R}^q, E), L^2(\mathbb{R}^q, \tilde{E}))$. The map

$$S^{(\mu_1,\mu_2)}_{\rho,\delta}(\mathbb{R}^q \times \mathbb{R}^q; E, \tilde{E})_{\mathcal{C}_1} \ni a \mapsto \mathrm{Op}_0(a) \in \mathcal{C}_1(L^2(\mathbb{R}^q, E), L^2(\mathbb{R}^q, \tilde{E}))$$

is continuous. Moreover, if $E = \tilde{E}$, then

$$\mathrm{Tr}_{L^2(\mathbb{R}^q,E)}[\mathrm{Op}_0(a)] = \frac{1}{(2\pi)^q}\iint_{\mathbb{R}^{2q}} \mathrm{Tr}_E(a(y,\eta))dy d\eta.$$

Proof Consider first $a(y,\eta) \in \mathcal{S}(\mathbb{R}^q \times \mathbb{R}^q, \mathcal{C}_1(E, \tilde{E}))$. We have

$$Au(y) = (2\pi)^{-q}\int_{\mathbb{R}^q} e^{i(y\eta} a(y,\eta)\int_{\mathbb{R}^q} e^{-iy'\eta} u(y')\mathrm{d}y'\mathrm{d}\eta$$

$$= (2\pi)^{-q} \int_{\mathbb{R}^q} e^{iy\eta} a(y,\eta)(1-\Delta_\eta)^p \int_{\mathbb{R}^q} e^{-iy'\eta} \langle y' \rangle^{-2p} u(y') dy' d\eta$$

$$= (2\pi)^{-q} \int_{\mathbb{R}^q} e^{iy\eta} \langle y \rangle^{-2p} \underbrace{[\langle y \rangle^{2p} (\mathcal{L}^p a)(y,\eta)]}_{=b(y,\eta)} \int_{\mathbb{R}^q} e^{-iy'\eta} \langle y' \rangle^{-2p} u(y') dy' d\eta,$$

where

$$\mathcal{L} = e^{-iy\eta}(1-\Delta_\eta)e^{iy\eta} = 1 + \sum_{j=1}^q (y_j + D_{\eta_j})^2.$$

By Lemma 9, $A \in \mathcal{C}_1(L^2(\mathbb{R}^q, E), L^2(\mathbb{R}^q, \tilde{E}))$, and for any $\mu' < -q$ the $\mathcal{C}_1$-norm of A can be estimated by a continuous seminorm of $(1-\Delta_y)^{2p} b(y,\eta)$ in $S_{\rho,\delta}^{(\mu',-4p)}(\mathbb{R}^q \times \mathbb{R}^q; E, \tilde{E})_{\mathcal{C}_1}$. Because the map $a(y,\eta) \mapsto (1-\Delta_y)^{2p} b(y,\eta)$ is continuous in

$$S_{\rho,\delta}^{\mu'-4p\delta,-8p}(\mathbb{R}^q \times \mathbb{R}^q; E, \tilde{E})_{\mathcal{C}_1,(0)} \to S_{\rho,\delta}^{\mu',-4p}(\mathbb{R}^q \times \mathbb{R}^q; E, \tilde{E})_{\mathcal{C}_1,(0)},$$

we see that the $\mathcal{C}_1$-norm of A can be estimated by a continuous seminorm of $a(y,\eta)$ in $S_{\rho,\delta}^{(\mu'-4p\delta,-8p)}(\mathbb{R}^q \times \mathbb{R}^q; E, \tilde{E})_{\mathcal{C}_1}$. Therefore, the map

$$S_{\rho,\delta}^{(\mu'-4p\delta,-8p)}(\mathbb{R}^q \times \mathbb{R}^q; E, \tilde{E})_{\mathcal{C}_1,(0)} \ni a \mapsto \mathrm{Op}_0(a) \in \mathcal{C}_1(L^2(\mathbb{R}^q, E), L^2(\mathbb{R}^q, \tilde{E}))$$

is well defined and continuous. Because $\mu' < -q$ is arbitrary, we obtain the first part of the proposition by applying Lemma 3.

It remains to show the asserted trace formula in the case that $E = \tilde{E}$. Since both functionals

$$S_{\rho,\delta}^{(\mu_1,\mu_2)}(\mathbb{R}^q \times \mathbb{R}^q; E, E)_{\mathcal{C}_1,(0)} \ni a \mapsto [\mathrm{Tr}_{L^2(\mathbb{R}^q,E)} \circ \mathrm{Op}_0](a) \in \mathbb{C}$$

$$S_{\rho,\delta}^{(\mu_1,\mu_2)}(\mathbb{R}^q \times \mathbb{R}^q; E, E)_{\mathcal{C}_1,(0)} \ni a \mapsto \frac{1}{(2\pi)^q} \iint_{\mathbb{R}^{2q}} \mathrm{Tr}_E(a(y,\eta)) dy d\eta \in \mathbb{C}$$

are continuous, it suffices to show that they are equal on the dense subset

$$[C_c^\infty(\mathbb{R}_y^q) \otimes \mathcal{F}_{y' \to \eta}(C_c^\infty(\mathbb{R}_{y'}^q))] \otimes \mathcal{C}_1(E),$$

but this is evident in view of the multiplicativity of the trace functional for tensor products of trace class operators. By Lemma 3, we can drop the subscript (0) from the symbol spaces. □

Theorem 6 *Let $p = \lceil (q+1)/4 \rceil$. Every $A \in \Psi_{\rho,\delta}^{\vec{\mu}}(\mathbb{R}^q; (E,\kappa), (\tilde{E},\tilde{\kappa}))_{\mathcal{C}_1}$ induces a $\mathcal{C}_1$-operator*

$$A : \mathcal{W}^{\vec{s}}(\mathbb{R}^q, E) \to \mathcal{W}^{\vec{s}'}(\mathbb{R}^q, \tilde{E})$$

for all $\vec{s}, \vec{s}' \in \mathbb{R}^2$ such that $\vec{\mu} + (q + 4p\delta, 8p) < \vec{s} - \vec{s}'$, where this inequality is to hold componentwise.

Proof Arguing as in the proof of Theorem 4 reduces the claim to the case that $\vec{s} = \vec{s}' = (0,0)$ and trivial group actions $\kappa_\rho \equiv \mathrm{Id}_E$ and $\tilde{\kappa}_\rho \equiv \mathrm{Id}_{\tilde{E}}$ (note that eliminating the group actions is based on Proposition 2). But this case is discussed in Proposition 3. □

Remark 1 Theorem 6 is certainly not optimal, especially regarding the growth condition on the variable $y \in \mathbb{R}^q$ as $|y| \to \infty$. However, in the case that $\delta = 0$, the condition on the order μ_1 is the best possible as the scalar case shows. In applications concerning compact manifolds with edge singularities, the abstract edge calculus is used to describe the local structure of operators near the edge. In this case, one may generally assume without loss of generality that the symbol $a(y, \eta)$ of an operator $A = \mathrm{Op}_0(a)$ has compact support in y, so the growth condition of Theorem 6 with respect to $y \in \mathbb{R}^q$ is satisfied in such applications.

6 Appendix: Banach Operator Ideals in Hilbert Spaces

We refer to [12, 13, 23] for background on the theory of operator ideals. We remind the reader that all Hilbert spaces are assumed to be complex and separable.

Definition 6 Suppose that for every pair of Hilbert spaces E and $\tilde{E}$ one is given a subset $\mathcal{I}(E, \tilde{E}) \subset \mathcal{L}(E, \tilde{E})$ with the following properties:

1. All finite-rank operators $F : E \to \tilde{E}$ belong to $\mathcal{I}(E, \tilde{E})$.
2. Whenever $A, B \in \mathcal{I}(E, \tilde{E})$, then $A + B \in \mathcal{I}(E, \tilde{E})$.
3. We have $GAH \in \mathcal{I}(E_0, E_1)$ whenever $G \in \mathcal{L}(\tilde{E}, E_1)$, $A \in \mathcal{I}(E, \tilde{E})$, and $H \in \mathcal{L}(E_0, E)$.

Then the collection

$$\mathcal{I} = \bigcup_{E,\tilde{E}} \mathcal{I}(E, \tilde{E})$$

is called an operator ideal (in the category of Hilbert spaces).

An operator ideal $\mathcal{I}$ is called normed if there is a function $\|\cdot\|_{\mathcal{I}} : \mathcal{I} \to \mathbb{R}$ that restricts to a norm $\|\cdot\|_{\mathcal{I}(E,\tilde{E})}$ on $\mathcal{I}(E, \tilde{E})$ for all Hilbert spaces E and $\tilde{E}$ such that the following holds:

1. $\|e' \otimes \tilde{e}\|_{\mathcal{I}(E,\tilde{E})} = \|e'\|_{e'} \|\tilde{e}\|_{\tilde{E}}$ for all $e' \in E'$ and all $\tilde{e} \in \tilde{E}$.
2. $\|GAH\|_{\mathcal{I}(E_0,E_1)} \leq \|G\|_{\mathcal{L}(\tilde{E},E_1)} \|A\|_{\mathcal{I}(E,\tilde{E})} \|H\|_{\mathcal{L}(E_0,E)}$ for all $G \in \mathcal{L}(\tilde{E}, E_1)$, $A \in \mathcal{I}(E, \tilde{E})$, and $H \in \mathcal{L}(E_0, E)$.

A normed operator ideal $\mathcal{I}$ is a Banach operator ideal if $\mathcal{I}(E, \tilde{E})$ is complete with the norm $\|\cdot\|_{\mathcal{I}(E,\tilde{E})}$ for all Hilbert spaces E and $\tilde{E}$.

We remark that it is a consequence of the axioms that

$$\|A\|_{\mathcal{L}(E,\tilde{E})} \leq \|A\|_{\mathcal{I}(E,\tilde{E})}$$

for all $A \in \mathcal{I}(E, \tilde{E})$, see [12, Proposition 6.1.4]. In particular, the embedding

$$\big(\mathcal{I}(E, \tilde{E}), \|\cdot\|_{\mathcal{I}}\big) \hookrightarrow \big(\mathcal{L}(E, \tilde{E}), \|\cdot\|_{\mathcal{L}(E,\tilde{E})}\big)$$

is continuous.

There are two trivial Banach operator ideals, $\mathcal{I} = \mathcal{L}$ and $\mathcal{I} = \mathcal{K}$: The first consists of $\mathcal{I}(E, \tilde{E}) = \mathcal{L}(E, \tilde{E})$ (all bounded operators), the second one has $\mathcal{I}(E, \tilde{E}) = \mathcal{K}(E, \tilde{E})$ (all compact operators) for Hilbert spaces E and $\tilde{E}$. Both $\mathcal{L}$ and $\mathcal{K}$ are Banach operator ideals with respect to the usual operator norm.

In applications of spectral and index theory, the Schatten–von Neumann classes $\mathcal{C}_p$, $1 \le p < \infty$, occur frequently. These are Banach operator ideals with respect to the Schatten p-norms. For $p = 1$ and $E = \tilde{E}$, the class $\mathcal{C}_1(E, E)$ coincides with the trace class operators, and for $p = 2$ the class $\mathcal{C}_2$ specializes to the Hilbert–Schmidt operators.

Let $A \in \mathcal{L}(E, \tilde{E})$. For $r \in \mathbb{N}$, the rth approximation number of A is defined as

$$\alpha_r(A) = \inf\{\|A - F\|_{\mathcal{L}(E,\tilde{E})} : F \in \mathcal{L}(E, \tilde{E}),\ \dim R(F) < r\}.$$

Definition 7 For $1 \le p < \infty$ and Hilbert spaces E and $\tilde{E}$, define the Schatten–von Neumann class $\mathcal{C}_p(E, \tilde{E})$ as the space of all operators $A \in \mathcal{L}(E, \tilde{E})$ such that

$$\|A\|_{\mathcal{C}_p} = \Big(\sum_{r=1}^{\infty} \alpha_r(A)^p\Big)^{1/p} < \infty.$$

The following theorem summarizes some of the properties of this class.

Theorem 7

1. *An operator $A : E \to \tilde{E}$ belongs to the class $\mathcal{C}_p(E, \tilde{E})$, $1 \le p < \infty$, if and only if*

$$\sup\Big\{\Big(\sum |\langle Ae_k, \tilde{e}_k\rangle|^p\Big)^{1/p} : (e_k) \subset E,\ (\tilde{e}_k) \subset \tilde{E} \text{ are finite orthonormal}\Big\} < \infty.$$

In this case,

$$\|A\|_{\mathcal{C}_p} = \sup\Big\{\Big(\sum |\langle Ae_k, \tilde{e}_k\rangle|^p\Big)^{1/p} : (e_k) \subset E,\ (\tilde{e}_k) \subset \tilde{E} \text{ are finite orthonormal}\Big\}.$$

2. *$\mathcal{C}_p$ is a Banach operator ideal with norm $\|\cdot\|_{\mathcal{C}_p}$.*
3. *Composition of operators induces a continuous map*

$$\mathcal{C}_p(E_1, E_2) \times \mathcal{C}_{p'}(E_0, E_1) \to \mathcal{C}_1(E_0, E_2),$$

where $1 < p, p' < \infty$ with $\frac{1}{p} + \frac{1}{p'} = 1$. More precisely, we have

$$\|AB\|_{\mathcal{C}_1} \le \|A\|_{\mathcal{C}_p}\|B\|_{\mathcal{C}_{p'}}.$$

Theorem 7 implies that the classes $\mathcal{C}_p$ are invariant under taking Hilbert space adjoints. More precisely, we have $\|A^*\|_{\mathcal{C}_p(E,\tilde{E})} = \|A\|_{\mathcal{C}_p(\tilde{E},E)}$.

Acknowledgement This material is based in part upon work supported by the National Science Foundation under Grant No. DMS-0901202.

References

1. H. Amann, Operator-valued Fourier multipliers, vector-valued Besov spaces, and applications. Math. Nachr. **186**, 5–56 (1997)
2. C. Dorschfeldt, B.-W. Schulze, Pseudo-differential operators with operator-valued symbols in the Mellin-Edge-Approach. Ann. Global Anal. Geom. **12**, 135–171 (1994)
3. C. Dorschfeldt, U. Grieme, B.-W. Schulze, Pseudo-differential Calculus in the Fourier-Edge-Approach on Non-compact Manifolds. In: Spectral Theory, Microlocal Analysis, Singular Manifolds, Math. Top. vol. 14 (Akademie Verlag, Berlin, 1997), pp. 249–299
4. K.-J. Engel, R. Nagel, *One-parameter Semigroups for Linear Evolution Equations*. Graduate Texts in Mathematics, vol. 194 (Springer, New York, 2000)
5. T. Hirschmann, Functional analysis in cone and edge Sobolev spaces. Ann. Global Anal. Geom. **8**, 167–192 (1990)
6. L. Hörmander, in *The Analysis of Linear Partial Differential Operators III*, Grundlehren der mathematischen Wissenschaften, vol. 274, Corrected 2nd printing (Springer, Berlin, 1994)
7. T. Krainer, G. Mendoza, The kernel bundle of a holomorphic Fredholm family. Comm. Partial Differential Equations **38**, 2107–2125 (2013)
8. T. Krainer, G. Mendoza, Elliptic systems of variable order. Rev. Mat. Iberoam. **31**, 127–160 (2015)
9. T. Krainer, G. Mendoza, Boundary value problems for first order elliptic wedge operators. Preprint arXiv:1307.2398 on arXiv.org. (Accepted for publication in the American Journal of Mathematics)
10. H. Kumano-go, *Pseudodifferential Operators* (MIT Press, Cambridge, 1981)
11. G. Luke, Pseudodifferential operators on Hilbert bundles, J. Differential Equations **12**, 566–589 (1972)
12. A. Pietsch, *Operator Ideals*. North-Holland Mathematical Library vol. 20, (North-Holland Publishing Co., Amsterdam-New York, 1980)
13. A. Pietsch, *Eigenvalues and s-numbers*. Cambridge Studies in Advanced Mathematics vol. 13, (Cambridge University Press, Cambridge, 1987)
14. B. Scharf, H.-J. Schmeißer, W. Sickel, Traces of vector-valued Sobolev spaces. Math. Nachr. **285**, 1082–1106 (2012)
15. H.-J. Schmeißer, Vector-valued Sobolev and Besov spaces. In: Seminar Analysis of the Karl-Weierstraß-Institute of Mathematics 1985/86, Teubner-Texte Math. vol. 96 (Teubner, Leipzig, 1987), pp. 4–44
16. E. Schrohe, *Fréchet Algebras of Pseudodifferential Operators and Boundary Value Problems* (1998, unpublished manuscript)
17. B.-W. Schulze, Pseudo-differential operators on manifolds with edges. In: Symposium Partial Differential Equations, Holzhau 1988, Teubner-Texte Math. vol. 112 (Teubner, Leipzig, 1989), pp. 259–287
18. B.-W. Schulze, *Pseudo-differential Operators on Manifolds with Singularities* (North Holland, Amsterdam, 1991)
19. B.-W. Schulze, *Boundary Value Problems and Singular Pseudo-differential Operators*. Pure and Applied Mathematics (John Wiley & Sons, Ltd., Chichester, 1998)
20. J. Seiler, *Pseudodifferential Calculus on Manifolds with Non-Compact Edges*. Doctoral Dissertation, University of Potsdam, Germany, 1997
21. J. Seiler, Continuity of edge and corner pseudodifferential operators. Math. Nachr. **205**, 163–182 (1999)
22. M. Shubin, *Pseudodifferential Operators and Spectral Theory*, 2nd edn. (Springer, Berlin, 2001)
23. B. Simon, *Trace Ideals and Their Applications*. Mathematical Surveys and Monographs, vol. 120, 2nd edn. (American Mathematical Society, Providence, 2005)
24. H. Triebel, *Fractals and Spectra. Related to Fourier Analysis and Function Spaces*. Monographs in Mathematics, vol. 91 (Birkhäuser, Basel, 1997)
25. H. Widom, On a class of integral operators on a half-space with discontinuous symbol. J. Funct. Anal. **88**, 166–193 (1990)

Boundary Value Problems for Elliptic Wedge Operators: The First-Order Case

Thomas Krainer and Gerardo A. Mendoza

Abstract This chapter is a description of some of the results obtained by the authors in connection with the problem in the title. These, discussed following a summary of background material concerning wedge differential operators, consist of the notion of trace bundle, an extension of the Douglis–Nirenberg calculus to handle spaces of anisotropic varying regularity and associated pseudodifferential operators, and boundary value problems proper, the latter in the first-order case. The concepts concerning the main results are illustrated with simple examples.

1 Introduction

The present chapter is an account of results published in a series of papers [11–13] in connection with boundary value problems for elliptic wedge operators on a manifold with fibered boundary. Briefly, in [11] we address the fundamental issue of boundary values, in [12] we construct an extension of the Douglis–Nirenberg calculus (see for instance [3, 9] for the role of this calculus in the classical context), while in [13] we address elliptic boundary value problems for first-order wedge operators and prove, in particular, sufficient conditions for well-posedness of such problems. Here, we shall address the main aspects of each of these papers in subsequent sections, after dealing with background information and some notation.

Our point of view, properly translated, closely parallels that of regular elliptic boundary value problems, and indeed our approach, restricted in [13] to first-order operators, allows a full analysis of the classical problem as a special case.

Of course, the work described here does not exist in a vacuum. However, we shall cite only the most directly pertinent work, and refer the reader to the papers indicated above for a more representative listing of research in the general area of

G. A. Mendoza (✉)
Department of Mathematics, Temple University, Philadelphia, PA 19122, USA
e-mail: gmendoza@temple.edu

T. Krainer
Department of Mathematics and Statistics, Pennsylvania State University,
3000 Ivyside Park, Altoona Campus, Altoona, PA 16601, USA
e-mail: krainer@psu.edu

J. Escher et al., (eds.), *Elliptic and Parabolic Equations,* Springer Proceedings in Mathematics & Statistics 119, DOI 10.1007/978-3-319-12547-3_9

elliptic problems for elliptic operators on manifolds with singularities. The preceding notwithstanding, we call attention to the papers of Mazzeo and Vertman [18] treating higher-order problems under certain assumptions, as well as to the extensive work of Schulze and collaborators (for example [10, 25]) on algebras of pseudodifferential operators on manifolds with singularities.

2 Setup

The differential-topological setup is that of Mazzeo [15]. Namely, a compact manifold $\mathcal{M}$ whose boundary $\mathcal{N}$ is the total space of a locally trivial fibration $\wp : \mathcal{N} \to \mathcal{Y}$ with typical fiber $\mathcal{Z}$. (For an extension of this kind of structure, the reader is directed to the paper by Albin et al. [1], which gives a very clear description of the process of resolving singularities of an arbitrary stratified pseudomanifold through a series of blowups, keeping track of boundary fibrations.) The base space $\mathcal{Y}$, the edge, may have several components which are manifolds of possibly different dimensions and the typical fibers over different components may not be diffeomorphic, but we will ignore this for notational simplicity.

The analytic objects are wedge operators, i.e., elements of $x^{-m}\operatorname{Diff}^m_e(\mathcal{M}; E, F)$, where $\operatorname{Diff}^m_e(\mathcal{M}; E, F)$ is the class of edge differential operators of order m defined by Mazzeo, that is, linear differential operators on $\mathcal{M}$ of order m with smooth coefficients which along the boundary differentiate only in directions tangent to the fibers; E and F are Hermitian vector bundles over $\mathcal{M}$, and x is a defining function for $\mathcal{N}$, positive in $\mathring{\mathcal{M}}$. An element of $\operatorname{Diff}^m_e(\mathcal{M}; E, F)$ is thus a regular differential operator on $\mathcal{M}$ which in local coordinates x, y_j, z_μ near any point of the boundary, with x as just indicated and the y_j restricted to $\partial\mathcal{M}$ being constant on fibers, has the form

$$\sum_{k+|\alpha|+|\beta|\le m} a_{k\alpha\beta}(xD_x)^k(xD_y)^\alpha D_z^\beta \tag{1}$$

with respect to local trivializations of E and F. The coefficients $a_{k\alpha\beta}$ are smooth up to the boundary. A regular differential operator A of order m on $\mathcal{M}$ is an element of $x^{-m}\operatorname{Diff}^m_e$ by way of the cheap trick $A = x^{-m}(x^m A)$. In this case, $\mathcal{Y} = \mathcal{N}$ and the fibers are just the points of $\mathcal{N}$. One of the initial motivations for the structural specification of the elements of Diff^m_e comes from what results when writing a regular differential operator in cylindrical coordinates.

The functional analytic component enters through a choice of a b-density $\mathfrak{m}_b = x^{-1}\mathfrak{m}$ as in Melrose [20]; $\mathfrak{m}$ is a smooth density. With the b-density and the Hermitian structures of E and F, one gets weighted L^2 spaces, e.g., $x^{-\gamma}L^2_b(\mathcal{M}; E)$.

3 Some Considerations

Any elliptic element of $\mathrm{Diff}_e^m(\mathcal{M}; E, F)$ (we review the intrinsic notion of ellipticity in Sect. 4), viewed initially as an operator

$$C_c^\infty(\mathring{\mathcal{M}}; E) \subset x^{-\gamma} L_b^2(\mathcal{M}; E) \to x^{-\gamma} L_b^2(\mathcal{M}; F), \tag{2}$$

admits only one closed extension (γ is a real number; the spaces are L^2 spaces with respect to the measure $x^{2\gamma}\mathfrak{m}_b$), whereas generically elliptic elements of the space $x^{-m}\mathrm{Diff}_e^m(\mathcal{M}; E, F)$ admit infinitely many such extensions: This is the reason why boundary value problems make sense for wedge operators but not for edge operators. Having a unique closed extension is a property shared by other classes of operators such as the differential operators in the Θ-calculus of Epstein et al. [6], and more generally, those associated to Lie structures at infinity of Ammann et al. [2].

Let A be an elliptic element in $x^{-m}\mathrm{Diff}_e^m(\mathcal{M}; E, F)$. Recall that the domain of the maximal extension of A, initially as an operator (2), is the space

$$\mathcal{D}_{\max}(A) = \{u \in x^{-\gamma} L_b^2(\mathcal{M}; E) : Au \in x^{-\gamma} L_b^2(\mathcal{M}; F)\}$$

and that the minimal domain, $\mathcal{D}_{\min}(A)$, is the domain of the closure of A starting with (2). In the case of a regular elliptic operator of order m, the minimal domain is $H_0^m(\mathcal{M}; E)$ (we take $\gamma = 1/2$ in this case because $L^2(\mathcal{M}; \mathfrak{m}) = x^{-1/2} L^2(\mathcal{M}, \mathfrak{m}_b)$).

One seeks among other things to establish the existence of a split exact sequence

$$0 \to \mathcal{D}_{\min}(A) \to H_A^m \to \mathcal{S}_A \to 0 \tag{3}$$

in which H_A^m is a conveniently chosen subspace of the maximal domain. For regular elliptic differential operators, this sequence is analogous, and in a natural general theory should reduce, to the classical sequence

$$0 \to H_0^m(\mathcal{M}, E) \to H^m(\mathcal{M}, E) \to \bigoplus_{j=0}^{m-1} H^{m-j-1/2}(\partial\mathcal{M}; E_{\partial\mathcal{M}}) \to 0$$

that is associated with taking Cauchy data on the boundary. On the face of it, in (3) one could take $H_A^m = \mathcal{D}_{\max}(A)$ and $\mathcal{S}_A = \mathcal{D}_{\max}(A)/\mathcal{D}_{\min}(A)$; viewing the quotient as the orthogonal of $\mathcal{D}_{\min}(A)$ in $\mathcal{D}_{\max}(A)$ with respect to the inner product

$$(u, v)_A = (u, v) + (Au, Av) \tag{4}$$

on the maximal domain gives a natural splitting. However, this choice is generally bad, because one should also require that the inclusion $H_A^m \hookrightarrow x^{-\gamma} L_b^2$ be compact. The following example shows that this need not be the case when H_A^m is taken to be $\mathcal{D}_{\max}(A)$ endowed with the norm defined by the inner product (4).

Example 1 Let $\mathcal{M}$ be the closed unit disk in $\mathbb{R}^2$ and let Δ be the Euclidean Laplacian. We claim that the inclusion of $\mathcal{D}_{\max}(\Delta)$ in $L^2(\mathcal{M})$ is not compact. If it were, then also the inclusion of $\ker\Delta$ in L^2 is compact. But with the norm defined by (4) in $\mathcal{D}_{\max}(\Delta)$ we have

$$\|u\|_\Delta^2 = \|u\|^2 + \|\Delta u\|^2, \tag{5}$$

so the Δ-norm and the L^2 norm are the same on $\ker\Delta$. But the compactness of the inclusion map now implies that the unit sphere of $\ker\Delta$ is compact in the L^2 norm, a contradiction since $\ker\Delta$ is infinite-dimensional. Thus, the inclusion of $\mathcal{D}_{\max}(\Delta)$ in $L^2(\mathcal{M})$ is not compact.

We now discuss briefly the role of the weight $x^{2\gamma}$ and the use of a b-density rather than a regular density. The weight $x^{2\gamma}$ connects with geometric information such as what appears when introducing cylindrical coordinates along a submanifold $\mathcal{Y}$ of codimension k in a smooth manifold: A smooth measure near $\mathcal{Y}$ becomes essentially $x^{k-1}dx\,dy\,dz$, x being the radial variable and dz representing the measure on the sphere S^{k-1}. Both the factor x^{-1} making up the b-density and the weight $x^{2\gamma}$ can be removed by conjugating the operator with multiplication by an appropriate power of x. We eventually take advantage of this and pick $\gamma = m/2$, but keep the b-density since this brings to the foreground the multiplicative structure of $\mathbb{R}_+$ (for which $x^{-1}dx$ is a Haar measure). The class $x^{-m}\mathrm{Diff}_e^m$ is invariant under conjugation as described; however not so the class Diff^m of regular differential operators, which under such operations end up subsumed in the more general class of wedge operators; depending on the particularities of the problem, this may be advantageous.

4 Ellipticity, the Wedge Cotangent Bundle and the Structure Ring

Ellipticity of an element $A \in x^{-m}\mathrm{Diff}_e^m(\mathcal{M}; E, F)$ means ellipticity of $P = x^m A$ in the sense of [15]. This is ellipticity of P over the interior of $\mathcal{M}$ in the usual sense, and, near a boundary point where P is written as in (1), invertibility of

$$\sum_{k+|\alpha|+|\beta|=m} a_{k\alpha\beta}\xi^k\eta^\alpha\zeta^\beta.$$

While this is a perfectly good practical definition of ellipticity, it disregards basic information of the manifold-with-boundary-fibration and the class of w-differential operators. Still, fixing x allows a definition of principal symbol of A by way of declaring it to be the edge-principal symbol of $x^m A$, which is what the above expression is in coordinates. This is, however, not quite satisfactory since it does depend, albeit mildly, on the choice of defining function.

The natural structure bundle for the class of wedge differential operators is the wedge cotangent bundle ${}^wT^*\mathcal{M}$. It is constructed in a fashion similar to Melrose's b-tangent bundle in [19, 20] or Mazzeo's edge tangent bundle in [15], as follows (see [7] for details). The space of continuous differential 1-forms on $\mathcal{M}$, whose pullback to the fibers of $\mathcal{N} \to \mathcal{Y}$ vanishes, is a finitely generated projective module,

${}^{w}C(\mathcal{M}; T^*\mathcal{M})$, over the ring of continuous functions on $\mathcal{M}$, and is therefore, by a theorem of Swan [26], (isomorphic to) the space of sections of a vector bundle over $\mathcal{M}$ which we denote by ${}^{w}T^*\mathcal{M}$. This vector bundle is easily seen to be a C^∞ bundle, and is the natural structure bundle to the same extent as ${}^{e}T\mathcal{M}$ is the natural structure bundle in the case of edge operators and ${}^{b}T\mathcal{M}$ is in the case of b-operators. The inclusion map ${}^{w}\mathrm{ev}_* : {}^{w}C(\mathcal{M}; T^*\mathcal{M}) \to C(\mathcal{M}; T^*\mathcal{M})$ determines a (smooth) bundle homomorphism ${}^{w}\mathrm{ev} : {}^{w}T^*\mathcal{M} \to T^*\mathcal{M}$ covering the identity; ${}^{w}\mathrm{ev}$ is an isomorphism over the interior while over the boundary its kernel is the conormal bundle to the fibers of $\mathcal{Z} \to \mathcal{Y}$.

The naturality of ${}^{w}T^*\mathcal{M}$ is further justified by it being the domain of a principal symbol map for elements $A \in x^{-m}\mathrm{Diff}^m_e(\mathcal{M}; E, F)$: There is a smooth homomorphism

$$ {}^{w}\!\sigma(A) \in C^\infty({}^{w}T^*\mathcal{M}\backslash 0; \mathrm{Hom}({}^{w}\pi^* E,\, {}^{w}\pi^* F)), $$

the wedge symbol of A, related to the standard principal symbol of A over $\mathring{\mathcal{M}}$ by

$$ {}^{w}\!\sigma(A) = \sigma(A) \circ {}^{w}\mathrm{ev}. $$

Naturally, ellipticity is defined as invertibility of ${}^{w}\!\sigma(A)$.

The section ${}^{w}\!\sigma(A)$ can also be obtained by an oscillatory test using real-valued functions in the ring

$$ \mathcal{R} = \{ f \in C^\infty(\mathcal{M}) : f|_{\mathcal{N}} \text{ is constant on the fibers of } \wp \}, $$

see [13]. The fundamental role of this ring can be seen from the observation that it determines the boundary structure of $\mathcal{M}$. It can also be used to define the spaces $x^{-m}\mathrm{Diff}^m_e$ without resorting to coordinates, as described in the introduction of the just cited paper. From another point of view, the differentials of real-valued elements of $\mathcal{R}$ generate ${}^{w}C^\infty(\mathcal{M}; T^*\mathcal{M})$ as a module over $C^\infty(\mathcal{M}; \mathbb{R})$. Finally, observe that if the configuration of $\mathcal{M}$ with its boundary fibration comes from blowing up a smooth manifold along a smooth submanifold (that is, from cylindrical coordinates), then $\mathcal{R}$ is, to first order along $\mathcal{N}$, the pullback of the ring of smooth functions on the original manifold.

Incidentally, when a regular elliptic operator on a smooth manifold is written in cylindrical coordinates with the given submanifold as the axis, the result is not just a wedge operator as already pointed out, but it is also a w-elliptic operator. Along the same vein, a smooth Riemannian metric on the cotangent bundle of the original manifold becomes a smooth metric on the wedge cotangent bundle.

5 Indicial and Normal Families

The operator $P = x^m A$ is in particular a b-operator (by way of replacing the fine structure of the fibration $\mathcal{N} \to \mathcal{Y}$ with the one in which each connected component of $\mathcal{N}$ is one fiber): $P \in \mathrm{Diff}^m_b(\mathcal{M}; E, F)$. The operator P has the property that if $\phi \in C^\infty(\mathcal{M}; E)$, then $(P\phi)|_{\mathcal{N}}$ depends only on $\phi|_{\mathcal{N}}$, thus giving an operator $P|_{\mathcal{N}} : C^\infty(\mathcal{N}; E_{\mathcal{N}}) \to C^\infty(\mathcal{N}; F_{\mathcal{N}})$. By $E_{\mathcal{N}}$, we mean the part of E over $\mathcal{N}$. Also, $x^{-i\sigma} P x^{-i\sigma} \in \mathrm{Diff}_b^m$, and the indicial family of A (or P) is defined as

$$^b\widehat{P}(\sigma) = (x^{-i\sigma} P x^{i\sigma})|_{\mathcal{N}}, \quad \sigma \in \mathbb{C}.$$

Due to the factor x occurring with each derivative in y, this operator does not differentiate in y: It gives a family of differential operators $\widehat{P}_y(\sigma) : C^\infty(\mathcal{Z}_y; E_{\mathcal{Z}_y}) \to C^\infty(\mathcal{Z}_y; F_{\mathcal{Z}_y})$ depending smoothly on $(\sigma, y) \in \mathbb{C} \times \mathcal{Y}$ and holomorphically (polynomially) in σ. Here, $\mathcal{Z}_y$ is the fiber of $\wp : \mathcal{N} \to \mathcal{Y}$ over y.

Let $\pi_\wedge : \mathcal{N}^\wedge \to \mathcal{N}$ be the closed inward-pointing part of the normal bundle $N_{\mathcal{M}}\mathcal{N}$ of $\mathcal{N}$ in $\mathcal{M}$. Its boundary, being the zero section of $N_{\mathcal{M}}\mathcal{N}$, is identified with $\mathcal{N}$ and so carries the same fibration structure as $\partial\mathcal{M}$. The part of $\mathcal{N}^\wedge$ over $\mathcal{Z}_y$ is denoted $\mathcal{Z}_y^\wedge$.

The infinitesimal generator of the radial action on $\mathcal{N}^\wedge$ is $x_\wedge \partial_{x_\wedge}$ where $x_\wedge$ is any linear function $N_{\mathcal{M}}\mathcal{N}$, positive on $\overset{\circ}{\mathcal{N}^\wedge}$, for example, the one induced by dx. We shall write also x for $x_\wedge$, since there is little risk of confusion. The family $^b\widehat{P}(\sigma)$ is polynomial in σ, so replacing σ by $-ix\partial_x$ (more properly, by $-i\nabla_{x\partial_x}$) gives an operator $^bP \in \mathrm{Diff}_b^m(\mathcal{N}^\wedge; E^\wedge, F^\wedge)$ (which does not differentiate in y), where, for example, $E^\wedge$ is the pullback of E to $\mathcal{N}^\wedge$ (and ∇ the pullback of some (fixed) Hermitian connection on E). The indicial operator of A is then defined to be the operator

$$^bA = \frac{1}{x^m}\, {}^bP \in \frac{1}{x^m}\mathrm{Diff}_b^m(\mathcal{N}^\wedge; E^\wedge, F^\wedge).$$

In local coordinates, if $x^m A$ is given by (1) near some boundary point, then

$$^bA = \frac{1}{x^m} \sum_{k+|\beta| \le m} a_{k0\beta}(0, y, z)(xD_x)^k D_z^\beta.$$

Note that because ^bP does not differentiate in y, the indicial operator may be viewed as a family

$$\mathcal{Y} \ni y \mapsto {}^bA_y = \frac{1}{x^m}\, {}^bP \in \frac{1}{x^m}\mathrm{Diff}_b^m(\mathcal{Z}_y^\wedge; E_{\mathcal{Z}_y^\wedge}, F_{\mathcal{Z}_y^\wedge}).$$

We wrote $E_{\mathcal{Z}_y^\wedge}$ to mean $E^\wedge$ along $\mathcal{Z}_y^\wedge$ with a slight abuse of the notation.

One defines an $\mathbb{R}_+$-action $\varrho \mapsto \kappa_\varrho$ on sections u of $E^\wedge$ by letting $(\kappa_\varrho u)(\nu)$ be the result of parallel transport of $u(\varrho\nu)$ from $\varrho\nu$ to ν along the fiber of $\mathcal{N}^\wedge$ followed by multiplication by ϱ^γ (i.e., on functions, $(\kappa_\varrho f)(\nu) = \varrho^\gamma f(\varrho\nu)$). Thus, κ_ϱ is a bundle homomorphism covering the radial action $\nu \mapsto \tau_\varrho \nu = \varrho\nu$. The factor ϱ^γ ensures that κ_ϱ is unitary on $x^{-\gamma}L_b^2(\mathcal{N}^\wedge; E^\wedge)$. The b-density for the latter space is defined with the aid of a tubular neighborhood map:

$$\mathcal{N}^\wedge \supset V \xrightarrow{\varphi} W \subset \mathcal{M}$$

as

$$\mathfrak{m}_b^\wedge = \lim_{\rho\to 0} (\varphi \circ \tau_\rho)^* \mathfrak{m}_b.$$

The normal family of A associates to each element $\boldsymbol{\eta} \in T^*Y\backslash 0$ an element of $x^{-m}\mathrm{Diff}_b^m(\mathcal{Z}_y^\wedge; E_{\mathcal{Z}_y^\wedge}, F_{\mathcal{Z}_y^\wedge})$ by means of the formula:

$$A_\wedge(\boldsymbol{\eta})u = \big(\lim_{\rho\to\infty} \rho^{-m}\kappa_\rho^{-1} e^{-i\rho\wp_\wedge^* g}\Phi^* A \Phi_* e^{i\rho\wp_\wedge^* g}\kappa_\rho \tilde{u}\big)\big|_{\mathcal{Z}_y^\wedge},\ u \in C_c^\infty(\overset{\circ}{\mathcal{Z}_y^\wedge}; E_{\mathcal{Z}_y^\wedge}), \quad (6)$$

with $g \in C^\infty(\mathcal{Y})$ real valued with $dg(y) = \boldsymbol{\eta}$, and $\tilde{u}$ a C_c^∞ extension of u. The maps Φ_* (and its inverse Φ^*) are defined using φ, the radial action on $\mathcal{N}^\wedge$, and parallel transport on E and F as needed. See [7, Proposition 2.10] for details on this; a local argument shows that $A_\wedge$ is independent of φ. In local coordinates, if $x^m A$ is given by (1) near some boundary point, then

$$A_\wedge(\boldsymbol{\eta}) = \frac{1}{x^m} \sum_{k+|\alpha|+|\beta|\le m} a_{k\alpha\beta}(0, y, z)(xD_x)^k (x\eta)^\alpha D_z^\beta, \quad \boldsymbol{\eta} = \eta\cdot dy.$$

This expression makes sense also at $\boldsymbol{\eta} = 0$, where it becomes canonically equal to ${}^bA$.

6 The Kernel Bundle of a Holomorphic Fredholm Family

The argument leading to the definition of ${}^bP$ can be extended to give a formal power series expansion

$$P = \sum_{\ell=0}^{N} {}^bP_\ell x^\ell + {}^b\widetilde{P}_{N+1} x^{N+1}$$

in which the ${}^bP_\ell$ and ${}^b\widetilde{P}_{N+1}$ are all b-operators on $\mathcal{N}^\wedge$, the ${}^bP_\ell$ commute with $\nabla_{x\partial_x}$ and ${}^b\widetilde{P}_{N+1}$ is defined only near $\mathcal{N}$. The operator ${}^bP_0$ is equal to ${}^bP$, each operator ${}^bP_\ell$ includes derivatives in y of at most order $\max\{\ell, m\}$, similarly the remainder ${}^b\widetilde{P}_{N+1}$.

If $u \in x^{-\gamma}L_b^2(\mathcal{M}; E)$, its Mellin transform is

$$\widehat{u}(\sigma, p) = \int_{\pi_\wedge^{-1}(p)} x^{-i\sigma}\omega(x)\Phi^* u \,\frac{dx}{x}.$$

This may be viewed variously as an element of L^2 on $\{\Im\sigma = \gamma\}\times\mathcal{N}$, as a holomorphic function on $\Im\sigma > \gamma$ with values in $L^2(\mathcal{N}, E|_\mathcal{N})$, and so on. Changing the cutoff function ω (which is supported in a small neighborhood of $\mathcal{N}$) changes $\widehat{u}(\sigma, p)$ by an entire additive term.

One gets a hold on boundary values of elements $u \in \mathcal{D}_{\max}(A)$ by exploiting an idea from [21],here, applied to the fact that for such u one has that $\widehat{u}(\sigma, p)$ is holomorphic in $\Im\sigma > \gamma$, whereas $(x^m Au)^\wedge = \widehat{f}(\sigma)$ is holomorphic in $\Im\sigma > \gamma - m$, so

$$ {}^b\widehat{P}(\sigma)\widehat{u}(\sigma) = \widehat{f}(\sigma) - \sum_{\ell=1}^{N} {}^b\widehat{P}_\ell(\sigma)\widehat{u}(\sigma + i\ell) - ({}^b\widetilde{P}_{N+1}x^{n+1}u)^\wedge(\sigma), $$

where the right-hand side is holomorphic in $\Im\sigma > \gamma - 1$ (but in principle less regular in y than u since it includes derivatives in y). It follows that

$$ \widehat{u}(\sigma) = {}^b\widehat{P}(\sigma)^{-1}[\widehat{f}(\sigma) - \sum_{\ell=1}^{N} {}^b\widehat{P}_\ell(\sigma)\widehat{u}(\sigma + i\ell) - ({}^b\widetilde{P}_{N+1}x^{n+1}u)^\wedge(\sigma)], \tag{7} $$

for $\Im\sigma > \gamma - 1$ such that ${}^b\widehat{P}(\sigma)^{-1}$ exists.

Assume henceforth that A is elliptic. Then, $\widehat{P}_y(\sigma)$ is elliptic for each (σ, y) and invertible in regions $|\Im\sigma| < a$ for any a when $|\Re\sigma|$ is large enough, uniformly for y in compact sets (so on $\mathcal{Y}$ itself). This implies in particular that the set

$$ \operatorname{spec}_{b,y}(A) = \{\sigma : \widehat{P}_y(\sigma) \text{ is not invertible}\}, $$

the boundary spectrum of A (or P) at y is discrete (see [20, 21]). This set may vary with y, but in a number of important geometric situations in the edge (complete) setting it does not, see for example Mazzeo and Melrose [16], Mazzeo and Phillips [17], Epstein et al. [6] to name but a few. The set

$$ \operatorname{spec}_e(A) = \{(\sigma, y) \in \mathbb{C} \times \mathcal{Y} : \sigma \in \operatorname{spec}_{b,y}(A)\} $$

is the edge spectrum of A.

Thus, with the assumed ellipticity of A, (7) holds for sure in

$$ \{(\sigma, p) \in \mathbb{C} \times \mathcal{N} : \Im\sigma > \gamma - 1,\ (\sigma, \wp(p)) \notin \operatorname{spec}_e(A)\}. $$

Using this in the right-hand side of (7) gives now the information about the meromorphic structure of $\widehat{u}(\sigma)$ in $\Im\sigma > \gamma - 2$. Iterating, one gets information on $\Im\sigma > \gamma - m$.

The caveat is that the right-hand side has to be treated as a distribution (at least in the y variable) so one has to proceed with much care.

Clearly, the meromorphic invertibility of ${}^b\widehat{P}_y(\sigma)$ for σ in the set

$$ \Sigma = \{\sigma \in \mathbb{C} : \gamma - m < \Im\sigma < \gamma\} $$

for each $y \in \mathcal{Y}$ does play an important role. If K is a Hilbert space and $V \subset \mathbb{C}$ is open, we write $\mathfrak{M}(V, K)$ for the space of meromorphic K-valued functions on V and $\mathfrak{H}(V, K)$ for the subspace of holomorphic elements. Thus, $f \in \mathfrak{M}(V, K)$ if there is, for each $\sigma_0 \in V$, a number $\mu_0 \in \mathbb{N}_0$ such that $\sigma \mapsto (\sigma - \sigma_0)^{\mu_0} f(\sigma)$ is holomorphic near σ_0. With this notation, we have that the holomorphic (polynomial) family ${}^b\widehat{P}_y$ defines a map

$$ {}^b\widehat{P}_y : \mathfrak{M}(\Sigma; H^m(\mathcal{Z}_y; E_{\mathcal{Z}_y})) \to \mathfrak{M}(\Sigma; L^2(\mathcal{Z}_y; F_{\mathcal{Z}_y})) $$

which in turn gives a map

$$ \begin{aligned} [{}^b\widehat{P}_y] : \mathfrak{M}(\Sigma; H^m(\mathcal{Z}_y; E_{\mathcal{Z}_y}))/\mathfrak{H}(\Sigma; H^m(\mathcal{Z}_y; E_{\mathcal{Z}_y})) \\ \to \mathfrak{M}(\Sigma; L^2(\mathcal{Z}_y; F_{\mathcal{Z}_y}))/\mathfrak{H}(\Sigma; L^2(\mathcal{Z}_y; F_{\mathcal{Z}_y})) \end{aligned} $$

It is the kernel of this last map that is of interest, as y varies in $\mathcal{Y}$. This space is more conveniently expressed with the space of singular parts of its elements. The latter form a finite-dimensional space of meromorphic $H^m(\mathcal{Z}_y; E_{\mathcal{Z}_y})$-valued functions on $\mathbb{C}$ with poles in $\operatorname{spec}_{b,y}(A) \cap \Sigma$. To get an analytic hold on these elements, we note that the singular part of an element $[\widehat{\tau}] \in \ker[{}^b\widehat{P}_y]$ represented by $\widehat{\tau} \in \mathfrak{M}(\Sigma; H^m(\mathcal{Z}_y; E_{\mathcal{Z}_y}))$ can be obtained by computing the integral

$$ \mathfrak{s}_\Omega(\widehat{\tau})(\sigma) = \frac{i}{2\pi} \oint_{\partial\Omega} \frac{\widehat{\tau}(\zeta)}{\zeta - \sigma}\, d\zeta, \qquad |\sigma| \gg 1 $$

with the positive (counterclockwise) orientation, where $\Omega \Subset \Sigma$ is open, has smooth (or rectifiable) boundary and contains $\operatorname{spec}_{b,y}(A)$. The integral is of course independent of the representative of $[\widehat{\tau}]$. Write

$$ \widehat{\mathscr{T}}_y = \mathfrak{s}(\ker[{}^b\widehat{P}_y]). $$

If $\widehat{\tau} \in \widehat{\mathscr{T}}_y$ and Γ is a simple closed smooth (or rectifiable) curve surrounding $\operatorname{spec}_{b,y}(A) \cap \Sigma$, then

$$ \tau = -\frac{1}{2\pi} \oint_\Gamma x^{i\sigma} \widehat{\tau}(\sigma)\, d\sigma, \tag{8} $$

with the positive orientation viewed as a section of $E_{\mathcal{Z}_y^\wedge}$, is an element of $\ker {}^bA_y$ (τ is essentially the inverse Mellin transform of $\widehat{\tau}$). Indeed,

$$ \begin{aligned} x^m\, {}^bA_y \tau &= -\frac{1}{2\pi} \oint_\Gamma x^{i\sigma}\, {}^b\widehat{P}(\sigma) u(\sigma)\, d\sigma \\ &= 0 \end{aligned} $$

since ${}^b\widehat{P}(\sigma)u(\sigma)$ is entire. Evidently τ has the from

$$ \tau = \sum_{\substack{\sigma \in \Sigma \\ \sigma \in \operatorname{spec}_{b,y}(A)}} \sum_{\ell=0}^{N_\sigma} \tau_{\sigma\ell}\, x^{i\sigma} \log^\ell x. \tag{9} $$

The coefficients $\tau_{\sigma\ell}$ are sections (*a fortiori* smooth) of $E_{\mathcal{Z}_y}$. We let $\mathscr{T}_y$ be the space of all elements obtained from $\widehat{\mathscr{T}}_y$ as just described.

Theorem 1 ([11, Theorem 3.2]) *Assume*

$$\{(\sigma, y) \in \mathbb{C} \times \mathcal{Y} : \Im\sigma = \gamma,\ \gamma - m\} \cap \operatorname{spec}_e(A) = \emptyset. \tag{10}$$

Define

$$\mathscr{T} = \bigsqcup_{y \in \mathcal{Y}} \mathscr{T}_y, \quad \pi : \mathscr{T} \to \mathcal{Y} \text{ the canonical map.}$$

Then $\mathscr{T} \to \mathcal{Y}$ *is a smooth vector bundle. A smooth section of* $\mathscr{T}$ *is a map*

$$\mathcal{Y} \ni y \mapsto \tau(y) \in \mathscr{T}_y$$

which viewed as a section of E *over* $\overset{\circ}{\mathcal{N}}{}^\wedge$ *is smooth in the usual sense.*

We call $\mathscr{T}$ the trace bundle of A. In this note, we will always assume that (10) holds.

Example 2 We pause to illustrate some of the ideas. Let $\mathcal{Y}$ be a closed orientable surface, $\mathcal{Z}$ an arbitrary closed manifold, $L^+ \to \mathcal{Y}$ a nontrivial complex line bundle, L^- its dual, and

$$\Phi = \begin{bmatrix} \Phi_{11} & \Phi_{12} \\ \Phi_{21} & \Phi_{22} \end{bmatrix} : \begin{matrix} L^+ \\ \oplus \\ L^- \end{matrix} \to \begin{matrix} L^+ \\ \oplus \\ L^- \end{matrix}$$

some vector bundle homomorphism covering the identity $\mathcal{Y} \to \mathcal{Y}$. The homomorphisms Φ_{11} and Φ_{22} are required to be isomorphisms. Let $\mathcal{M} = [0, 1) \times \mathcal{Y} \times \mathcal{Z}$ and let E be the pullback of $L^+ \oplus L^-$ to $\mathcal{M}$, a trivial rank 2 bundle: $E = \pi^* L^+ \oplus \pi^* L^-$, $\pi : \mathcal{M} \to \mathcal{Y}$ the projection. Let Δ_{L^+} be a Laplacian on sections of L^+. For example, fix a Riemannian metric on $T^*\mathcal{Y}$, a connection and a Hermitian metric on L^+ and let Δ_{L^+} be the Bochner Laplacian, $\Delta_{L^+} = \nabla^*\nabla$. Let Δ_{L^-} be a Laplacian on sections of L^-. Exploiting the product structure of $\mathcal{M}$, we let these Laplacians act on the factors of E in the natural way. Next we pick a Laplacian $\Delta_{\mathcal{Z}}$ (acting on functions on $\mathcal{Z}$) and let $Q_{\mathcal{Z}} = \Delta_{\mathcal{Z}} + c$, $c \geq 0$, act on sections of $\pi^* L^+$ or $\pi^* L^-$ in the canonical way, again exploiting the product structure of $\mathcal{M}$ and the pullback nature of these bundles; we specify the constant c later. Finally, writing x for the standard coordinate in $[0, 1)$ we define

$$P = \begin{bmatrix} (xD_x)^2 & 0 \\ 0 & (xD_x)^2 \end{bmatrix} + x^2 \begin{bmatrix} \Delta_{L^+} & 0 \\ 0 & \Delta_{L^-} \end{bmatrix} + \begin{bmatrix} Q_{\mathcal{Z}}\Phi_{11} & \Phi_{12} \\ \Phi_{21} & Q_{\mathcal{Z}}\Phi_{22} \end{bmatrix}$$

acting on $C^\infty(\mathcal{M}; E)$. This is an elliptic edge operator with respect to the obvious boundary fibration of $\mathcal{M}$, acting on sections of $\pi^* L^+ \oplus \pi^* L^-$, so $A = x^{-2} P \in$

$x^{-2}\mathrm{Diff}^2_e(\mathcal{M}; E)$ is an elliptic wedge operator. We compute its boundary spectrum. The indicial family of P is

$$ {}^b\widehat{P}(\sigma) = \begin{bmatrix} \sigma^2 & 0 \\ 0 & \sigma^2 \end{bmatrix} + \begin{bmatrix} Q_{\mathcal{Z}}\Phi_{11} & \Phi_{12} \\ \Phi_{21} & Q_{\mathcal{Z}}\Phi_{22} \end{bmatrix}. $$

Let $\{\psi_k\}$ be a complete orthonormal system of eigenvectors of $Q_{\mathcal{Z}}$, $\{\lambda_k^2\}$ the corresponding eigenvalues. The λ_k are nonnegative and assumed to form a nondecreasing sequence. Let ν^+ be a frame of L^+ in a neighborhood $U \subset \mathcal{Y}$ of some $y_0 \in \mathcal{Y}$, let ν^- be the dual frame. Keeping the same notation for the lifted frames, they make up a frame for E over $\pi^{-1}(U)$. In terms of the resulting frame, a section of $E_{\mathcal{N}}$ over $\wp^{-1}(U)$ has the form

$$ \sum_{k=0}^{\infty} \psi_k(z) \begin{bmatrix} c_k^+(y) \\ c_k^-(y) \end{bmatrix} $$

Using formulas such as $\Phi_{12}(\nu^-) = \varphi_{12}\nu^+$ with smooth $\varphi_{ij} : U \to \mathbb{C}$, we get

$$ \sum_{k=0}^{\infty} \psi_k(z) \begin{bmatrix} \sigma^2 + \lambda_k^2\varphi_{11}(y) & \varphi_{12}(y) \\ \varphi_{21}(y) & \sigma^2 + \lambda_k^2\varphi_{22}(y) \end{bmatrix} \begin{bmatrix} c_k^+(y) \\ c_k^-(y) \end{bmatrix} \tag{11} $$

for ${}^b\widehat{P}(\sigma)u$ with respect to the same frame. The boundary spectrum at y consists of the roots σ of all polynomials

$$ \sigma^4 + \lambda_k^2(\varphi_{11}(y) + \varphi_{22}(y))\sigma^2 + \lambda_k^4\varphi_{11}(y)\varphi_{22}(y) - \varphi_{12}(y)\varphi_{21}(y), \quad k = 0, 1, \dots $$

The functions $\varphi_{11}, \varphi_{12}$ are globally defined and independent of choice of frames. The product $\varphi_{12}\varphi_{21}$ is independent of the choice of frame so it is also globally defined. Evidently, as y varies, these roots can have very complicated behavior.

We illustrate the simplest possibilities assuming $\Phi_{21} = 0$, in which case

$$ \mathrm{spec}_{b,y}(A) = \{\pm i\lambda_k\sqrt{\varphi_{11}(y)},\ \pm i\lambda_k\sqrt{\varphi_{22}(y)} : k \in \mathbb{N}_0\}. $$

Take $\gamma = 1$. Pick $c > 0$ in the definition of $Q_{\mathcal{Z}}$ so $\lambda_0 > 0$. Suppose $1/\lambda_1^2 < \sup_{\mathcal{Y}} |\varphi_{jj}| < 1/\lambda_0^2$. Then, (10) holds and the part of the b-spectrum that matters, the part in $\Sigma = \{\sigma \in \mathbb{C} : -1 < \Im\sigma < 1\}$, consists of the points

$$ \pm i\lambda_0\sqrt{\varphi_{11}(y)},\ \pm i\lambda_0\sqrt{\varphi_{22}(y)}. $$

Each of these is locally well defined in a manner that gives a locally smooth function of y, since the φ_{jj} vanishes nowhere. That the roots can be arranged to depend locally smoothly on y cannot be guaranteed if $\varphi_{12}\varphi_{21} \neq 0$.

1. If $\varphi_{11}(y) - \varphi_{22}(y) \neq 0$ for all y then these four roots are all different from each other for each y , and $\widehat{\mathscr{T}}_y$ is the span of

$$\frac{\psi_1(z)}{\sigma \mp i\lambda_0\sqrt{\varphi_{11}(y)}}\nu^+, \qquad \frac{\psi_1(y)}{\sigma \mp i\lambda_0\sqrt{\varphi_{22}(y)}}\Big(\frac{-\varphi_{12}}{\lambda_0^2(\varphi_{11}-\varphi_{22})}\nu^+ + \nu^-\Big).$$

The fibers of $\mathscr{T}$ are four-dimensional, the trace bundle is isomorphic to $L^+ \oplus L^+ \oplus L^- \oplus L^-$ which is trivial since $c_1(L^+) + c_1(L^-) = 0$, while neither individual summand, nor $L^+ \oplus L^+$ and $L^- \oplus L^-$ is trivial since $2c_1(L^+) \neq 0$. In particular, if $\operatorname{spec}_{b,y}(A)$ is independent of y (which happens when φ_{11} and φ_{22} are constant), the part of $\mathscr{T}$ associated with a single pole in $\operatorname{spec}_e(A) \cap \Sigma \times \mathcal{Y}$ is not trivial despite the fact that E itself is a trivial bundle.
2. Suppose now that $\varphi_{11}(y_0) = \varphi_{22}(y_0)$ for some y_0. For y near y_0 define $\sigma_1^{\pm}(y) = \pm i\lambda_0\sqrt{\phi_{11}(y)}$, likewise $\sigma_2^{\pm}(y)$. Then, $\widehat{\mathscr{T}}_{y_0}$ is spanned by

$$\chi_1^{\pm}(y_0) = \frac{1}{2\sigma_1^{\pm}(y_0)(\sigma - \sigma_1^{\pm}(y_0))}\begin{bmatrix}1\\0\end{bmatrix}$$

and

$$\chi_2^{\pm}(y_0) = \Big(\frac{1}{(\sigma - \sigma_1^{\pm}(y_0))^2} - \frac{1}{\sigma_1^{\pm}(y_0)(\sigma - \sigma_1^{\pm}(y_0))}\Big)\begin{bmatrix}1\\0\end{bmatrix} + \frac{2\sigma_1^{\pm}(y_0)^2}{\sigma_1^{\pm}(y_0)(\sigma - \sigma_1^{\pm}(y_0))}\begin{bmatrix}0\\1\end{bmatrix}.$$

In the notation for the $\chi_j^{\pm}(y_0)$, we are taking advantage of the fact that $\varphi_{11}(y_0) = \varphi_{22}(y_0)$ and identify ν^+ and ν^- with the respective column matrices. These formulas are obtained by applying the inverse of

$$M(\sigma, y) = \begin{bmatrix}\sigma^2 + \lambda_0^2\varphi_{11}(y) & \varphi_{12}(y)\\ 0 & \sigma^2 + \lambda_0^2\varphi_{22}(y)\end{bmatrix}$$

(the matrix in (11) with $k = 0$ and $\varphi_{21} = 0$) at $y = y_0$ to the column with components f^+, f^- and then computing the singular part of the resulting expression at $\sigma = \sigma_1^{\pm}(y_0)$, which gives $f^+\chi_1^{\pm}(y_0) + f^-\chi_2^{\pm}(y_0)$. One now obtains a frame for $\widehat{\mathscr{T}}$ near y_0 by taking the singular part of each of $M^{-1}(\sigma, y)M(\sigma, y_0)\chi_j^{\pm}$:

$$\chi_j^{\pm}(\sigma, y) = \mathfrak{s}_\Omega(\chi_j^{\pm}(\cdot, y_0))(\sigma)$$

where Ω is a disk containing the $\sigma_{jj}^{\pm}(y)$ in its interior (y is kept in a neighborhood V of y_0). Using the components of these four vectors as coefficients, one rewrites the result in terms of ν^+ and ν^-. The resulting expressions for the $\chi_j^{\pm}(\sigma, y)$ are smooth in the complement of $(\Sigma \times V) \cap \operatorname{spec}_e(A)$, and so applying (8) to each of them gives a smooth local section of $\mathscr{T}$ over V, altogether making up a local frame.

In general, the construction of local frames of $\mathscr{T}$ near some y_0 as described at the end of the example yields pointwise bases for $\mathscr{T}_y$ that are smooth in the sense described in Theorem 1. In [11], we also prove that if $A^\star$ is the formal adjoint of A and $\mathscr{T}^\star$ its trace bundle, then, taking $\gamma = m/2$ for convenience,

$$\mathscr{T}_y \times \mathscr{T}_y^\star \ni (u, v) \mapsto [u, v]_y^\flat = ({}^bA\omega u, \omega v)_{x^{-m/2}L_b^2} - (\omega u, {}^bA^\star \omega v)_{x^{-m/2}L_b^2} \in \mathbb{C} \quad (12)$$

is nondegenerate and gives a smooth Hermitian pairing of $\mathscr{T}$ and $\mathscr{T}^\star$. This is Theorem 5.3 of [11]. The pairing is independent of the specific cutoff function ω, but one needs to be included because the L^2 spaces are over $\mathcal{Z}_y^\wedge$. What is behind nondegeneracy is the general fact that if A is an arbitrary elliptic operator on some open $\mathring{\mathcal{M}}$, then the Hilbert space adjoint of A with its minimal domain is $A^\star$ with its maximal domain. The smoothness of the pairing follows from writing the pointwise pairing as a contour integral of Mellin transforms (as was in fact done in the just cited paper).

The proof in [11] of Theorem 1 above proceeds in two main steps. First, we show the existence, in a neighborhood of each $y_0 \in \mathcal{Y}$, of a system of sections τ_j that are smooth in the sense of the theorem and are a pointwise basis of each fiber. The existence of such local systems of solutions was also proved by Costabel and Dauge [4] and Schmutzler [24] using different methods. We then prove that two frames τ_j, τ_j' are related by smooth transition functions by observing that if $\tau_j^\star$ is a local frame for $\mathscr{T}^*$, then, first

$$\tau_j' = \sum_k a_{kj}\tau_k$$

for some functions a_{kj}, trivially since the τ_j give bases pointwise, and then that the a_{kj} must be smooth because they satisfy the system of equations

$$[\tau_j', \tau_\ell^\star]_y^\flat = a_{kj}(y)[\tau_k, \tau_\ell^\star]_y^\flat$$

in which the matrices with components $[\tau_j', \tau_\ell^\star]_y^\flat$ and $[\tau_k, \tau_\ell^\star]_y^\flat$ are smooth, since the τ_j', τ_k, and $\tau_\ell^\star$ are smooth, and the second matrix is invertible by the nondegeneracy of the pairing. It follows that the set of frames which are smooth in the sense of Theorem 1 admit smooth transition functions, so the condition defines a smooth structure for the trace bundle in the usual sense.

We close this section with one last observation whose relevancy will become apparent in each of the next two sections. The fiber of $\mathscr{T}$ at y consists of elements of the form (9) in the kernel of ${}^bA_y$. Since ${}^bA_y$ and $\nabla_{x\partial_x}$ commute, the latter defines a bundle homomorphism which we shall denote by

$$x\partial_x : \mathscr{T} \to \mathscr{T}.$$

This homomorphism is smooth because at the level of $\widehat{\mathscr{T}}_y$ it is just multiplication by $i\sigma$. The eigenvalues on the fiber $\mathscr{T}_y$, the numbers $i\sigma$ with $\sigma \in \operatorname{spec}_{b,y}(A) \cap \Sigma$, generally vary with the fiber, as will the Jordan canonical form of $x\partial_x$.

7 Elliptic Systems of Variable Order

To motivate the results described in this section, mostly coming from [12], it is useful to follow the construction of the trace bundle in the case of a regular elliptic differential operator on a manifold with boundary.

Example 3 Suppose A is such an operator on $\mathcal{M}$. In local coordinates x, y_j near a point of $\mathcal{N} = \mathcal{Y}$ (each point of $\mathcal{N}$ is a fiber of the boundary fibration, so there are no z_μ),

$$A = \sum_{k+|\alpha|\le m} a_{k\alpha}(x,y) D_x^k D_y^\alpha$$

with smooth $a_{k\alpha}$ up to $x = 0$. Using

$$x^m D_x^k D_y^\alpha = x^{m-k-|\alpha|} p_k(xD_x + i|\alpha|)(xD_y)^\alpha$$

one gets

$$x^m A = \sum_{k+|\alpha|\le m} a_{k\alpha}(x,y) x^{m-k-|\alpha|} p_k(xD_x + i|\alpha|)(xD_y)^\alpha,$$

so

$${}^b\widehat{P}_y(\sigma) = a_{m0}(0,y) p_m(\sigma)$$

which translates to ${}^bP_y = a_{m0}(0,y)p_m(xD_x)$, therefore

$${}^bA_y = a_{m0}(0,y) D_x^m.$$

Taking $\gamma = 1/2$, the relevant strip Σ in the complex plane is $1/2 - m < \Im\sigma < 1/2$, the set of poles of ${}^b\widehat{P}_y(\sigma)^{-1}$ is

$$\mathrm{spec}_{b,y}(A) = \{0, -i, -2i, \ldots, -(m-1)i\},$$

all poles are simple, and the elements of $\widehat{\mathscr{T}}_y$ have the form

$$\hat{\tau} = \sum_{\ell=0}^{m-1} \frac{\tau_\ell}{\sigma + i\ell}$$

where $\tau_\ell \in E_y$. Thus, $\mathscr{T}_y$ consists of all polynomials

$$\tau = -\frac{1}{2\pi} \sum_{\ell=0}^{m-1} \oint_\Gamma x^{i\sigma} \frac{\tau_\ell}{\sigma + i\ell}\, d\sigma = -\frac{1}{2\pi} \sum_{\ell=0}^{m-1} \tau_\ell x^\ell$$

as functions on $\mathbb{R}_+$ with values in the fiber E_y of E. These are exactly the elements in the kernel of ${}^bA_y$ since $a_{m0}(0, y)$ is invertible by ellipticity. Therefore, by our definition, the trace bundle of A is

$$\mathscr{T} = \bigsqcup_{y \in \mathcal{Y}} \{ \sum_{\ell=0}^{m-1} \tau_\ell x^\ell : \tau_\ell \in E_y \},$$

the direct sum of m copies of $E_{\mathcal{Y}}$. It is a particularity of regular elliptic operators of order m on sections of a bundle E that they all share the same trace bundle. As expected, the operator $x\partial_x$ discussed in the last paragraph of the previous section acts on $\mathscr{T}$. Its eigenvalues are the numbers $0, 1, \dots, m-1$, the eigenvectors in the fiber over $y \in \mathcal{Y}$ corresponding to the eigenvalue ℓ are the monomials $\tau_\ell x^\ell$, $\tau_\ell \in E_y$. The trace bundle splits globally into the direct sum of m subbundles, each isomorphic to E.

Connecting the above example with a boundary problem

$$\begin{cases} Au = f & \text{in } \mathring{\mathcal{M}} \\ B\gamma u = g & \text{on } \mathcal{Y} \end{cases}$$

for A (where the solution u is sought in $H^m(\mathcal{M}; E)$), note that the classical traces $\gamma_s(u) = D_x^s u|_{\mathcal{Y}}$, assembled into the Taylor polynomial

$$\gamma_A(u) = \sum_{\ell=0}^{m-1} \frac{i^\ell}{\ell!} \gamma_\ell(u)\, x^\ell$$

of u, yield a section of the trace bundle of A. The relation between the regularity of the coefficients $\gamma_\ell(u)$ and the power x^ℓ, namely that the component of $\gamma_A(u)$ in the eigenspace of $x\partial_x$ with eigenvalue ℓ lies in $H^{m-\ell-1/2}(\mathcal{Y}; \mathscr{T})$, is not to be viewed as an accident but as an expression of a tight link between these concepts. This assertion will be fully justified by the results to be described in the next section.

In the case of a general elliptic wedge operator A, the fiberwise action of $x\partial_x$ on its trace bundle will generically have eigenvalues and Jordan canonical form varying with the base point. This makes the issue of regularity of boundary values rather more complicated. We deal with this in [12] as a problem independently of the motivating example. In the rest of the section, we present some of the ideas going into that paper, mostly from the local point of view, rarely yielding to expressing things globally.

On account that the results are independent of those of [11] which were described in the previous section, we view $\pi : \mathscr{T} \to \mathcal{Y}$ as some vector bundle and consider an arbitrary smooth endomorphism $\mathfrak{a} : \mathscr{T} \to \mathscr{T}$ on which no conditions are placed. We continue to assume that $\mathcal{Y}$ is compact although this is not necessary in the general theory. We fix some Hermitian metric on $\mathscr{T}$, not necessarily related to $\mathfrak{a}$, and a smooth density $\mathfrak{m}_{\mathcal{Y}}$. With these we define $L^2(\mathcal{Y}; \mathscr{T})$.

The following example motivates the next step:

Example 4 Let $a : \mathbb{C}^m \to \mathbb{C}^m$ be diagonal with entries $\ell + 1/2$, $\ell = 0, \dots, m-1$. For $\eta \in \mathbb{R}^q$ define $\langle\eta\rangle = (1 + |\eta|^2)^{1/2}$ and set $\langle\eta\rangle^{m-a} = \exp(\log\langle\eta\rangle(m-a))$. Then $\langle\eta\rangle^{m-a}$ is diagonal with entries $\langle\eta\rangle^{m-\ell-1/2}$ and if $u \in C_c^{-\infty}(\mathbb{R}^q; \mathbb{C}^m)$ is such that

$$\frac{1}{(2\pi)^q} \int_{\mathbb{R}^q} e^{iy\cdot\eta} \langle\eta\rangle^{m-a} \widehat{u}(\eta)\, d\eta$$

belongs to $L^2(\mathbb{R}^d; \mathbb{C}^m)$, then the components $u_0, \dots, u_{m-1}$ satisfy

$$u_\ell \in H^{m-\ell-1/2}(\mathbb{R}^q).$$

Returning to the general case, define $\rho^{\mathfrak{a}}$ for $\rho > 0$ as expected:

$$\rho^{\mathfrak{a}(y)} = \frac{i}{2\pi} \oint_{\Gamma_y} \rho^\sigma (\mathfrak{a}(y) - \sigma)^{-1}\, d\sigma, \quad \rho > 0,$$

where Γ_y is a positively oriented contour enclosing $\operatorname{spec}(\mathfrak{a}(y))$. This defines a smooth isomorphism $\mathscr{T} \to \mathscr{T}$. Finally, let g be a Riemannian metric on $\mathcal{Y}$, define $\langle\boldsymbol{\eta}\rangle = (1 + g(\boldsymbol{\eta}, \boldsymbol{\eta}))^{1/2}$, then $p(\boldsymbol{\eta}) = \langle\boldsymbol{\eta}\rangle^{\mathfrak{a}(y)}$, for $\boldsymbol{\eta} \in T_y^*\mathcal{Y}$. We will prove in a moment that p is a symbol in the class $S^\infty_{1,\delta}$ for $\delta > 0$ arbitrarily small, locally in any sufficiently small neighborhood of any point of $\mathcal{Y}$. The order is locally bounded (by our assumption of compactness of $\mathcal{Y}$, also globally bounded).

To define a pseudodifferential operator with p as principal symbol, construct a global parametrization of the conormal to the diagonal $\Delta_{\mathcal{Y}} \subset \mathcal{Y} \times \mathcal{Y}$ as in [8, p. 381]. Namely, view $T\mathcal{Y}$ as the normal bundle to $\Delta_{\mathcal{Y}}$: $N\Delta_{\mathcal{Y}} = \{(v, -v) : v \in T\mathcal{Y}\}$, let V be a neighborhood of the zero section of $T\mathcal{Y}$ on which $\exp : V \to \mathcal{Y} \times \mathcal{Y}$ is a diffeomorphism onto its image W. If $(y, y') \in W$, let $(v, -v)$, $v \in T_{y''}\mathcal{Y}$ be the unique element mapped to (y, y') by exp and define $\varphi(y, y', \boldsymbol{\eta}) = \langle\boldsymbol{\eta}, v\rangle$ for $\boldsymbol{\eta} \in T^*_{y''}\mathcal{M}$. Finally, pick a smooth homogeneous fiberwise density ν on $T^*\mathcal{Y}$ (in local coordinates (y, η), $d\nu = f(y)|d\eta_1 \wedge \cdots \wedge d\eta_q|$, f smooth and positive). With these ingredients define $\Lambda^{\mathfrak{a}}$ through its Schwartz kernel:

$$K_{\Lambda^{\mathfrak{a}}}(y, y') = \frac{1}{(2\pi)^q} \int e^{i\varphi(y,y',\boldsymbol{\eta})} \sum_\alpha \chi_\alpha(y, y') p_\alpha(y, y', \eta)\, d\nu$$

where the χ_α are carefully chosen smooth functions with $\sum \chi_\alpha$ supported in W, equal to 1 in a neighborhood of $\Delta_{\mathcal{Y}}$, and the p_α are constructed using $\langle\eta\rangle_y^{\mathfrak{a}}$ using certain adapted trivializations as we explain below. We then define

$$H^{\mathfrak{a}+s}(\mathcal{Y}; E) = \{u \in C^{-\infty}(\mathcal{Y}; \mathscr{T}) : \Lambda^{\mathfrak{a}+s} u \in L^2(\mathcal{Y}; E)\}$$

for real s using the endomorphism $\mathfrak{a} + sI$ of $\mathscr{T}$.

It remains to explain the choice of functions χ_α and what the p_α are. The purpose is to gain some control on the behavior of $\mathfrak{a}$ through careful localization as follows. Fix $\delta \in (0, 1)$. Given $y_0 \in \mathcal{Y}$, choose for each eigenvalue σ_ℓ of $\mathfrak{a}(y_0) : \mathscr{T}_{y_0} \to \mathscr{T}_{y_0}$, a number $\delta_\ell \in (0, \delta)$ small enough that the closures of the disks $D_\ell = D(\sigma_\ell, \delta_\ell/2) \subset \mathbb{C}$ are disjoint. Then, there is a neighborhood U of y_0 such that $\operatorname{spec}(\mathfrak{a}(y)) \subset \bigcup D_\ell$ if $y \in U$. By way of the projections

$$\Pi_{\ell,y} = \frac{i}{2\pi} \oint_{\partial D_\ell} (\mathfrak{a}(y) - \sigma)^{-1}\, \mathrm{d}\sigma, \quad y \in U,$$

one gets a decomposition

$$\mathscr{T}_U = \bigoplus_\ell \mathscr{T}_U^\ell, \qquad \mathscr{T}_U^\ell = \Pi_\ell \mathscr{T}_U$$

of $\mathscr{T}$ over U into smooth subbundles, each invariant under $\mathfrak{a}$; the eigenvalues of $\mathfrak{a}|_{\mathscr{T}^\ell} : \mathscr{T}_U^\ell \to \mathscr{T}_U^\ell$ cluster within $D(\sigma_\ell, \delta_\ell/2)$. Picking U small enough allows us to assume additionally that the bundles $\mathscr{T}_U^\ell$ (hence also $\mathscr{T}_U$) are trivial. We refer to the above as a δ-admissible decomposition of $\mathscr{T}$ over U ([12, Definition 2.2]). We trivialize $\mathscr{T}_U$ through the trivializations of the $\mathscr{T}_U^\ell$: Let $\phi : \pi^{-1}(U) \to U \times \mathbb{C}^r$ be such an adapted trivialization ($r = \operatorname{rank}\mathscr{T}$), define $a_\phi : U \times \mathbb{C}^r \to U \times \mathbb{C}^r$ by $a_\phi = \phi \circ \mathfrak{a} \circ \phi^{-1}$. Finally, assume that U is also the domain of a local chart of $\mathcal{Y}$.

We now pick an open cover $\{U_\alpha\}$ of $\mathcal{Y}$ consisting of open sets as just described and build up the $\chi_\alpha \in C_c^\infty(U_\alpha \times U_\alpha)$ from a partition of unity near the diagonal subordinate to the cover $\{U_\alpha \times U_\alpha\}$ of $\Delta_\mathcal{Y}$ so that $\sum \chi_\alpha = 1$ near $\Delta_\mathcal{Y}$. Next, with adapted trivializations ϕ_α, we let

$$p_\alpha(y, y', \eta) = \phi_\alpha(y)^{-1} \langle\eta\rangle_y^{a_{\phi_\alpha}(y)} \phi_\alpha(y'),$$

so that $p_\alpha(y, y', \eta) : \mathscr{T}_{y'} \to \mathscr{T}_y$. Implicit in this definition is that we are using parallel transport on $\mathscr{T}|_{U_\alpha}$ with respect to a (flat) connection adapted to a δ-admissible decomposition. See Safarov [23] or Pflaum [22] for a systematic analysis of the role of connections in the definition of standard pseudodifferential operators acting on vector bundles.

It remains to show that the p_α are symbols. In the following lemma, we let $\langle\eta\rangle_y^2 = 1 + \sum g^{ij}(y)\eta_i\eta_j$ with smooth, positive definite g^{ij}. We will drop the reference to ϕ from the notation.

Lemma 1 ([12, Lemma 3.4]) *The function $(y, \eta) \mapsto \langle\eta\rangle_y^{a(y)}$ is, in each open subset $V \Subset U$, a symbol in the Hörmander class $S_{1,\delta}^M$ for some M depending on V.*

The proof of the lemma consists of establishing that

$$D_y^\alpha \partial_\eta^\beta \langle\eta\rangle_y^{a(y)} = p_{\alpha\beta}(y, \eta) \langle\eta\rangle_y^{a(y)}$$

with $\|p_{\alpha\beta}(y, \eta)\| \le C\langle\eta\rangle_y^{-|\beta|+\delta}$ (one can take $\delta = 0$ in the estimate if $\alpha = 0$), then observing that

$$\|\langle\eta\rangle_y^{a(y)}\| \le C(1 + |\eta|)^M, \quad (y, \eta) \in V \times \mathbb{R}^q$$

for some C if $M > \sup\{\Re\sigma : \sigma \in \operatorname{spec}(a(y)),\ y \in V\}$.

To get the estimate for $p_{\alpha\beta}$, note first that since $\rho\partial_\rho\rho^{a(y)} = a(y)\rho^{a(y)}$,

$$\partial_{\eta_j}\langle\eta\rangle_y^{a(y)} = \frac{\partial_{\eta_j}\langle\eta\rangle_y}{\langle\eta\rangle_y} a(y)\langle\eta\rangle_y^{a(y)},$$

and by induction,

$$\partial_\eta^\beta\langle\eta\rangle_y^{a(y)} = p_\beta(y,\eta)\langle\eta\rangle_y^{a(y)},$$

where $p_\beta(y,\eta)$ is a classical symbol of order $-|\beta|$. So we only need to deal with derivatives in y. Differentiating

$$\langle\eta\rangle_y^{a(y)} = \sum_j \frac{i}{2\pi}\oint_{\partial D_\ell}\langle\eta\rangle_y^\sigma (a(y)-\sigma)^{-1}\,\mathrm{d}\sigma$$

one gets that

$$D_y^\alpha\langle\eta\rangle_y^{a(y)} = \sum_\ell \frac{i}{2\pi}\oint_{\partial D_\ell}\langle\eta\rangle_y^\sigma R_\alpha(y,\sigma)\,\mathrm{d}\sigma$$

where $R_\alpha(y,\sigma) = \sum_{k=0}^{|\alpha|}\sigma^k R_{\alpha k}$ where $R_{\alpha k}$ is a sum of terms each being the product of a classical symbol of order zero followed by a product of at most $|\alpha| - k + 1$ factors $(a(y)-\sigma)^{-1}$, each of these factors separated by a factor $\partial_y^{\alpha'}a(y)$. Since $\mathrm{dist}(\mathrm{spec}(a(y)) \cap D_\ell, \partial D_\ell)$ is uniformly bounded from below when $y \in V$ (since $V \Subset U$), the norm of $R_\alpha(y,\sigma)$ is also uniformly bounded when $y \in V$ and $\sigma \in \bigcup_\ell \partial D_\ell$. Now use

$$(D_y^\alpha\langle\eta\rangle_y^{a(y)})\langle\eta\rangle_y^{-a(y)} = \sum_\ell\left(\frac{i}{2\pi}\right)^2\int_{\partial D_\ell\times\partial D_\ell}\langle\eta\rangle_y^{\sigma-\sigma'}R_\alpha(y,\sigma)(a(y)-\sigma')^{-1}\,\mathrm{d}\sigma\,\mathrm{d}\sigma'$$

to get the estimate

$$\|(D_y^\alpha\langle\eta\rangle_y^{a(y)})\langle\eta\rangle_y^{-a(y)}\| \le C\langle\eta\rangle^\delta$$

using that $\|\langle\eta\rangle_y^{\sigma-\sigma'}\| \le C(1+|\eta|)^\delta$ when $\sigma,\sigma' \in \partial D_\ell$. This completes the proof of the lemma.

In addition to defining the spaces $H^{s+\mathfrak{a}}(\mathcal{Y};\mathscr{T})$ to deal with the expected issue of varying regularity of traces of elements of the maximal domain of an elliptic wedge operator, one also needs to develop a theory of pseudodifferential operators able to deal with the varying regularity in order to express general boundary conditions. This of course is reminiscent of the relation between the Douglis–Nirenberg calculus [5], see also Chazarain and Piriou [3], and boundary conditions in the classical situation.

The ingredients going into the definition of such operators are a pair of vector bundles $\mathscr{T},\mathscr{S} \to \mathcal{Y}$ endowed with endomorphisms $\mathfrak{a}$, $\mathfrak{b}$. Symbol classes are defined locally on δ-admissible domains common to both $\mathfrak{a}$ and $\mathfrak{b}$ which are also domains of local charts. Passing to trivializations and local coordinates, we define ([12, Definition 3.1])

$$S^{\mu}_{1,\delta}(U \times \mathbb{R}^q; (\mathbb{C}^r, a), (\mathbb{C}^{r'}, b))$$

for any real μ to be the space of all $p(y,\eta) \in C^\infty(U \times \mathbb{R}^q, \mathrm{Hom}(\mathbb{C}^r, \mathbb{C}^{r'}))$ such that for every subset $K \Subset U$ and all $\alpha, \beta \in \mathbb{N}_0^q$ there exists a constant $C_{K,\alpha,\beta} > 0$ such that

$$\left\| \langle\eta\rangle^{b(y)} (D_y^\alpha \partial_\eta^\beta p(y,\eta)) \langle\eta\rangle^{-a(y)} \right\| \leq C_{K,\alpha,\beta} \langle\eta\rangle^{\mu-|\beta|+\delta|\alpha|}$$

for all $(y,\eta) \in K \times \mathbb{R}^q$. For example, the symbol in Lemma 1 belongs to the class $S^0_{1,\delta}(U \times \mathbb{R}^q; (\mathbb{C}^r, a), (\mathbb{C}^r, 0))$. It also belongs to $S^0_{1,\delta}(U \times \mathbb{R}^q; (\mathbb{C}^r, 0), (\mathbb{C}^r, -a))$, as can be seen by composing $D_y^\alpha \partial_\eta^\beta \langle\eta\rangle_y^{a(y)}$ with $\langle\eta\rangle^{-a}$ on the left rather than on the right.

The classes $S^{\mu}_{1,\delta}(U \times \mathbb{R}^q; (\mathbb{C}^r, a), (\mathbb{C}^{r'}, b))$ have all the usual properties such as invariance under changes of coordinates, and asymptotic summability. Ellipticity of $p \in S^{\mu}_{1,\delta}(U \times \mathbb{R}^q; (\mathbb{C}^r, a), (\mathbb{C}^{r'}, b))$ is defined as usual: The existence of $q \in S^{-\mu}_{1,\delta}(U \times \mathbb{R}^q; (\mathbb{C}^{r'}, b), (\mathbb{C}^r, a))$ such that $pq = 1$ modulo lower-order terms.

Elements in the class just defined are locally in $S^M_{1,\delta}$ for large enough M, so one can use them to define pseudodifferential operators; write $\Psi^{\mu}_{1,\delta}(U; (\mathbb{C}^r, a), (\mathbb{C}^{r'}, b))$ for the space with symbols as just defined. The expected properties hold: invariance under changes of local coordinates and under compatible composition, the latter meaning

$$\Psi^{\mu'}_{1,\delta}(U; (\mathbb{C}^{r'}, b), (\mathbb{C}^{r''}, c)) \circ \Psi^{\mu}_{1,\delta}(U; (\mathbb{C}^r, a), (\mathbb{C}^{r'}, b)) \subset \Psi^{\mu+\mu'}_{1,\delta}\big(U; (\mathbb{C}^r, a), (\mathbb{C}^{r''}, c)\big).$$

The reason for the good behavior at this level is that because the symbols are in a Hörmander class, all the properties of these classes are inherited. In particular, one gets asymptotic expansions for the symbol of a composition, from which one can directly determine the validity of the assertion about compositions in our class.

Consequently, the rough theory, including global definitions of classes of operators of varying order, follows largely as expected (even though some proofs had to be reworked completely) leading to the definition of the classes

$$\Psi^{\mu}_{1,\delta}\big(\mathcal{Y}; (\mathscr{T}, \mathfrak{a}), (\mathscr{S}, \mathfrak{b})\big)$$

associated with pairs of vector bundles with endomorphism. We single out a subclass having twisted homogeneous principal symbols (of order μ), meaning that they are locally constructed using symbols with the property

$$p(y, \rho\eta) = \rho^\mu \rho^{-b(y)} p(y,\eta) \rho^{a(y)} \quad \text{for all } \rho > 1 \text{ and } |\eta| > 1$$

modulo arbitrary symbols in $S^{\mu-1+\delta}_{1,\delta}(U; (\mathbb{C}^r, a), (\mathbb{C}^{r'}, b))$. An operator P in the class $\Psi^{\mu}_{1,\delta}(\mathcal{Y}; (\mathscr{T}, \mathfrak{a}), (\mathscr{S}, \mathfrak{b}))$ has, as expected, a globally defined principal symbol

$$\boldsymbol{\sigma}(P) : T^*\mathcal{Y}\backslash 0 \to \mathrm{Hom}(\pi_{\mathcal{Y}}^* \mathscr{T}, \pi_{\mathcal{Y}}^* \mathscr{S})$$

satisfying

$$\sigma(P)(\rho\eta) = \rho^{\mu}\rho^{-\mathfrak{b}(y)}\sigma(P)(\eta)\rho^{\mathfrak{a}(y)}, \quad \rho > 0,\ \eta \in T^*\mathcal{Y}\backslash 0.$$

8 Boundary Value Problems for First-Order Elliptic Wedge Operators

In this section, we describe some of the results contained [13]. As in that paper, we limit our discussion here to operators of order 1, which allows us to circumvent a number of technical complications proper to the higher-order case; these are being addressed in a forthcoming paper [14].

Henceforth A is an elliptic element of $x^{-1}\mathrm{Diff}^1_e(\mathcal{M}; E, F)$. We will base our discussion on the spaces $x^{-1/2}L^2_b$, i.e., we set $\gamma = 1/2$ (see the last paragraph of Sects. 2 and 3). We aim at describing our specific approach to setting up boundary value problems for A, then establishing sufficient conditions for well-posedness of these problems.

Recall that the trace bundle of A generally does depend on A and that it carries a natural endomorphism $x\partial_x$. In brief, we need to describe a restriction, or trace, operator

$$\gamma_A : \mathcal{D}_{\max}(A) \to C^{-\infty}(\mathcal{Y}; \mathscr{T}),$$

to state the problem, then also ancillary objects to state sufficient conditions for well-posedness.

Our first observation is that if u is a smooth section of $\mathscr{T}$ and ω is a cut-off function, then $\omega u \in \mathcal{D}_{\max}(A)$ if $u \in C^\infty(\mathcal{Y}; \mathscr{T})$; this is Lemma 5.6 of [13]. Strictly speaking, u is defined on $\mathcal{N}^\wedge$ and is a smooth section of $E^\wedge$ over the interior of $\mathcal{N}^\wedge$. Multiplication by ω produces an element of compact support which is transferred to $\mathcal{M}$ via a tubular neighborhood map (and parallel transport). Furthermore, see Lemma 5.6, op. cit., the map

$$C^\infty(\mathcal{Y}; \mathscr{T}) \ni u \mapsto \mathfrak{P}u = \omega u \in \mathcal{D}_{\max}(A)$$

is continuous.

We now use $\mathfrak{P}$ and the pairing (12) to define γ_{A^*}. For details see Sect. 5, op. cit. Let $\mathfrak{P}^\star : C^\infty(\mathcal{Y}; \mathscr{T}^\star) \to \mathcal{D}_{\max}(A^\star)$ be the corresponding map for the formal adjoint of A. Pick $v \in \mathcal{D}_{\max}(A^\star)$. Then

$$C^\infty(\mathcal{Y}; \mathscr{T}) \ni u \mapsto \langle \lambda_v, u\rangle = [\mathfrak{P}u, v]_A = (A\mathfrak{P}u, v)_{x^{-1/2}L^2} - (\mathfrak{P}u, A^\star v)_{x^{-1/2}L^2} \in \mathbb{C}$$

is continuous, so it defines a distribution with values in the bundle $\mathscr{T}^* \otimes |\bigwedge|\mathcal{Y}$. Here, $\mathscr{T}^*$ is the dual bundle of $\mathscr{T}$ with the opposite complex structure, not the trace bundle of $A^\star$, and $|\bigwedge|\mathcal{Y}$ is the density bundle of $\mathcal{Y}$, which we trivialize using the density $\mathfrak{m}_{\mathcal{Y}}$. We get an antilinear map $v \mapsto \lambda_v$ which is converted into a linear map $v \to \gamma_{A^\star} v$ using the nondegenerate pairing $[\cdot,\cdot]^\flat$ to identify $\mathscr{T}^*$ with $\mathscr{T}^\star$, so that

$$[\mathfrak{P}u, v]_A = \int_{\mathcal{Y}} [u, \gamma_{A^\star} v]_y^\flat \, d\mathfrak{m}_{\mathcal{Y}}.$$

We now have:

Theorem 2 *(Theorem 5.11, op. cit.) The trace map is a continuous operator*

$$\gamma_A : \mathcal{D}_{\max}(A) \to H^{-x\partial_x - 1/2}(\mathcal{Y}; \mathscr{T}).$$

Note that if A is a regular first-order elliptic operator, then $x\partial x$ acts as 0, since the traces are zeroth-order polynomials by Example 3, so in this case the space $H^{-x\partial_x - 1/2}(\mathcal{Y}; \mathscr{T})$ is the classical space of sections of $E_{\mathcal{Y}}$ of Sobolev regularity $-1/2$.

With $\mathfrak{P}$ and γ_A, we can now define a replacement for the classical Sobolev space $H^1(\mathcal{M}; E)$ in which solutions are sought when A is a regular first-order elliptic operator, see Sect. 8, op. cit. Namely, we let $H^1_{\mathscr{T}}(\mathcal{M}; E)$ be the completion of $C^\infty_{\mathscr{T}}(\mathcal{M}; E) = \dot{C}^\infty(\mathcal{M}; E) + \mathfrak{P}C^\infty(\mathcal{Y}; \mathscr{T})$, the latter being a replacement of $C^\infty(\mathcal{M}; E)$, with respect to the norm defined by

$$\|u\|^2_{H^1_{\mathscr{T}}} = \|u\|^2_A + \|\gamma_A u\|^2_{H^{1/2 - x\partial_x}}.$$

We recall that $\|u\|^2_A = \|u\|^2 + \|Au\|^2$ is the norm on $\mathcal{D}_{\max}(A)$ but that this norm is typically too weak to give a compact embedding $\mathcal{D}_{\max}(A) \hookrightarrow x^{-1/2} L^2_b$, as illustrated in Example 1.

The following example shows how this trick works.

Example 5 Consider the situation of Example 1, in which $\mathcal{M}$ is the closed unit disk in $\mathbb{R}^2$ and Δ the standard Laplacian. We claim that the norm defined by

$$\|u\|^2_{H^2_{\mathscr{T}}} = \|u\|^2_\Delta + \|\gamma_0 u\|^2_{H^{3/2}(\partial\mathcal{M})} + \|\gamma_1 u\|^2_{H^{1/2}(\partial\mathcal{M})}$$

on $C^\infty(\mathcal{M})$, in which $\gamma_\ell = \partial^\ell u|_{\partial\mathcal{M}}$, is equivalent to the H^2 norm on $\mathcal{M}$. Indeed, on the one hand, the continuity of

$$H^2(\mathcal{M}) \xrightarrow{\gamma_\ell} H^{2-\ell-1/2}(\partial\mathcal{M}), \quad \ell = 0, 1,$$

gives

$$\|u\|^2_{H^2_{\mathscr{T}}} \leq C\|u\|^2_{H^2(\mathcal{M})}$$

for some C. On the other hand,

$$\int_{\mathcal{M}} \|\nabla u\|^2 \, d\lambda = -\int_{\mathcal{M}} u \, \Delta u \, d\lambda - \int_{\partial\mathcal{M}} u \frac{\partial \overline{u}}{\partial \nu} \, ds$$

and the Cauchy–Schwarz inequality give the reverse estimate; ν is the inward pointing normal, $d\lambda$ is Lebesgue measure and ds arc length measure. Thus, while the Δ-graph norm (5) is not sufficiently strong to give compactness of the embedding

$\mathcal{D}_{\max}(\Delta) \to L^2(\mathcal{M})$, the norm $\|u\|^2_{H^2_{\mathscr{T}}}$ does. One gets automatically a split exact sequence

$$0 \to H_0^2(\mathcal{M}) \to H^2_{\mathscr{T}}(\mathcal{M}) \to H^{2-(x\partial_x+1/2)}(\partial\mathcal{M}, \mathscr{T}) \to 0$$

as we discussed in Sect. 3.

In the general case we are discussing, the space $H^1_{\mathscr{T}}(\mathcal{M}; E)$ satisfies the basic requirements mentioned in Sect. 3. First, (3) takes on the form

$$0 \to \mathcal{D}_{\min}(A) \to H^1_{\mathscr{T}}(\mathcal{M}; E) \xrightarrow{\gamma_A} H^{1-(x\partial_x+1/2)}(\partial\mathcal{M}, \mathscr{T}) \to 0.$$

This is an exact sequence which splits through a continuous extension operator

$$\mathfrak{E} : H^{1-(x\partial_x+1/2)}(\partial\mathcal{M}, \mathscr{T}) \to H^1_{\mathscr{T}}(\mathcal{M}; E),$$

a right inverse of γ_A. Second, the inclusion

$$H^1_{\mathscr{T}}(\mathcal{M}; E) \hookrightarrow x^{-1/2} L^2_b(\mathcal{M}; E)$$

is compact. We shall omit here the discussion of $\mathfrak{E}$ and this last statement, referring the reader to [13] for the details.

In order to discuss ellipticity of a boundary value problem, we need to revisit the normal family, see (6). For each y, the operators $A_\wedge(\eta)$, $\eta \in T^*_y\mathcal{Y}$, are cone operators, elements of $x^{-1}\mathrm{Diff}^1_b(\mathcal{Z}^\wedge_y; E_{\mathcal{Z}^\wedge_y}, F_{\mathcal{Z}^\wedge_y})$. As such they have maximal and minimal domains, and each its own trace space, which is just the fiber $\mathscr{T}_y$ lifted via $\pi_{\mathcal{Y}} : T^*\mathcal{Y} \to \mathcal{Y}$ to η because the indicial operator of $A_\wedge(\eta)$ is ${}^bA_y$. They also have each a trace map $\gamma_{A_\wedge(\eta)} : \mathcal{D}_{\max}(A_\wedge(\eta)) \to \pi^*_{\mathcal{Y}}\mathscr{T}_y$. The vector bundle $\mathfrak{K} \to T^*\mathcal{Y}\backslash 0$ whose fiber at η is the kernel of

$$A_\wedge(\eta) : \mathcal{D}_{\max}(A_\wedge(\eta)) \subset x^{-1/2}L^2_b(\mathcal{Z}^\wedge_y; E_{\mathcal{Z}^\wedge_y}) \to x^{-1/2}L^2_b(\mathcal{Z}^\wedge_y; F_{\mathcal{Z}^\wedge_y}), y = \pi_{\mathcal{Y}}(\eta),$$

is a smooth vector bundle. The image $\gamma_{A_\wedge}(\mathfrak{K}) = \mathscr{K} \subset \pi^*_{\mathcal{Y}}\mathscr{T}$ is a subbundle, see Theorem 6.2, op. cit.

Boundary conditions will be, as generically in the classical case, conditions of the form $B\gamma_A u = g$ where

$$B : C^\infty(\mathcal{Y}; \mathscr{T}) \to C^\infty(\mathcal{Y}; \mathscr{G}), \quad B \in \Psi^\mu_{1,\delta}(\mathcal{Y}; (\mathscr{T}, -(x\partial_x + 1/2)), (\mathscr{G}, \mathfrak{a}))$$

for some $\mu \in \mathbb{R}$. As the notation indicates, the vector bundle $\mathscr{G} \to \mathcal{Y}$ comes equipped with a smooth endomorphism $\mathfrak{a}$. We assume that B has twisted homogeneous principal symbol $\boldsymbol{\sigma}(B)$ (see the end of Sect. 7 above). To account for the possible necessity of Atiyah–Patodi–Singer conditions, let

$$\Pi \in \Psi^0_{1,\delta}(\mathcal{Y}; (\mathscr{G}, \mathfrak{a}), (\mathscr{G}, \mathfrak{a}))$$

be a projection with twisted homogeneous principal symbol $\boldsymbol{\sigma}(\Pi)$. On account that Π is a continuous projection, its range, $\Pi H^{1-\mu+\mathfrak{a}}(\mathcal{Y}; \mathscr{G})$, is a closed subspace of

$H^{1-\mu+\mathfrak{a}}(\mathcal{Y};\mathscr{G})$. And because $\boldsymbol{\sigma}(\Pi)=\boldsymbol{\sigma}(\Pi)^2$, the range of $\boldsymbol{\sigma}(\Pi)$ is a subbundle $\mathscr{G}_\Pi$ of $\pi_{\mathcal{Y}}^*\mathscr{G}$ over $T^*\mathcal{Y}\backslash 0$.

We are now ready to state the central result of [13]:

Theorem 3 *(Theorem 9.6, op. cit.). Let* $A\in x^{-1}\mathrm{Diff}_e^1(\mathcal{M};E,F)$ *be w-elliptic. Assume that (10) holds with* $\gamma=1/2$ *and* $m=1$*, so there is a well-defined trace bundle* $\mathscr{T}\to\mathcal{Y}$*. Assume further that*

$$A_\wedge(\boldsymbol{\eta}):\mathcal{D}_{\min}(A_\wedge(\boldsymbol{\eta}))\subset x^{-1/2}L_b^2(\mathcal{M};E)\to x^{-1/2}L_b^2(\mathcal{M};F)\ \textit{is injective,}$$

$$A_\wedge(\boldsymbol{\eta}):\mathcal{D}_{\max}(A_\wedge(\boldsymbol{\eta}))\subset x^{-1/2}L_b^2(\mathcal{M};E)\to x^{-1/2}L_b^2(\mathcal{M};F)\ \textit{is surjective}$$

for all $\boldsymbol{\eta}\in T^*\mathcal{Y}\backslash 0$*. Finally, assume*

$$\boldsymbol{\sigma}(\Pi B):\mathscr{K}\to\mathscr{G}_\Pi\ \textit{on}\ T^*\mathcal{Y}\backslash 0\ \textit{is an isomorphism.}$$

Then, the operator

$$\begin{bmatrix} A \\ \Pi B\gamma_A \end{bmatrix}: H^1_{\mathscr{T}}(\mathcal{M};E)\to \begin{matrix} x^{-1/2}L_b^2(\mathcal{M};F) \\ \oplus \\ \Pi H^{1-\mu+\mathfrak{a}}(\mathcal{Y};\mathscr{G}) \end{matrix}$$

is a Fredholm operator.

In other words, the boundary value problem:

$$\begin{cases} Au=f\in x^{-1/2}L_b^2(\mathcal{M};F) \\ \Pi B(\gamma_A u)=g\in \Pi H^{1-\mu+\mathfrak{a}}(\mathcal{Y};\mathcal{G}), \end{cases}$$

in which u is sought in $H^1_{\mathscr{T}}(\mathcal{M};E)$, is well-posed.

Acknowledgments This work was partially supported by the National Science Foundation, Grants DMS-0901202 (TK) and DMS-0901173 (GAM).

References

1. P. Albin, É. Leichtnam, R. Mazzeo, P. Piazza, The signature package on Witt spaces. Ann. Sci. Éc. Norm. Supér (4). **45**, 241–310 (2012)
2. B. Ammann, R. Lauter, V. Nistor, Pseudodifferential operators on manifolds with a Lie structure at infinity, Ann. of Math. (2) **165**(3), 717–747 (2007)
3. J. Chazarain, A. Piriou, in *Introduction to the Theory of Linear Partial Differential Equations*. Studies in Mathematics and its Applications, vol. 14 (North-Holland, Amsterdam-New York, 1982)
4. M. Costabel, M. Dauge, in *Edge Asymptotics on a Skew Cylinder*. Symposium "Analysis on Manifolds with Singularities" (Breitenbrunn, 1990). Teubner-Texte Math., vol. 131 (Teubner, Stuttgart, 1992), p. 2842

5. A. Douglis, L. Nirenberg, Interior estimates for elliptic systems of partial differential equations. Comm. Pure Appl. Math. **8**, 503–538 (1955)
6. C.L. Epstein, R.B. Melrose, G.A. Mendoza, Resolvent of the Laplacian on strictly pseudoconvex domains. Acta Math. **167**(1–2), 1–106 (1991)
7. J. B. Gil, T. Krainer, G.A. Mendoza, On the closure of elliptic wedge operators. J. Geom. Anal. **23**(4), 2035–2062 (2013)
8. V. Guillemin, S. Sternberg, in *Geometric Asymptotics*. Mathematical Surveys, vol. 14 (AMS, Providence, 1977)
9. L. Hörmander, Pseudo-differential operators and non-elliptic boundary problems. Ann. of Math. **83**, 129–209 (1966)
10. D. Kapanadze, B.-W. Schulze, J. Seiler, Operators with singular trace conditions on a manifold with edges. Integral Equ. Operator Theory. **61**(2), 241–279 (2008)
11. T. Krainer, G. Mendoza, The kernel bundle of a holomorphic Fredholm family. Comm. Partial Differential Equ.. **38**(12), 2107–2125 (2013)
12. T. Krainer, G. Mendoza, Elliptic systems of variable order. Rev. Mat. Iberoam. arXiv:1301.5820 (to appear)
13. T. Krainer, G. Mendoza, Boundary value problems for first order elliptic wedge operators. Preprint 1307.2398 on arXiv.org (Accepted for publication in the American Journal of Mathematics)
14. T. Krainer, G. Mendoza, Boundary value problems for general elliptic wedge operators (in preparation)
15. R. Mazzeo, Elliptic theory of differential edge operators I. Comm. Partial Differential Equations. **16** (1991), 1615–1664
16. R. Mazzeo, R.B. Melrose, Meromorphic extension of the resolvent on complete spaces with asymptotically constant negative curvature. J. Funct. Anal. **75**(2), 260–310 (1987)
17. R. Mazzeo, R.S. Phillips, Hodge theory on hyperbolic manifolds. Duke Math. J. **60**(2), 509–559 (1990)
18. R. Mazzeo, B. Vertman, Elliptic theory of differential edge operators, II: Boundary value problems. arXiv:1307.2266
19. R.B. Melrose, Transformation of boundary value problems. Acta Math. **147**, 149–236 (1981)
20. R.B. Melrose, in *The Atiyah-Patodi-Singer Index Theorem*. Research Notes in Mathematics, vol. 4 (A K Peters, Ltd., Wellesley, MA, 1993)
21. R.B. Melrose, G.A. Mendoza, Elliptic operators of totally characteristic type. MSRI Preprint (1983)
22. M.J. Pflaum, The normal symbol on Riemannian manifolds. New York J. Math. **4**, 97–125 (1998)
23. Y. Safarov, Pseudodifferential operators and linear connections. Proc. London Math. Soc. (3). **74**(2), 379–416 (1997)
24. B. Schmutzler, The structure of branching asymptotics for elliptic bondary value problems in domains with edges. Symposium "Analysis on Manifolds with Singularities" (Breitenbrunn, 1990). Teubner-Texte Math., vol. 131 (Teubner, Stuttgart, 1992), pp. 201–207
25. B.-W. Schulze, *Pseudo-differential Operators on Manifolds with Singularities* (North Holland, Amsterdam, 1991)
26. R. Swan, Vector bundles and projective modules. Trans. Amer. Math. Soc. **105**, 264–277 (1962)

The Time Singular Limit for a Fourth-Order Damped Wave Equation for MEMS

Philippe Laurençot and Christoph Walker

Abstract We consider a free boundary problem modeling electrostatic microelectromechanical systems. The model consists of a fourth-order damped wave equation for the elastic plate displacement which is coupled to an elliptic equation for the electrostatic potential. We first review some recent results on existence and nonexistence of steady states as well as on local and global well-posedness of the dynamical problem, the main focus being on the possible touchdown behavior of the elastic plate. We then investigate the behavior of the solutions in the time singular limit, when the ratio between inertial and damping effects decays to zero.

1 Introduction

An idealized electrostatically actuated microelectromechanical system (MEMS) consists of a fixed horizontal ground plate held at zero potential above which an elastic plate (or membrane) held at potential V is suspended, see Fig. 1. A Coulomb force is generated by the potential difference across the device and results in a displacement of the elastic plate, thereby converting electrostatic energy into mechanical energy, see [4, 19] for a more detailed account and further references. After a suitable scaling and assuming homogeneity in the transversal horizontal direction (i.e., no y-dependence in Fig. 1), the ground plate is assumed to be located at $z = -1$ and the plate displacement $u = u(t,x) \in (-1, \infty)$ evolves according to

$$\begin{aligned} \gamma^2 \partial_t^2 u + \partial_t u + \beta \partial_x^4 u - \tau \partial_x^2 u \\ = -\lambda \left(\varepsilon^2 \, |\partial_x \psi(t,x,u(t,x))|^2 + |\partial_z \psi(t,x,u(t,x))|^2\right) \end{aligned} \tag{1}$$

P. Laurençot (✉)
Toulouse Institute of Mathematics, UMR 5219, University of Toulouse, CNRS, 31062 Toulouse Cedex 9, France
e-mail: laurenco@math.univ-toulouse.fr

C. Walker
Institute for Applied Mathematics, Leibniz University of Hannover, 30167 Hannover, Germany
e-mail: walker@ifam.uni-hannover.de

J. Escher et al., (eds.), *Elliptic and Parabolic Equations,* Springer Proceedings in Mathematics & Statistics 119, DOI 10.1007/978-3-319-12547-3_10

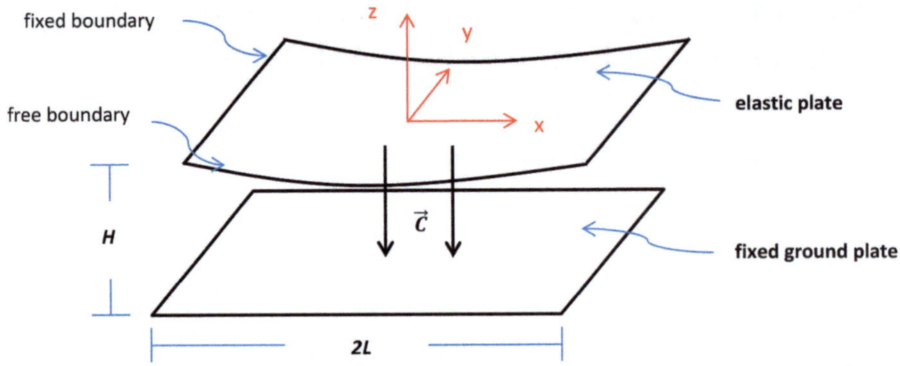

Fig. 1 Sketch of an idealized electrostatic MEMS device

for $t > 0$ and $x \in I := (-1, 1)$ with clamped boundary conditions

$$u(t, \pm 1) = \partial_x u(t, \pm 1) = 0, \quad t > 0, \tag{2}$$

and initial conditions

$$u(0, x) = u^0(x), \quad \gamma^2 \partial_t u(0, x) = \gamma^2 u^1(x), \quad x \in I. \tag{3}$$

In (1), $\gamma^2 \geq 0$ measures the ratio of inertial and damping forces which are given by the second- and first-order time derivatives, respectively, while $\beta \partial_x^4 u$ with $\beta \geq 0$ and $-\tau \partial_x^2 u$ with $\tau \geq 0$ account for bending and stretching of the elastic plate, respectively. The right-hand side of (1) reflects the electrostatic forces exerted on the elastic plate, where the parameter $\lambda > 0$ is proportional to the square of the voltage difference between the two components, and the parameter $\varepsilon > 0$ denotes the aspect ratio of the device (that is, the ratio of height to length). The boundary conditions (2) describe an elastic plate being clamped at its fixed boundary. Finally, the electrostatic potential $\psi = \psi(t, x, z)$ satisfies a rescaled Laplace equation in the time-varying region

$$\Omega(u(t)) := \{(x, z) \in I \times (-1, \infty) \,:\, -1 < z < u(t, x)\}$$

between the ground plate and the elastic plate which reads

$$\varepsilon^2 \, \partial_x^2 \psi + \partial_z^2 \psi = 0, \qquad (x, z) \in \Omega(u(t)), \quad t > 0, \tag{4}$$

$$\psi(t, x, z) = \frac{1 + z}{1 + u(t, x)}, \qquad (x, z) \in \partial\Omega(u(t)), \quad t > 0. \tag{5}$$

Note that, when $\gamma > 0$, Eq. (1) is a hyperbolic nonlocal semilinear fourth-order equation for the plate displacement u, which is coupled to the second-order elliptic equation (4) in the moving domain $\Omega(u(t))$ for the electrostatic potential ψ. If damping effects dominate over inertia effects, one may set $\gamma = 0$ in (1)–(3) and thus obtains a parabolic equation for u.

A noteworthy feature of the above model is that it is only meaningful as long as the elastic plate does not touch down on the ground plate, that is, the deflection u satisfies $u > -1$. From a physical point of view, it is expected that above a certain critical threshold value of λ, the elastic plate "pulls in" and smashes down on the ground plate. Obviously, the stable operating conditions of a given MEMS device heavily depend on the possible occurrence of this so-called *"pull in" instability*. Mathematically, the touchdown singularity manifests in the definition of $\Omega(u(t))$ which becomes disconnected if $u(t,x)$ reaches the value -1 at some point x, and also in the right-hand side of (1) as $\partial_z \psi$ becomes singular at such points since $\psi = 1$ along $z = u$ while $\psi = 0$ along $z = -1$.

1.1 State of the Art

According to the previous discussion, the mathematical investigation aims at showing that the parameter λ indeed governs the dynamics of (1)–(5), in particular the touchdown behavior and the closely related issues of global well-posedness and existence of steady states. More precisely, above a certain threshold value of λ it is conjectured that solutions to (1)–(5) cease to exist globally in time and that there are no steady states, while for λ below this critical value, solutions are global and there are at least two steady states. Moreover, if solutions do not exist globally, then the elastic plate pulls in at some finite time $T_c < \infty$, i.e.,

$$\lim_{t \to T_c} \min_{x \in I} \{u(t,x)\} = -1. \tag{6}$$

In the special situation of the so-called *small aspect ratio model*, which corresponds to setting $\varepsilon = 0$ in (1)–(5), it turns out that the electrostatic potential ψ is explicitly given by $\psi(t,x,z) = (1+z)/(1+u(t,x))$ in $\Omega(u(t))$ and the full system (1)–(5) reduces to a singular evolution equation only involving u. In this case, a quite complete characterization of the expected dynamics—confirming almost all of these conjectures—is obtained in [8, 16], see also [7, 17] for further information as well as [4] and the references therein for the small aspect ratio model in general.

In contrast, due to the present coupling, the free boundary problem with $\varepsilon > 0$ turns out to be even more involved and the literature is far more sparse in this case. A series of recent papers, however, addresses these questions for the free boundary problem when $\gamma = \beta = 0$: see [12] for steady-state solutions, [1] for the corresponding parabolic problem, and [2, 3] for a quasilinear version thereof. The fourth-order case $\beta > 0$ with $\gamma \geq 0$ is investigated in [13] and we shall review its main results below (see also [14] for a quasilinear version for $\beta > 0$ and $\gamma = 0$).

In all the just-cited references on the free boundary problem, a very crucial ingredient in the analysis is the understanding of the elliptic problem (4)–(5) in the domain $\Omega(u)$ in dependence of a given (free) boundary described by a function $u : [-1,1] \to (-1,\infty)$ for a fixed time t (being suppressed for the moment). In particular, precise information on the gradient trace of the potential $\psi = \psi_u$ on the elastic plate is required as a function of u. For this, one can transform the Laplace equation (4)–(5) for ψ_u to an elliptic problem in the fixed rectangle $I \times (0,1)$ (with coefficients depending on u and its x-derivatives up to second order and being singular in case that u approaches -1) for a transformed electrostatic potential ϕ_u given by

$$\phi_u(x,\eta) := \psi_u(x,(1+u(x))\eta - 1), \quad (x,\eta) \in I \times (0,1). \tag{7}$$

Using then elliptic regularity theory and pointwise multiplications in Sobolev spaces, the following key result can be shown [1, Proposition 5]:

Proposition 1 *Given* $2\alpha \in (0,1/2)$ *and* $\kappa \in (0,1)$ *define an open subset of* $H^{2+2\alpha}(I)$ *by*

$$S_\alpha(\kappa) := \{v \in H^{2+2\alpha}(I);\ v(\pm 1) = \partial_x v(\pm 1) = 0,\ \|v\|_{H^{2+2\alpha}(I)} < 1/\kappa,$$
$$\text{and } v(x) > -1+\kappa \text{ for } x \in I\}.$$

Then, for each $u \in S_\alpha(\kappa)$, *there is a unique solution* $\psi = \psi_u \in H^2(\Omega(u))$ *to (4)–(5) and*

$$\|\psi_u\|_{H^2(\Omega(u))} \le C_0(\kappa), \quad u \in S_\alpha(\kappa), \tag{8}$$

with

$$\|\phi_{u_1} - \phi_{u_2}\|_{H^2(I\times(0,1))} \le C_0(\kappa)\|u_1 - u_2\|_{H^{2+2\alpha}(I)}, \quad u_1,u_2 \in S_\alpha(\kappa), \tag{9}$$

for some positive constant $C_0(\kappa)$ *depending on* κ *and also on* α *and* ε, *but not on* $u \in S_\alpha(\kappa)$. *Moreover, the mapping*

$$g : S_\alpha(\kappa) \to H^{2\alpha}(I), \quad u \mapsto \varepsilon^2\left|\partial_x\psi_u(x,u(x))\right|^2 + \left|\partial_z\psi_u(x,u(x))\right|^2$$

is analytic, bounded, and uniformly Lipschitz continuous.

In fact, estimate (8) follows from [1, Lemma 6] while (9) is shown in [1, Eq. (38)]. Importantly, the minimal regularity required in order to control the potential ψ_u in terms of suitable norms of u appears to be that the latter belongs to $W_q^2(I)$ for some $q > 2$, whence necessarily $2\alpha > 0$ above (see [1, Proposition 5] for a more precise result). The regularity properties of the map g are stated in this form for simplicity but are still sufficient in the fourth-order case $\beta > 0$ considered in the following.

As pointed out above, the system (1)–(5) is investigated in [13]. In particular, existence and nonexistence of steady states are derived in [13] in dependence of the voltage value λ. While the former is a rather immediate consequence of the implicit function theorem (once Proposition 1 is established), the latter is based on a nonlinear version of the eigenfunction method which involves a positive eigenfunction in $H^4(I)$ associated to a positive eigenvalue of the fourth-order operator $\beta\partial_x^4 - \tau\partial_x^2$ subject to clamped boundary conditions [6, 15, 18]. The result reads [13, Theorem 1.7]:

Theorem 1 *(Steady states)*

(i) **Existence:** *There is* $\lambda_s > 0$ *such that for each* $\lambda \in (0, \lambda_s)$*, there exists an asymptotically stable steady-state solution* $(U_\lambda, \Psi_\lambda)$ *to (1)–(5) with* $U_\lambda \in H^4(I)$ *satisfying* $-1 < U_\lambda < 0$ *in* I *and* $\Psi_\lambda \in H^2(\Omega(U_\lambda))$.

(ii) **Nonexistence:** *There is* $\lambda_c \geq \lambda_s$ *such that there is no (sufficiently smooth) steady-state solution* (u, ψ) *to (1)–(5) for* $\lambda > \lambda_c$.

Yet, open problems are whether $\lambda_c = \lambda_s$ and whether there is a second (unstable) steady state for $\lambda < \lambda_s$ as in the small aspect ratio model, see [8, 16].

In [13], also the well-posedness of the dynamical problem is addressed. Due to Proposition 1, one may write (1)–(5) as a single semilinear Cauchy problem for the plate displacement u (and its time derivative) that one can then solve by means of semigroup theory. We recall here the main statements from [13, Propositions 3.1 and 3.2, Corollaries 5.7 and 5.10] and indicate an explicit dependence on the parameter γ for future purposes:

Theorem 2 (Well-posedness). *Let* $\gamma \geq 0$ *and* $2\alpha \in (0, 1/2)$. *Consider an initial condition* (u^0, u^1) *in* $H^{4+2\alpha}(I) \times H^{2+2\alpha}(I)$ *satisfying* $u^0(\pm 1) = \partial_x u^0(\pm 1) = u^1(\pm 1) = \partial_x u^1(\pm 1) = 0$ *and such that* $u^0 > -1$ *in* I. *Then the following hold:*

(i) **Local existence:** *For each* $\lambda > 0$*, there is a unique solution* (u_γ, ψ_γ) *to (1)–(5) on the maximal interval of existence* $[0, T_\gamma)$ *in the sense that*

$$u_\gamma \in C([0, T_\gamma), H^{2+2\alpha}(I)) \cap C^1([0, T_\gamma), H^{2\alpha}(I)) \quad \text{if} \quad \gamma > 0,$$

$$u_0 \in C([0, T_0), H^4(I)) \cap C^1([0, T_0), L_2(I)) \quad \text{if} \quad \gamma = 0,$$

with

$$\partial_t^k u_\gamma \in L_1((0, T), H^{4+2\alpha-2k}(I)), \quad k = 0, 1, 2, \quad T \in (0, T_\gamma), \quad \gamma > 0,$$

satisfies (1)–(3) together with

$$u_\gamma(t, x) > -1, \quad (t, x) \in [0, T_\gamma) \times I,$$

while $\psi_{u_\gamma}(t) \in H^2(\Omega(u_\gamma(t)))$ *solves (4)–(5) in* $\Omega(u_\gamma(t))$ *for each* $t \in [0, T_\gamma)$.

(ii) **Touchdown:** *There is* $\gamma_1 > 0$ *such that, if* $\gamma \in [0, \gamma_1]$*, then the solution* (u_γ, ψ_γ) *to (1)–(5) obeys the following criterion for global existence: if for each* $T > 0$ *there is* $\kappa(T) \in (0, 1)$ *such that*

$$u_\gamma(t) \geq -1 + \kappa(T) \quad \text{in } I$$

for $t \in [0, T_\gamma) \cap [0, T]$*, then* $T_\gamma = \infty$.

(iii) **Global existence:** *Given $\kappa \in (0,1)$, there are numbers $\lambda_* = \lambda_*(\gamma,\kappa) > 0$ and $N_* = N_*(\gamma,\kappa) > 0$ such that $T_\gamma = \infty$ provided that*

$$\left\|(u^0, u^1)\right\|_{H^{4+2\alpha}(I)\times H^{2+2\alpha}(I)} \leq N_*, \quad u^0 \geq -1+\kappa \ \text{ in } I,$$

and $\lambda \leq \lambda_*$. In this case, $u_\gamma \in L_\infty((0,\infty), H^{2+2\alpha}(I))$ with

$$\inf_{(t,x)\in[0,\infty)\times I} u_\gamma(t,x) > -1.$$

Actually, in case of the damping dominated limit $\gamma = 0$, less regularity on the initial data is required while more regularity on the solution (u_0, ψ_0) may be obtained, see [13, Propositions 3.1 and 3.6] for details. The global existence result for small λ values stated in part (iii) of the above theorem is based on the exponential decay of the associated semigroup which stems from the damping term. This fact will be exploited further in Sect. 2. Note that for small voltage values λ, touchdown is impossible, even in infinite time. An interesting, but still lacking salient feature of the physical model is a relation between $\lambda_c \geq \lambda_s$ from Theorem 1 and (an optimally chosen) λ_*.

The probably most important contribution to be brought forward by Theorem 2 is the global existence criterion stated in part (ii) which implies that touchdown is the only singularity preventing global existence. This is in clear contrast to the second-order case $\beta = 0$ considered in [1–3], where—in principle—a finite existence time T_γ may also be due to a blowup of some Sobolev norm of $u_\gamma(t,\cdot)$ as $t \to T_\gamma$. Roughly speaking, this physically most relevant feature is achieved by fully exploiting the additional information coming from the fourth-order term as well as the underlying gradient flow structure of (1)–(5), the latter seeming to have been unnoticed so far though being inherent in the model derivation. Indeed, introducing the total energy

$$\mathcal{E}(u_\gamma) := \mathcal{E}_{\mathrm{m}}(u_\gamma) - \lambda \mathcal{E}_{\mathrm{e}}(u_\gamma)$$

involving the mechanical energy

$$\mathcal{E}_{\mathrm{m}}(u_\gamma) := \frac{\beta}{2}\left\|\partial_x^2 u_\gamma\right\|_{L_2(I)}^2 + \frac{\tau}{2}\left\|\partial_x u_\gamma\right\|_{L_2(I)}^2$$

and the electrostatic energy

$$\mathcal{E}_{\mathrm{e}}(u_\gamma) := \int_{\Omega(u_\gamma)} \left[\varepsilon^2|\partial_x \psi_{u_\gamma}(x,z)|^2 + |\partial_z \psi_{u_\gamma}(x,z)|^2\right] \mathrm{d}(x,z), \tag{10}$$

the following energy equality holds [13, Propositions 1.3 and 1.6]:

Proposition 2 *(Energy equality). Under the assumptions of Theorem 2(i),*

$$\mathcal{E}(u_\gamma(t)) + \frac{\gamma^2}{2}\left\|\partial_t u_\gamma(t)\right\|_{L_2(I)}^2 + \int_0^t \left\|\partial_t u_\gamma(s)\right\|_{L_2(I)}^2 \mathrm{d}s = \mathcal{E}(u^0) + \frac{\gamma^2}{2}\left\|u^1\right\|_{L_2(I)}^2 \tag{11}$$

for $t \in [0, T_\gamma)$.

Note, however, that the energy $\mathcal{E}$ is the sum of terms with different signs and is thus *not* coercive. The main difficulty in the proof of Proposition 2 is the computation of the derivative of $\mathcal{E}_e(u_\gamma)$ with respect to u_γ since its dependence on u_γ is somehow implicit and involves the domain $\Omega(u_\gamma)$. Nevertheless, the derivative can be interpreted as the shape derivative of the Dirichlet integral of $\psi_\gamma = \psi_{u_\gamma}$, which can be computed and shown to be equal to the right-hand side of (1)—except for the sign—by shape optimization arguments [13]. An additional difficulty stems from the fact that the time regularity of u_γ as stated in part (i) of Theorem 2 is not sufficient for a direct computation and one rather has to use an approximation argument.

Then, to prove the global existence criterion from part (ii) of Theorem 2, one may proceed as follows: As long as $u_\gamma(t,\cdot)$ stays away from -1, one may control the electrostatic energy $\mathcal{E}_e(u_\gamma(t))$ by the mechanical energy $\mathcal{E}_m(u_\gamma(t))$ and then derive from the time decrease of $\mathcal{E}(u_\gamma(t))$ implied by Proposition 2 first a bound on the $H^2(I)$-norm of $u_\gamma(t)$ and subsequently also on higher Sobolev norms by a bootstrapping argument which yields global existence.

1.2 The Time Singular Limit

In many research papers, mostly dedicated to the small aspect ratio model with $\varepsilon = 0$, inertial effects are neglected from the outset as damping effects may be predominant, a few exceptions being [7, 10]. In this note, we now shall investigate the behavior of the solutions in the damping dominated limit $\gamma^2 \to 0$. Obviously, considering such a time singular limit from a mathematical point of view requires in particular a common interval of existence, independent of γ, that is, a lower bound on the maximal existence time T_γ. This is provided by the first result of this chapter:

Proposition 3 *(Minimal existence time). Let $2\alpha \in (0,1/2)$, $\gamma \in (0,\gamma_1]$, $\lambda > 0$. Consider an initial condition $(u^0,u^1) \in H^{4+2\alpha}(I) \times H^{2+2\alpha}(I)$ such that $u^0 \in S_\alpha(\kappa)$ for some $\kappa \in (0,1)$ and $u^1(\pm 1) = \partial_x u^1(\pm 1) = 0$. Let (u_γ,ψ_γ) be the unique solution to (1)–(5) defined on the maximal interval of existence $[0,T_\gamma)$. There are $\hat{\gamma} := \hat{\gamma}(\kappa,\gamma_1,\|u^1\|_{H_D^{2\alpha}(I)}) \in (0,\gamma_1]$, $N := N(\kappa,\gamma_1) > 0$, and $\Lambda := \Lambda(\kappa,\gamma_1) > 0$ such that:*

(i) There is $\hat{T} := \hat{T}(\lambda,\kappa,\gamma_1,\|u^0\|_{H^{4+2\alpha}(I)}) \in (0,\infty)$ such that, for all $\gamma \in (0,\hat{\gamma})$, $T_\gamma > \hat{T}$ and $u_\gamma(t) \in S_\alpha(\kappa/2)$ for $t \in [0,\hat{T}]$.

(ii) If $\lambda \in (0,\Lambda)$ and $\|u^0\|_{H^{4+2\alpha}(I)} \le N$, then $T_\gamma = \infty$ and $u_\gamma(t) \in S_\alpha(\kappa/2)$ for $t \ge 0$ and $\gamma \in (0,\hat{\gamma})$.

The proof of this proposition is given in Sect. 2. It relies on an exponential decay of the energy associated to the damped wave equation being independent of $\gamma \in [0,\gamma_1]$.

As a consequence we are in a position to investigate the damping dominated limit and prove that (u_γ,ψ_γ) converges toward (u_0,ψ_0) in a suitable sense as $\gamma^2 \to 0$.

Theorem 3 (Damping dominated limit). *Under the assumptions of Proposition 3(i) and as* $\gamma^2 \longrightarrow 0$,

$$u_\gamma \longrightarrow u_0 \ \ in \ \ C([0,\hat{T}], H^{2+2\xi}(I)) \tag{12}$$

for each $\xi \in (0,\alpha)$ *and*

$$\phi_{u_\gamma} \longrightarrow \phi_{u_0} \ \ in \ \ C\big([0,\hat{T}], H^2(I \times (0,1))\big), \tag{13}$$

where ϕ_{u_γ} *is the transformed electrostatic potential given by (7) (with* u *replaced by* u_γ*). In addition, if* $u^0 = u^1 = 0$*, then*

$$\partial_t u_\gamma \longrightarrow \partial_t u_0 \ \ in \ \ L_p(0,\hat{T}; H^{2\alpha}(I)) \tag{14}$$

for each $p \in (1,\infty)$*. Under the assumptions of Proposition 3(ii), statements (12)-(14) are true for each* $T > 0$ *instead of* $\hat{T}$.

The proof of Theorem 3 is performed in Sect. 3. It is based on compactness properties of $(u_\gamma, \psi_\gamma)_{\gamma\in(0,\gamma_1)}$ being provided by the energy functional $\mathcal{E}$.

2 A Lower Bound on the Maximal Existence Time

In order to prove Proposition 3, we consider an initial condition (u^0, u^1) belonging to $H_D^{4+2\alpha}(I)\times H_D^{2+2\alpha}(I)$ and such that $u^0 \in S_\alpha(\kappa)$ for some $\kappa \in (0,1)$ and $2\alpha \in (0,1/2)$, where

$$H_D^\theta(I) := \begin{cases} \{v \in H^\theta(I)\,;\, v(\pm 1) = \partial_x v(\pm 1) = 0\}, & \theta > \dfrac{3}{2}, \\ \{v \in H^\theta(I)\,;\, v(\pm 1) = 0\}, & \dfrac{1}{2} < \theta < \dfrac{3}{2}, \\ H^\theta(I), & \theta < \dfrac{1}{2}. \end{cases}$$

We fix $\gamma \in (0,\gamma_1]$ with $\gamma_1 > 0$ introduced in Theorem 2(ii) and let (u_γ, ψ_γ) with

$$u_\gamma \in C([0,T_\gamma), H_D^{2+2\alpha}(I)) \cap C^1([0,T_\gamma), H_D^{2\alpha}(I))$$

and

$$\partial_t^k u_\gamma \in L_1((0,T), H_D^{4+2\alpha-2k}(I)), \quad k = 0,1,2, \quad T \in (0,T_\gamma),$$

be the unique solution to (1)–(5) on the maximal interval of existence $[0,T_\gamma)$ as provided by Theorem 2. Then, introducing the operator

$$A_\alpha := \beta\partial_x^4 - \tau\partial_x^2 \in \mathcal{L}(H_D^{4+2\alpha}(I), H_D^{2\alpha}(I)),$$

we have

$$\gamma^2 \frac{\mathrm{d}^2}{\mathrm{d}t^2} u_\gamma + \frac{\mathrm{d}}{\mathrm{d}t} u_\gamma + A_\alpha u_\gamma = -\lambda g(u_\gamma), \quad t \in (0,T_\gamma),$$

in $H_D^{2\alpha}(I)$, the function g being defined in Proposition 1.

We want to control a suitable norm of $u_\gamma(t)$ for which we basically use an idea from [9, Sect. 2], the difference mainly being the focus on estimates which are uniform with respect to $\gamma \in (0, \gamma_1]$. To this end, define

$$v(t) := u_\gamma(t) - u^0 \quad \text{and} \quad f(t) := -\lambda g(u_\gamma(t)) - A_\alpha u^0$$

for $t \in [0, T_\gamma)$. Then v solves the equation

$$\gamma^2 \frac{\mathrm{d}^2}{\mathrm{d}t^2} v + \frac{\mathrm{d}}{\mathrm{d}t} v + A_\alpha v = f, \quad t \in (0, T_\gamma), \tag{15}$$

in $H_D^{2\alpha}(I)$ with initial condition $(v(0), \partial_t v(0)) = (0, u^1)$. Recall that there are real numbers $c_2 \geq c_1 \geq c_0 \geq 1$ such that

$$\|z\|^2_{H_D^{2\alpha}(I)} \leq c_0 \|z\|^2_{H_D^{2+2\alpha}(I)} \leq c_1 \left\| A_\alpha^{1/2} z \right\|^2_{H_D^{2\alpha}(I)} \leq c_2 \|z\|^2_{H_D^{2+2\alpha}(I)} \tag{16}$$

for all $z \in H_D^{2+2\alpha}(I)$. Then, defining

$$E(t) := \left\| A_\alpha^{1/2} v(t) \right\|^2_{H_D^{2\alpha}(I)} + \gamma^2 \left\| \partial_t v(t) \right\|^2_{H_D^{2\alpha}(I)}, \quad t \in (0, T_\gamma),$$

and

$$F(t) := \gamma \left\langle v(t), \partial_t v(t) \right\rangle_{H_D^{2\alpha}(I)}, \quad t \in (0, T_\gamma),$$

we deduce from (15), (16), and the self-adjointness of $A_\alpha^{1/2}$ in $H_D^{2\alpha}(I)$ that

$$\frac{\mathrm{d}}{\mathrm{d}t} E(t) = -2 \left\| \partial_t v(t) \right\|^2_{H_D^{2\alpha}(I)} + 2 \left\langle f(t), \partial_t v(t) \right\rangle_{H_D^{2\alpha}(I)} \tag{17}$$

and

$$|F(t)| \leq \frac{c_1}{2} E(t) \tag{18}$$

for a. e. $t \in (0, T_\gamma)$. Next, let

$$b := \min \left\{ \frac{2}{2\gamma_1^2 + c_1 + 1}, \frac{1}{2c_1}, \frac{1}{\gamma_1 c_1} \right\}$$

and introduce $G(t) := E(t) + b\gamma F(t)$ for $t \in (0, T_\gamma)$. According to (16)–(18) and Young's inequality,

$$\begin{aligned}
\tfrac{\mathrm{d}}{\mathrm{d}t} G(t) &= \left(-2 + b\gamma^2\right) \left\| \partial_t v(t) \right\|^2_{H_D^{2\alpha}(I)} - b \left\langle v(t), \partial_t v(t) + A_\alpha v(t) - f(t) \right\rangle_{H_D^{2\alpha}(I)} \\
&\quad + 2 \left\langle f(t), \partial_t v(t) \right\rangle_{H_D^{2\alpha}(I)} \\
&\leq \left(-2 + b\gamma^2\right) \left\| \partial_t v(t) \right\|^2_{H_D^{2\alpha}(I)} - b \left\| A_\alpha^{1/2} v(t) \right\|^2_{H_D^{2\alpha}(I)} \\
&\quad + b \left(\tfrac{1}{4c_1} \left\| v(t) \right\|^2_{H_D^{2\alpha}(I)} + c_1 \left\| \partial_t v(t) \right\|^2_{H_D^{2\alpha}(I)} \right) \\
&\quad + \tfrac{b^2}{2} \left\| v(t) \right\|^2_{H_D^{2\alpha}(I)} + \tfrac{1}{2} \left\| f(t) \right\|^2_{H_D^{2\alpha}(I)} + b \left\| \partial_t v(t) \right\|^2_{H_D^{2\alpha}(I)} + \tfrac{1}{b} \left\| f(t) \right\|^2_{H_D^{2\alpha}(I)} \\
&\leq \left(-2 + b\gamma^2 + bc_1 + b\right) \left\| \partial_t v(t) \right\|^2_{H_D^{2\alpha}(I)} - \tfrac{b}{2} \left\| A_\alpha^{1/2} v(t) \right\|^2_{H_D^{2\alpha}(I)} \\
&\quad + \tfrac{b}{2} \left(\left(\tfrac{1}{2c_1} + b \right) \left\| v(t) \right\|^2_{H_D^{2\alpha}(I)} - \left\| A_\alpha^{1/2} v(t) \right\|^2_{H_D^{2\alpha}(I)} \right) \\
&\quad + \left(\tfrac{1}{2} + \tfrac{1}{b} \right) \left\| f(t) \right\|^2_{H_D^{2\alpha}(I)}
\end{aligned}$$

for a. e. $t \in (0, T_\gamma)$. Since the choice of b ensures that

$$\frac{1}{2c_1} + b \le \frac{1}{c_1} \quad \text{and} \quad -2 + b(\gamma^2 + c_1 + 1) \le -b\gamma^2,$$

the third term on the right-hand side is nonpositive by (16) and we obtain

$$\frac{\mathrm{d}}{\mathrm{d}t} G(t) \le -\frac{b}{2} E(t) + \frac{b+2}{2b} \|f(t)\|^2_{H_D^{2\alpha}(I)} \quad \text{for a.e. } t \in (0, T_\gamma).$$

Observe that (18) and the choice of b also ensure

$$\frac{1}{2} E(t) \le \left(1 - \frac{c_1 b\gamma}{2}\right) E(t) \le G(t) \le \left(1 + \frac{c_1 b\gamma}{2}\right) E(t) \le \left(1 + \frac{c_1 b\gamma_1}{2}\right) E(t)$$

for $t \in (0, T_\gamma)$, whence

$$\frac{\mathrm{d}}{\mathrm{d}t} G(t) \le -\frac{b}{2 + c_1 b\gamma_1} G(t) + \frac{b+2}{2b} \|f(t)\|^2_{H_D^{2\alpha}(I)} \quad \text{for a.e. } t \in (0, T_\gamma).$$

Consequently, setting $\omega := b/(2 + c_1 b\gamma_1)$,

$$E(t) \le 2G(t) \le \tfrac{b}{\omega} e^{-\omega t} E(0) + \tfrac{b+2}{b\omega}\left(1 - e^{-\omega t}\right) \sup_{s\in(0,t)} \left\{ \|f(s)\|^2_{H_D^{2\alpha}(I)} \right\}$$

for $t \in (0, T_\gamma)$. Now, owing to (16) and the definitions of $E(t)$ and $f(t)$, there is a constant $M := M(\gamma_1) > 0$ such that

$$\begin{aligned} &\|u_\gamma(t) - u^0\|^2_{H_D^{2+2\alpha}(I)} \\ &\le M\gamma^2 \|u^1\|^2_{H_D^{2\alpha}(I)} + M\left(1 - e^{-\omega t}\right) \left[\lambda^2 \sup_{s\in(0,t)} \left\{ \|g(u_\gamma(s))\|^2_{H_D^{2\alpha}(I)} \right\} + \|u^0\|^2_{H_D^{4+2\alpha}(I)} \right] \end{aligned}$$

for $t \in (0, T_\gamma)$. Since u^0 belongs to $S_\alpha(\kappa)$, it follows from the time continuity of u_γ in $H_D^{2+2\alpha}(I)$ and the continuous embedding of $H_D^{2+2\alpha}(I)$ in $L_\infty(I)$ that

$$\hat{T}_\gamma := \sup\left\{ t_0 \in (0, T_\gamma)\,;\, u_\gamma(t) \in S_\alpha(\kappa/2) \ \text{ for all } \ t \in [0, t_0) \right\} > 0.$$

Then, by Proposition 1,

$$\|g(u_\gamma(t))\|_{H_D^{2\alpha}(I)} \le c_3(\kappa) := \sup_{w\in S_\alpha(\kappa/2)} \left\{ \|g(w)\|_{H_D^{2\alpha}(I)} \right\}, \quad t \in [0, \hat{T}_\gamma),$$

and we conclude that

$$\begin{aligned} \|u_\gamma(t) - u^0\|^2_{H_D^{2+2\alpha}(I)} &\le M\gamma^2 \|u^1\|^2_{H_D^{2\alpha}(I)} \\ &\quad + M\left[\lambda^2 c_3(\kappa)^2 + \|u^0\|^2_{H_D^{4+2\alpha}(I)} \right] \left(1 - e^{-\omega t}\right) \end{aligned}$$

for $t \in [0, \hat{T}_\gamma)$. Since $u^0 \in S_\alpha(\kappa)$ and $H_D^{2+2\alpha}(I)$ embeds continuously in $L_\infty(I)$ with constant, say, $c_4 \geq 1$, the previous inequality ensures that

$$\|u_\gamma(t)\|_{H_D^{2+2\alpha}(I)} \leq \left\|u_\gamma(t) - u^0\right\|_{H_D^{2+2\alpha}(I)} + \left\|u^0\right\|_{H_D^{2+2\alpha}(I)} < \frac{2}{\kappa}$$

when

$$M\gamma^2\left\|u^1\right\|^2_{H_D^{2\alpha}(I)} + M\left[\lambda^2 c_3(\kappa)^2 + \left\|u^0\right\|^2_{H_D^{4+2\alpha}(I)}\right]\left(1 - e^{-\omega t}\right) < \frac{1}{\kappa^2}$$

and

$$u_\gamma(t) \geq u^0 - \left\|u_\gamma(t) - u^0\right\|_{L_\infty(I)} \geq \kappa - 1 - c_4\left\|u_\gamma(t) - u^0\right\|_{H_D^{2+2\alpha}(I)} > \frac{\kappa}{2} - 1$$

when

$$M\gamma^2\left\|u^1\right\|^2_{H_D^{2\alpha}(I)} + M\left[\lambda^2 c_3(\kappa)^2 + \left\|u^0\right\|^2_{H_D^{4+2\alpha}(I)}\right]\left(1 - e^{-\omega t}\right) < \frac{\kappa^2}{4c_4^2}.$$

Thus, as $\kappa^2/(4c_4^2) \leq 1 \leq 1/\kappa^2$, we deduce from the above analysis that $u_\gamma(t)$ belongs to $S_\alpha(\kappa/2)$ provided $t \in [0, \hat{T}_\gamma)$ and $\gamma \in (0, \gamma_1]$ satisfy

$$\gamma^2\left\|u^1\right\|^2_{H_D^{2\alpha}(I)} < \frac{\kappa^2}{8Mc_4^2} \tag{19}$$

and

$$\left[\lambda^2 c_3(\kappa)^2 + \left\|u^0\right\|^2_{H_D^{4+2\alpha}(I)}\right]\left(1 - e^{-\omega t}\right) < \frac{\kappa^2}{8Mc_4^2}. \tag{20}$$

Therefore, there exist

$$\hat{\gamma} := \hat{\gamma}(\kappa, \gamma_1, \|u^1\|_{H_D^{2\alpha}(I)}) \in (0, \gamma_1) \quad \text{and} \quad \hat{T} := \hat{T}(\lambda, \kappa, \gamma, \|u^0\|_{H_D^{4+2\alpha}(I)}) > 0$$

such that

$$u_\gamma(t) \in S_\alpha(\kappa/2) \quad \text{for} \quad t \in [0, \hat{T}_\gamma) \cap [0, \hat{T}] \quad \text{and} \quad \gamma \in (0, \hat{\gamma}).$$

Recalling the definition of $\hat{T}_\gamma$, the previous statement implies in particular that $\hat{T}_\gamma \geq \hat{T}$. Finally, owing to the positivity of ω, it is clear that if one requires that

$$\lambda^2 c_3(\kappa)^2 + \left\|u^0\right\|^2_{H_D^{4+2\alpha}(I)} < \frac{\kappa^2}{8Mc_4^2}$$

instead of (20), there exist $\Lambda := \Lambda(\kappa, \gamma_1) > 0$ and $N := N(\kappa, \gamma_1) > 0$ such that $u_\gamma(t)$ belongs to $S_\alpha(\kappa/2)$ for all $t \in [0, T_\gamma)$ and $\gamma \in (0, \hat{\gamma})$ provided that $\lambda \in (0, \Lambda)$ and $\|u^0\|_{H_D^{4+2\alpha}(I)} \leq N$, whence $T_\gamma = \infty$ due to Theorem 2(ii). This proves Proposition 3.

3 The Time Singular Limit $\gamma^2 \longrightarrow 0$

In order to prove Theorem 3, we stick to the notation from the previous section. Recall that

$$u_\gamma(t) \in S_\alpha(\kappa/2), \quad t \in [0, \hat{T}], \quad \gamma \in (0, \hat{\gamma}). \tag{21}$$

It then follows from (8) that

$$\|\psi_{u_\gamma}(t)\|_{H^2(\Omega(u_\gamma(t)))} \le C_0(\kappa), \quad t \in [0, \hat{T}], \quad \gamma \in (0, \hat{\gamma}).$$

This gives a uniform bound on the electrostatic energy $\mathcal{E}_e(u_\gamma(t))$ defined in (10) so that (11) implies

$$\frac{\gamma^2}{2}\|u_\gamma(t)\|_{L_2(I)}^2 + \int_0^t \left\|\partial_t u_\gamma(s)\right\|_{L_2(I)}^2 \,\mathrm{d}s \le c(\kappa), \quad t \in [0, \hat{T}], \quad \gamma \in (0, \hat{\gamma}). \tag{22}$$

Now, let $\xi \in (0, \alpha)$. Owing to (21) and (22), the set $\{u_\gamma\,;\, \gamma \in (0, \hat{\gamma})\}$ is bounded in $L_\infty(0, \hat{T}; H^{2+2\xi}(I))$ with $\{\partial_t u_\gamma;\, \gamma \in (0, \hat{\gamma})\}$ bounded in $L_2((0, \hat{T}) \times I)$. We then infer from the compactness of the embedding of $H^{2+2\alpha}(I)$ in $H^{2+2\xi}(I)$ and [20, Corollary 4] that there are a subsequence of $\gamma^2 \longrightarrow 0$ (not relabeled) and $\bar{u}_0$ in $C([0, \hat{T}], H_D^{2+2\xi}(I))$ such that

$$u_\gamma \longrightarrow \bar{u}_0 \ \text{ in } \ C([0, \hat{T}], H_D^{2+2\xi}(I)). \tag{23}$$

Clearly, $\bar{u}_0(t) \in S_\alpha(\kappa/4)$ for $t \in [0, \hat{T}]$ by (21) and (23). The latter and (9) also imply

$$\|\phi_{u_\gamma(t)} - \phi_{\bar{u}_0(t)}\|_{H^2(I\times(0,1))} \le C_0(\kappa/4)\|u_\gamma(t) - \bar{u}_0(t)\|_{H^{2+2\xi}(I)}$$

for $t \in [0, \hat{T}]$ and $\gamma \in (0, \hat{\gamma})$. Consequently, Theorem 3 follows if we can show that $\bar{u}_0$ and u_0 coincide. To this end, recall that the function $g : S(\kappa/4) \to H_D^{2\alpha}(I)$, defined in Proposition 1, is uniformly Lipschitz continuous. In particular, from (23) we deduce that for each $p \in (1, \infty)$,

$$g(u_\gamma) \longrightarrow g_0 := g(\bar{u}_0) \ \text{ in } \ L_p(0, \hat{T}; H_D^{2\alpha}(I)). \tag{24}$$

Thus, if v_γ denotes the solution to the linear Cauchy problem

$$\gamma^2 \frac{\mathrm{d}^2}{\mathrm{d}t^2} v + \frac{\mathrm{d}}{\mathrm{d}t} v + A_\alpha v = -\,\lambda g(u_\gamma), \quad t \in [0, \hat{T}],$$

subject to zero initial conditions

$$v(0) = \gamma^2 \partial_t v(0) = 0,$$

for $\gamma \in (0, \hat{\gamma})$ (with v_0 denoting accordingly the solution with $\gamma = 0$), it follows from (24), the fact that $-A_\alpha$ generates a strongly continuous cosine family in $H_D^{2\alpha}(I)$ as pointed out in [13, Sect. 3.2], and [5, VI.Theorem 7.6] that

$$v_\gamma \longrightarrow v_0 \ \text{ in } \ C([0, \hat{T}], H_D^{2\alpha}(I)), \quad \partial_t v_\gamma \longrightarrow \partial_t v_0 \ \text{ in } \ L_p(0, \hat{T}; H_D^{2\alpha}(I)). \tag{25}$$

On the other hand, if w_γ denotes the solution to the homogeneous Cauchy problem

$$\gamma^2 \frac{d^2}{dt^2} w + \frac{d}{dt} w + A_\alpha w = 0, \quad t > 0,$$

subject to the initial conditions

$$w(0) = u^0, \quad \gamma^2 \partial_t w(0) = \gamma^2 u^1,$$

for $\gamma \in (0, \hat{\gamma})$ (with w_0 denoting accordingly the solution with $\gamma = 0$), then

$$w_\gamma \longrightarrow w_0 \text{ in } C([0, \hat{T}], H_D^{2\alpha}(I)) \tag{26}$$

owing to [11, Theorem 3.2]. Clearly, by uniqueness of solutions to linear wave equations, we have $u_\gamma = v_\gamma + w_\gamma$, and consequently, from (23), (25), and (26) we derive that $\bar{u}_0 = v_0 + w_0$ solves

$$\frac{d}{dt} \bar{u}_0 + A_\alpha \bar{u}_0 = -\lambda g(\bar{u}_0), \quad t \in (0, \hat{T}], \qquad \bar{u}_0(0) = u^0.$$

Since the above Cauchy problem has a unique solution according to Theorem 2, namely u_0 (restricted to $[0, \hat{T}]$), we conclude that $\bar{u}_0 = u_0$ and since this limit is independent of the subsequence $\gamma^2 \longrightarrow 0$, Theorem 3 is proven.

References

1. J. Escher, Ph. Laurençot, Ch. Walker, A parabolic free boundary problem modeling electrostatic MEMS. Arch. Ration. Mech. Anal. **211**, 389–417 (2014)
2. J. Escher, Ph. Laurençot, Ch. Walker, Dynamics of a free boundary problem with curvature modeling electrostatic MEMS. Trans. Amer. Math. Soc. (to appear)
3. J. Escher, Ph. Laurençot, Ch. Walker, Finite time singularity in a free boundary problem modeling MEMS. C. R. Acad. Sci. Paris Sér. I Math. **351**, 807–812 (2013)
4. P. Esposito, N. Ghoussoub, and Y. Guo, *Mathematical Analysis of Partial Differential Equations Modeling Electrostatic MEMS*. Courant Lecture Notes in Mathematics, vol. 20 (Courant Institute of Mathematical Sciences, New York, 2010)
5. H.O. Fattorini, *Second Order Linear Differential Equations in Banach Spaces* (North-Holland, Amsterdam, 1985)
6. H.-Ch. Grunau, Positivity, change of sign and buckling eigenvalues in a one-dimensional fourth order model problem. Adv. Diff. Equat. **7**, 177–196 (2002)
7. Y. Guo, Dynamical solutions of singular wave equations modeling electrostatic MEMS. SIAM J. Appl. Dyn. Syst. **9**, 1135–1163 (2010)
8. Z. Guo, B. Lai, D. Ye, Revisiting the biharmonic equation modelling electrostatic actuation in lower dimensions. Proc. Amer. Math. Soc. **142**, 2027–2034 (2014)
9. A. Haraux, E. Zuazua, Decay estimates for some semilinear damped hyperbolic problems. Arch. Ration. Mech. Anal. **100**, 191–206 (1988)
10. N.I. Kavallaris, A.A. Lacey, C.V. Nikolopoulos, D.E. Tzanetis. A hyperbolic non-local problem modelling MEMS technology. Rocky Mountain J. Math. **41**, 505–534 (2011)
11. J. Kisyński, Sur les équations hyperboliques avec petit paramètre. Colloq. Math. **10**, 331–343 (1963)
12. Ph. Laurençot, Ch. Walker, A stationary free boundary problem modeling electrostatic MEMS. Arch. Ration. Mech. Anal. **207**, 139–158 (2013)

13. Ph. Laurençot, Ch. Walker, A free boundary problem modeling electrostatic MEMS: I. Linear bending effects. Math. Ann. **360**, 307–349 (2014)
14. Ph. Laurençot, Ch. Walker, A free boundary problem modeling electrostatic MEMS: II. Nonlinear bending effects. Math. Models Methods Appl. Sci. **24**, 2549–2568 (2014)
15. Ph. Laurençot, Ch. Walker, Sign-preserving property for some fourth-order elliptic operators in one dimension or in radial symmetry. J. Anal. Math. (to appear)
16. Ph. Laurençot, Ch. Walker, A fourth-order model for MEMS with clamped boundary conditions. Proc. Lond. Math. Soc. **109**, 1435–1464 (2014)
17. A.E. Lindsay, J. Lega, Multiple quenching solutions of a fourth order parabolic PDE with a singular nonlinearity modeling a MEMS capacitor. SIAM J. Appl. Math. **72**, 935–958 (2012)
18. M.P. Owen, Asymptotic first eigenvalue estimates for the biharmonic operator on a rectangle. J. Differ. Equat. **136**, 166–190 (1997)
19. J.A. Pelesko, D.H. Bernstein, *Modeling MEMS and NEMS* (Chapman & Hall/CRC, Boca Raton, FL, 2003)
20. J. Simon, Compact sets in the space $L^p(0,T;B)$. Ann. Mat. Pura Appl. **146**, 65–96 (1987)

Compositions in the Edge Calculus

B.-Wolfgang Schulze

Abstract We study the symbolic calculus of edge-degenerate operators from the point of view of compositions. The underlying manifold with edge is interpreted as a stratified space with two strata. These generate a two-component principal symbolic hierarchy. It reflects other features of the edge symbolic calculus, including Mellin plus Green and trace and potential symbols, belonging to the block matrix algebra of edge problems, analogously as boundary value problems. We confine ourselves to upper-left corners. However, since trace and potential entries are contained in the concept of Green symbols, the desired composition behaviour in general also follows including trace and potential symbols. The ideas of our proof rely to a large extent on operator-valued symbols with twisted symbolic estimates and on recovering procedures of subordinate conormal symbols of the cone algebra on the infinite stretched cone, modelling local wedges.

1 Introduction

This article is aimed at studying the edge pseudo-differential calculus, first established in [22]. We update here a proof on compositions in that calculus which is desirable in iterative constructions for higher corner and edge singularities. The approach in a joint article with Gil and Seiler [7]. is still very instructive and shows features that are of independent interest for the nature of edge quantisation. Recall that the edge calculus has several sources from boundary value problems with and without the transmission property at the boundary. In this context, a manifold with smooth boundary is interpreted as a manifold with edge. The boundary is interpreted as an edge and the inner normal $\overline{\mathbb{R}}_+$ as the model cone of the respective local wedges, in this case the half space. Motivated by the work of Boutet de Monvel [1], the boundary symbolic calculus has been recognised as an aspect of the cone calculus, in the sense of Eskin [5] and later on deepened in joint works with Rempel [15], see also [14, 16, 17], or the monograph [25]. Recall that the various approaches deal with different quantisations. Both Eskin and Boutet de Monvel employed the truncation

B.-W. Schulze (✉)
Institut für Mathematik, Universität Potsdam, 14469 Potsdam, Germany
e-mail: schulze@math.uni-potsdam.de.address

J. Escher et al., (eds.), *Elliptic and Parabolic Equations*, Springer Proceedings in Mathematics & Statistics 119, DOI 10.1007/978-3-319-12547-3_11

quantisation where a pseudo-differential operator P on an open smooth manifold M is truncated to an embedded smooth manifold X with boundary by forming $\mathrm{r}^+ P \mathrm{e}^+$ with e^+ being the extension by zero from $\mathrm{int}\, X$ to $M \setminus X$ and r^+ the operator of restriction to $\mathrm{int}\, X$. On the other hand, there are reasons to look at cone and edge degenerate operators and to alternative quantisations close to the singularity, namely, based on the Mellin transform in the distance variable $r \in \mathbb{R}_+$ to the singularity. In this framework, it is natural to realise operators in weighted spaces and subspaces with asymptotics. At the same time, it became clear that the boundary symbolic calculus is an aspect of the cone calculus also when the base of the cone is a, say, closed smooth manifold of any dimension, and that there is also a natural connection to operators on the cone with a control up to the conical exit to infinity in the sense of Shubin [30], Parenti [12], Cordes [2], Schrohe [18], Egorov and Schulze [4, 26], in different terminology concerning symbolic phenomena. The behaviour of (principal) edge symbols both at $r = 0$ and $r = \infty$ has to be observed, also with respect to some rescaling properties which are reflected by twisted homogeneity and, on the level of complete symbols, twisted symbolic estimates. Moreover, the symbols have a rich internal structure also with respect to other aspects, e.g., the nature of Green, trace, and potential operators, or asymptotic phenomena in r as $r \to 0$, connected with the expected asymptotics of solutions to elliptic equations, observed by Kondratyev [11] and many other authors of this period.

2 Edge Quantisation

2.1 *Mellin Operators and the Cone Algebra*

We first recall some tools from the cone algebra. This refers to the infinite straight cone $X^{\Delta} := (\overline{\mathbb{R}}_+ \times X)/(\{0\} \times X)$ with a base X which is assumed to be smooth, compact, Riemannian. The operators themselves live on the open stretched cone $X^{\wedge} := \mathbb{R}_+ \times X$. We systematically employ here *Kegel spaces* $\mathcal{K}^{s,\gamma}(X^{\wedge})$ of smoothness $s \in \mathbb{R}$ and weight $\gamma \in \mathbb{R}$ for $r \to 0$, or, more generally, $\mathcal{K}^{s,\gamma;e}(X^{\wedge})$ with an extra *exit weight* $e \in \mathbb{R}$ for $r \to \infty$. These spaces together with corresponding subspaces $\mathcal{K}_P^{s,\gamma}(X^{\wedge}), \mathcal{K}_P^{s,\gamma;e}(X^{\wedge})$ with asymptotics of type P for $r \to 0$ have been first employed in joint work with Rempel, see, e.g., [17 and the references there]. In order to keep this exposition more accessible in Sect. 4.1 below we list some constructions and notation in connection with such spaces. Other material on the cone calculus can be found in [16, 24, 26], or the joint monograph with Egorov [4].

First let X be a Riemannian manifold of dimension n. By $L^{\mu}_{(\mathrm{cl})}(X; \mathbb{R}^l)$, $\mu \in \mathbb{R}$, we denote the space of pseudo-differential operators on X depending on a parameter $\lambda \in \mathbb{R}^l$. By (cl), we indicate spaces of operators with classical or general symbols when considerations concern both cases. The local symbols $a(x, \xi, \lambda)$ depend on $(\xi, \lambda) \in \mathbb{R}^{n+l}$, treated as covariables. The space of parameter-dependent smoothing operators is denoted by $L^{-\infty}(X; \mathbb{R}^l) := \mathcal{S}(\mathbb{R}^l, L^{-\infty}(X))$. Here $L^{-\infty}(X)$ is identified with the Fréchet space $C^{\infty}(X \times X)$ via the fixed Riemannian metric on X. Modulo

$L^{-\infty}(X;\mathbb{R}^l)$ the operator families in $L^{\mu}_{(\mathrm{cl})}(X;\mathbb{R}^l)$ have the form $\mathrm{Op}_x(a)(\lambda)$,

$$\mathrm{Op}_x(a)(\lambda)u(x) = \iint e^{i(x-x')\xi}a(x,\xi,\lambda)u(x')\mathrm{d}x'\text{đ}\xi,$$

$\text{đ}\xi = (2\pi)^{-n}\mathrm{d}\xi$. If we say nothing else the elements $A(\lambda) \in L^{\mu}_{(\mathrm{cl})}(X;\mathbb{R}^l)$ are regarded as families of continuous operators

$$A(\lambda) : C_0^{\infty}(X) \to C^{\infty}(X).$$

Parameter-dependent pseudo-differential operators are very common; the conditions in our definition are slightly different from those in Shubin's book [32].
If $U \subseteq \mathbb{C}$ is an open set, E a Fréchet space, then $\mathcal{A}(U,E)$ denotes the space of holomorphic functions in U with values in E. We now assume that X is a smooth compact manifold. We employ the fact that the spaces of pseudo-differential operators are Fréchet in a natural way.

Definition 1 Let $M^{\mu}_{\mathcal{O}}(X;\mathbb{R}^l_{\lambda})$ for $\mu \in \mathbb{R}$ denote the space of all

$$h(z,\lambda) \in \mathcal{A}(\mathbb{C}_z, L^{\mu}_{\mathrm{cl}}(X;\mathbb{R}^l))$$

such that $h(\beta+i\rho,\lambda) \in L^{\mu}_{\mathrm{cl}}(X;\mathbb{R}^{1+l}_{\rho,\lambda})$ for every $\beta \in \mathbb{R}$, uniformly in compact β-intervals. For $l=0$ we simply write $M^{\mu}_{\mathcal{O}}(X)$.

We will also employ pseudo-differential operators based on the Mellin transform. Let us briefly recall the shape of Mellin operators that are known from the cone and edge calculus, i.e., of singularitiy order 1. In the simplest case we have

$$\mathrm{op}^{\gamma}_M(f)u(r) := \int_{\Gamma_{1/2-\gamma}}\int_{\mathbb{R}_+}(r/r')^{-z}f(z)u(r')\mathrm{d}'/r'\text{đ}z \tag{1}$$

$$= \int_{-\infty}^{\infty}\int_0^{\infty}(r/r')^{-(1/2-\gamma+\mathrm{i}\rho)}f(1/2-\gamma+\mathrm{i}\rho)u(r')\mathrm{d}r'/r'\text{đ}\rho,$$

$\text{đ}z = (2\pi\mathrm{i})^{-1}\mathrm{d}z$, for a symbol $f(z) \in S^{\mu}(\Gamma_{1/2-\gamma})$, cf. formula (54). In this notation $\rho = \mathrm{Im}\, z$ plays the role of the covariable. The expression M is interpreted as a Mellin oscillatory integral, first for $u \in C_0^{\infty}(\mathbb{R}_+)$ and then extended to more general distribution spaces, e.g., $\mathcal{H}^{s,\gamma}(\mathbb{R}_+)$, cf. notation in Sect. 4.1 below. There are Mellin operators in numerous variants, e.g., with symbols

$$f(r,r',z) \in C^{\infty}(\mathbb{R}_+\times\mathbb{R}_+, S^{\mu}(\Gamma_{1/2-\gamma}))$$

with variable coefficients, or taking values in several operator classes, analogously as those with the twisted symbolic estimates. In particular, we consider operators (1) whith f extending to the complex z-plane as a meromorphic function.
For any $f(z) \in L^{\mu}_{(\mathrm{cl})}(X;\Gamma_{(n+)1/2-\gamma})$, regarded as an operator-valued Mellin symbol, the operator $\mathrm{op}_M^{\gamma-n/2}(f)$, $n=\dim X$, induces continuous operators

$$\mathrm{op}_M^{\gamma-n/2}(f) : \mathcal{H}^{s,\gamma}(X^{\wedge}) \to \mathcal{H}^{s-\mu,\gamma}(X^{\wedge})$$

for all $s \in \mathbb{R}$; concerning notation, cf. Sect. 4.1.

Let us briefly mention an aspect of the kernel cut-off construction, originally introduced in [21, Sect. 2, Proposition 12], and then employed in different variants in the singular pseudo-differential calculus, cf. also [24, Sect. 2.1.3; 26, Theorems 1.1.2, 1.1.47, 1.2.47] which yields the following theorem.

Theorem 1 *For every* $f(z) \in L^{\mu}_{cl}(X;\Gamma_\beta)$, *there exists an* $h(z) \in M^{\mu}_{\mathcal{O}}(X)$ *such that* $h|_{\Gamma_\beta} = f$ *modulo* $L^{-\infty}(X;\Gamma_\beta)$.

In this chapter a cut-off function on the half axis is a real-valued element $\omega \in C^\infty(\overline{\mathbb{R}}_+)$ with bounded support such that $\omega = 1$ in a neighbourhood of $r = 0$. In the following definition, ω, ω', etc. are arbitrary cut-off functions on the r half axis. We write $\varphi \prec \varphi'$ or $\varphi' \succ \varphi$ for functions φ, φ' if $\varphi' = 1$ on $\operatorname{supp}\varphi$. Concerning other notation in the following definition, we refer to Sects. 4.1 and 4.3.

Definition 2 The cone algebra over $X^\wedge$ consists of operator spaces of order $\mu \in \mathbb{R}$

$$L^{\mu}(X^\wedge, \mathrm{g}) \quad \text{and subspaces} \quad L_{M+G}(X^\wedge, \mathrm{g}) \supseteq L_G(X^\wedge, \mathrm{g}) \tag{2}$$

of Mellin plus Green and Green operators, associated with weight data

$$\begin{aligned} \mathrm{g} := (\gamma, \gamma - \mu, \Theta),\ \gamma \in \mathbb{R},\ \text{for a weight interval } \Theta \\ := (-(k+1), 0],\ k \in \mathbb{N} \cup \{+\infty\}, \end{aligned} \tag{3}$$

furnished with operators

$$A := H + P_{\mathrm{int}} + M + G + E \tag{4}$$

that are defined as follows:

(i) Green operators $G \in L_G(X^\wedge, \mathrm{g}) \subset \bigcap_{s,e \in \mathbb{R}} \mathcal{L}(\mathcal{K}^{s,\gamma;e}(X^\wedge), \mathcal{K}^{\infty,\gamma-\mu;\infty}(X^\wedge))$ are defined by the property to induce continuous operators

$$G : \mathcal{K}^{s,\gamma;e}(X^\wedge) \to \mathcal{K}_P^{\infty,\gamma-\mu;\infty}(X^\wedge),\ G^* : \mathcal{K}^{s,-\gamma+\mu;e}(X^\wedge) \to \mathcal{K}_Q^{\infty,-\gamma;\infty}(X^\wedge) \tag{5}$$

for all $s, e \in \mathbb{R}$ and G-dependent asymptotic types P and Q, associated with the weight data $(\gamma - \mu, \Theta)$ and $(-\gamma, \Theta)$, respectively; in this subclass $\Theta = (\vartheta, 0]$ for $-\infty \le \vartheta < 0$ may be arbitrary. The formal adjoint G^* refers to the $\mathcal{K}^{0,0}(X^\wedge)$-scalar product.

(ii) Smoothing Mellin plus Green operators $M+G \in L_{M+G}(X^\wedge, \mathrm{g})$ are first defined for $\Theta = (-(k+1), 0]$, $k \in \mathbb{N}$, where $G \in L_G(X^\wedge, \mathrm{g})$ and

$$M := \omega r^{-\mu} \left\{ \sum_{j=0}^{k} r^j \operatorname{op}_M^{\gamma_j - n/2}(f_j) \right\} \omega' \tag{6}$$

for Mellin symbols $f_j(z) \in M^{-\infty}_{R_j}(X)$, with Mellin asymptotic types R_j, $n = \dim X$, where the weights satisfy the conditions

$$\gamma - j \le \gamma_j \le \gamma \quad \text{and} \quad \pi_{\mathbb{C}} R_j \bigcap \Gamma_{(n+1)/2-\gamma_j} = \emptyset$$

for $j = 0, \dots, k$. For $\Theta = (-\infty, 0]$ and $\mathrm{g} = (\gamma, \gamma - \mu, (-\infty, 0])$, we define

$$L_{M+G}(X^\wedge, \mathrm{g}) := \bigcap_{k \in \mathbb{N}} L_{M+G}(X^\wedge, \mathrm{g}_k)$$

for $\mathrm{g}_k := (\gamma, \gamma - \mu, (-(k+1), 0])$.

(iii) The operators H, P_{int} and E in (4) are of the form

$$H := r^{-\mu} \omega \mathrm{op}_M^{\gamma - n/2}(h) \omega', \quad P_{\mathrm{int}} := (1 - \omega) \operatorname{Op}_r(p_{\mathrm{int}})(1 - \omega''), \tag{7}$$

for cut-off functions $\omega'' \prec \omega \prec \omega'$, arbitrary

$$h(r, z) \in C^\infty(\overline{\mathbb{R}}_+, M_{\mathcal{O}}^\mu(X)), \; p_{\mathrm{int}}(r, \rho) \in C^\infty(\mathbb{R}_+, L_{\mathrm{cl}}^\mu(X; \mathbb{R}_\rho)),$$

where $p_{\mathrm{int}}(r, \rho)$ has bounded support in r, and

$$E \in \chi_{\mathrm{e}} L_{\mathrm{cl}}^{\mu;0}(X^{\asymp}) \chi_{\mathrm{e}}' \tag{8}$$

for functions $\chi_{\mathrm{e}}, \chi_{\mathrm{e}}' \in C^\infty(\mathbb{R})$ that are equal to 0 for $r < \varepsilon_0$ and 1 for $r > \varepsilon_1$ for some $0 < \varepsilon_0 < \varepsilon_1$.

The notation $X^{\asymp}$ in formula (8) means $\mathbb{R} \times X$, interpreted as a manifold with conical exits to $r \to \pm\infty$, and $L_{\mathrm{cl}}^{\mu;0}(X^{\asymp})$ means the space of classical pseudo-differential operators of order μ and exit order 0 on $X^{\asymp}$, cf. [18] and [26, Sect. 1.4].

Theorem 2

(i) *We have*

$$L^\mu(X^\wedge, \mathrm{g}) \subset L_{cl}^\mu(X^\wedge).$$

(ii) *An operator $A \in L^\mu(X^\wedge, \mathrm{g})$ can be written in the form*

$$A = r^{-\mu} \operatorname{Op}_r(p) \mod L^{-\infty}(X^\wedge)$$

for a symbol $p(r, \rho) = \tilde{p}(r, r\rho)$, $\tilde{p}(r, \tilde{\rho}) \in C^\infty(\overline{\mathbb{R}}_+, L_{cl}^\mu(X; \mathbb{R}_{\tilde{\rho}}))$ and

$$A = r^{-\mu} \mathrm{op}_M^{\gamma - n/2}(h) \mod L^{-\infty}(X^\wedge)$$

for a Mellin symbol $h(r, z) \in C^\infty(\overline{\mathbb{R}}_+, M_{\mathcal{O}}^\mu(X))$.

This is a consequence of the Mellin quantisation, developed in [26, Theorem 3.2.7] that allows us to pass from $\operatorname{Op}_r(p)$ to $\mathrm{op}_M^{\gamma - n/2}(h)$, up to a smoothing remainder. From (i), we have for an $A \in L^\mu(X^\wedge, \mathrm{g})$ its homogeneous principal symbol of order μ that we denote by $\sigma_\psi(A)(r, x, \rho, \xi)$, referring to the local splitting of variables (r, x) and covariables $(\rho, \xi) \neq 0$, and we set $\tilde{\sigma}_\psi(A)(r, x, \rho, \xi) := r^\mu \sigma_\psi(A)(r, x, r^{-1}\rho, \xi)$ which is smooth up to $r = 0$.

Theorem 3 *We have*

(i)
$$L^{\mu}_{M+G}(X^{\wedge}, \mathrm{g}) \subset L^{-\infty}(X^{\wedge})$$

and

(ii)
$$A \in L^{\mu}(X^{\wedge}, \mathrm{g}) \bigcap \{L^{-\infty}(X^{\wedge}) \bigcup \chi_e L^{-\infty;-\infty}(X^{\asymp})\chi_e'\} = L_{M+G}(X^{\wedge}, \mathrm{g}).$$

While the first property is an immediate consequence of the definition, the second one follows from the assumption that the operators involved in $M^{\mu}_{\mathcal{O}}(X)$ are classical. The following theorem can be found in [24, Sect. 2.2.3] or [26, Sect. 2.4.2].

Theorem 4 *Let*

$$A \in L^{\mu}(X^{\wedge}, \mathrm{g}), \;\; B \in L^{\nu}(X^{\wedge}, \mathrm{g})$$

for $\mathrm{g} = (\gamma - \nu, \gamma - (\mu + \nu), \Theta), \;\; \mathrm{h} = (\gamma, \gamma - \nu, \Theta), \;\; \mu, \nu, \gamma \in \mathbb{R}$. *Then*

$$AB \in L^{\mu+\nu}(X^{\wedge}, \mathrm{g} \circ \mathrm{h})$$

for $\mathrm{g} \circ \mathrm{h} = (\gamma, \gamma - (\mu + \nu), \Theta)$, *and we have*

$$\sigma_{\psi}(AB) = \sigma_{\psi}(A)\sigma_{\psi}(B), \; \tilde{\sigma}_{\psi}(AB) = \tilde{\sigma}_{\psi}(A)\tilde{\sigma}_{\psi}(B),$$

and the Mellin translation product [24, Remark 2.2,1], *or* [25, Remark 2.1.14]

$$\sigma_M^{\mu+\nu-l}(AB) = \sum_{i+j=l} (T^{\nu-j}\sigma_M^{\mu-i}(A))\sigma_M^{\nu-j}(B),$$

where $(T^{\beta} f)(z) := f(z + \beta)$ *for any real* β. *Moreover, if* A *or* B *belong to* L_{M+G} *or* L_G *then the same is true of* AB.

2.2 Smoothing Mellin Plus Green Edge Symbols

We now recall the concept of smoothing Mellin plus Green symbols, belonging to the edge symbolic calculus. Green symbols in the edge algebra are of a similar meaning as the singular Green symbols of Boutet de Monvel's pseudo-differential calculus of boundary value problems with the transmission property at the boundary, cf. [1] or [13]. An idea of joint papers with Rempel, e.g., [14], and the monograph [17] was to extend Boutet de Monvel's structures to boundary value problems without the transmission property, including the index theorem for the case without the transmission property, proved in [14] by a homotopy argument and a reduction to [1]. In fact, Boutet de Monvel's algebra of operators of order and type zero has been recognised as a substructure of the edge algebra on the respective manifold with boundary, regarded as a manifold with edge. The case of operators of arbitrary order and type is of a similar kind, and has been deepened later on by other authors, see, in particular, joint works with Seiler [27], Kapanadze and Seiler [10], Wei [29]. The correspondence to Boutet de Monvel's boundary symbolic calculus is particularly visible for Green symbols, see [17, Sect. 3.3.1, Definition 1] including the relationship between Green, trace and potential symbols, see [17, Sect. 3.3.2, Definition 1], or also [25, Sect. 2.2.3]. In particular, it turned out that, in contrast to Boutet de Monvels boundary symbols, the case of violated transmission property generates extra operators, expressed by smoothing Mellin symbols, first observed on the half axis in Eskin's book [5], for order zero and with Hilbert–Schmidt operators instead of Green operators in the cone calculus. The edge algebra itself deals with operator families on X^{Δ} rather than $\overline{\mathbb{R}}_{+}$, depending on edge variables and covariables (y, η), more precisely, with symbols taking values in $L^{\mu}(X^{\wedge}, \mathrm{g})$, cf. Definition 2(ii). Mellin edge symbols in Definiton 3(ii) below have been introduced in [25, formula (2.1.279)]. While the step from the half axis to a cone with a link of dimension > 0 is not too complicated, the way of organising edge symbols with the "right" dependence on (y, η) belongs to the non-trivial achievements of the edge calculus.

Let $[\eta]$ be a strictly positive function in $C^{\infty}(\mathbb{R}^{q}_{\eta})$ such that $[\eta] = |\eta|$ for $|\eta| \geq C$ for some $C > 0$. Set

$$\psi_{\eta}(r) := \psi([\eta]r) \quad \text{and} \quad \psi_{|\eta|}(r) := \psi(|\eta|r) \quad \text{for any} \quad \psi \in C^{\infty}(\overline{\mathbb{R}}_{+}). \tag{9}$$

Definition 3 Let $\Omega \subseteq \mathbb{R}^{p}$ be an open set, $\mu \in \mathbb{R}$, and $\mathrm{g} := (\gamma, \gamma - \mu, \Theta)$ weight data as in Definition 2.

(i) Green edge symbols

$$g(y, \eta) \in \mathcal{R}^{\mu}_{G}(\Omega \times \mathbb{R}^{p}, \mathrm{g}) \subset \bigcap_{s,e \in \mathbb{R}} S^{\mu}_{\mathrm{cl}}(\Omega \times \mathbb{R}^{q}; \mathcal{K}^{s,\gamma;e}(X^{\wedge}), \mathcal{K}^{\infty,\gamma-\mu;\infty}(X^{\wedge}))$$

are defined by the properties

$$g(y, \eta) \in \bigcap_{s,e \in \mathbb{R}} S^{\mu}_{\mathrm{cl}}(\Omega \times \mathbb{R}^{q}; \mathcal{K}^{s,\gamma;e}(X^{\wedge}), \mathcal{K}^{\infty,\gamma-\mu;\infty}_{P}(X^{\wedge})), \tag{10}$$

$$g^{*}(y, \eta) \in \bigcap_{s,e \in \mathbb{R}} S^{\mu}_{\mathrm{cl}}(\Omega \times \mathbb{R}^{q}; \mathcal{K}^{s,-\gamma+\mu;e}(X^{\wedge}), \mathcal{K}^{\infty,-\gamma;\infty}_{Q}(X^{\wedge}))$$

for g-dependent asymptotic types P and Q, associated with the weight data $(\gamma - \mu, \Theta)$ and $(-\gamma, \Theta)$, respectively; in this subclass $\Theta = (\vartheta, 0]$ for $-\infty \leq \vartheta < 0$ may be arbitrary. The formal adjoint g^{*} is pointwise and refers to the $\mathcal{K}^{0,0}(X^{\wedge})$-scalar product.

(ii) Smoothing Mellin plus Green edge symbols are of the form

$$m(y, \eta) + g(y, \eta) \in \mathcal{R}^{\mu}_{M+G}(\Omega \times \mathbb{R}^{q}, \mathrm{g})$$

for $g(y, \eta) \in \mathcal{R}^{\mu}_{G}(\Omega \times \mathbb{R}^{q}, \mathrm{g})$, and (first for $\Theta = (-(k+1), 0]$, $k \in \mathbb{N}$)

$$m(y, \eta) := \omega_{\eta} r^{-\mu} \left\{ \sum_{j=0}^{k} r^{j} \sum_{|\alpha| \leq j} \mathrm{op}_{M}^{\gamma_{j,\alpha} - n/2}(f_{j,\alpha})(y) \eta^{\alpha} \right\} \omega'_{\eta} \tag{11}$$

for Mellin symbols $f_{j,\alpha}(z) \in M^{-\infty}_{R_{j,\alpha}}(X)$, $n = \dim X$, where the weights satisfy the conditions

$$\gamma - j \leq \gamma_{j,\alpha} \leq \gamma \quad \text{and} \quad \pi_{\mathbb{C}} R_{j,\alpha} \bigcap \Gamma_{(n+1)/2-\gamma_{j,\alpha}} = \emptyset$$

for $j = 0, \dots, k$, $|\alpha| \leq j$, and cut-off functions ω, ω'. For $\Theta = (-\infty, 0]$ and $\mathrm{g} = (\gamma, \gamma - \mu, (-\infty, 0])$, we define $\mathcal{R}^{\mu}_{M+G}(\Omega \times \mathbb{R}^p, \mathrm{g}) := \bigcap_{k \in \mathbb{N}} \mathcal{R}^{\mu}_{M+G}(\Omega \times \mathbb{R}^p, \mathrm{g}_k)$ for $\mathrm{g}_k := (\gamma, \gamma - \mu, (-(k+1), 0])$.

Remark 1
We have

$$\mathcal{R}^{\mu}_{M+G}(\Omega \times \mathbb{R}^q, \mathrm{g}) \subset S^{\mu}_{\mathrm{cl}}(\Omega \times \mathbb{R}^q; \mathcal{K}^{s,\gamma;e}(X^{\wedge}), \mathcal{K}^{\infty,\gamma-\mu;\infty}(X^{\wedge}))$$

and

$$\mathcal{R}^{\mu}_{M+G}(\Omega \times \mathbb{R}^q, \mathrm{g}) \subset S^{\mu}_{\mathrm{cl}}(\Omega \times \mathbb{R}^q; \mathcal{K}^{s,\gamma;e}_P(X^{\wedge}), \mathcal{K}^{\infty,\gamma-\mu;\infty}_Q(X^{\wedge}))$$

for every $s, e \in \mathbb{R}$ and every asymptotic type P, for some resulting Q.

Remark 2 For every $\mathcal{R}^{\mu}_{G}(\Omega \times \mathbb{R}^q, \mathrm{g})$ and cut-off functions σ, σ' on the r half axis, regarded as operators of multiplication in Kegel spaces, we have

$$g - \sigma g,\ g - g\sigma',\ g - \sigma g \sigma' \in \mathcal{R}^{-\infty}_{G}(\Omega \times \mathbb{R}^q, \mathrm{g}).$$

Definition 4 For $(m+g)(y,\eta) \in \mathcal{R}^{\mu}_{M+G}(\Omega \times \mathbb{R}^p, \mathrm{g})$ and $m(y,\eta)$ of the form (11), we set

$$\sigma^{\mu-j}_{M}(m+g)(y,z,\eta) := \sum_{|\alpha| \leq j} f_{j,\alpha}(y,z)\eta^{\alpha}, \quad j = 0, \dots, k, \tag{12}$$

and

$$\sigma^{\mu-l}_{\wedge}(m+g)(y,\eta) := \omega_{|\eta|} r^{-\mu+l} \left\{ \sum_{j+|\alpha|=l} \mathrm{op}^{\gamma_{j,\alpha}-n/2}_{M}(f_{j,\alpha})(y)\eta^{\alpha} \right\} \omega'_{|\eta|} + g_{(\mu-j)}(y,\eta), \tag{13}$$

$\eta \neq 0$, for $l = 0, \dots, k$, and $\sigma^{\mu-l}_{\wedge}(m+g)(y,\eta) = g_{(\mu-l)}(y,\eta)$ for $l \in \mathbb{N}$, $l > k$; here $g_{(\mu-l)}(y,\eta)$ is the (twisted) homogeneous component of $g(y,\eta)$ of order $\mu - l$ as a classical symbol.

The expression (13) has the meaning of the (twisted) homogeneous component of a corresponding classical operator-valued symbol. For $l = 0$, we also write

$$\sigma_{\wedge}(m+g)(y,\eta) := \sigma^{\mu}_{\wedge}(m+g)(y,\eta),$$

and we have

$$\sigma_{\wedge}(m+g)(y,\eta) = \lim_{\delta \to \infty} \delta^{-\mu} \kappa^{-1}_{\delta}(m+g)(y,\delta\eta)\kappa_{\delta}, \tag{14}$$

cf. also Remark 8 below.

2.3 Edge-Degenerate Operators and Mellin-Edge Quantisation

Edge-degenerate pseudo-differential operators appear in the construction of parametrices of elliptic differential operators of the form

$$A = r^{-\mu} \sum_{|\alpha| \leq \mu} a_{j,\alpha}(r,y)(-r\partial_r)^j (rD_y)^\alpha \tag{15}$$

for coefficients $a_{j,\alpha}(r,y) \in C^\infty(\overline{\mathbb{R}}_+ \times \Omega, \mathrm{Diff}^{\mu-(j+|\alpha|)}(X))$. Here, $\mathrm{Diff}^\nu(X)$ is the space of all differential operators on X of order ν with smooth coefficients.

Theorem 5 *Let $\Omega \subseteq \mathbb{R}^p$ be an open set. For a symbol*

$$p(r,y,\rho,\eta) := \tilde{p}(r,y,r\rho,r\eta), \tag{16}$$

$$\tilde{p}(r,y,\tilde{\rho},\tilde{\eta}) \in C^\infty\left(\overline{\mathbb{R}}_+ \times \Omega, L^\mu_{cl}(X;\mathbb{R}^{1+q}_{\tilde{\rho},\tilde{\eta}})\right) \tag{17}$$

there exists an

$$h(r,y,z,\eta) := \tilde{h}(r,y,z,r\eta), \tag{18}$$

$$\tilde{h}(r,y,z,\tilde{\eta}) \in C^\infty\left(\overline{\mathbb{R}}_+ \times \Omega, M^\mu_{\mathcal{O}}(X;\mathbb{R}^q_{\tilde{\eta}})\right) \tag{19}$$

such that

$$\mathrm{op}^\beta_M(h)(y,\eta) := \mathrm{Op}_r(p)(y,\eta) \mod C^\infty(\Omega, L^\mu_{cl}(X^\wedge;\mathbb{R}^q)), \tag{20}$$

for every $\beta \in \mathbb{R}$, and $\tilde{h}$ is unique mod $C^\infty(\overline{\mathbb{R}}_+ \times \Omega, M^{-\infty}_{\mathcal{O}}(X;\mathbb{R}^q_{\tilde{\eta}}))$.

A proof of this theorem is given in [26, Sect. 3.2.2]. Setting

$$p_0(r,y,\rho,\eta) := \tilde{p}(0,y,r\rho,r\eta), \quad h_0(r,y,z,\eta) := \tilde{h}(0,y,z,r\eta), \tag{21}$$

we have

$$\mathrm{op}^\beta_M(h_0)(y,\eta) := \mathrm{Op}_r(p_0)(y,\eta) \tag{22}$$

modulo $C^\infty(\Omega, L^\mu_{\mathrm{cl}}(X;\mathbb{R}^q))$, for every $\beta \in \mathbb{R}$.

Definition 5 Let $\mathcal{R}^\mu(\Omega \times \mathbb{R}^q, \mathrm{g})$ for an open set $\Omega \subseteq \mathbb{R}^p$ and weight data $\mathrm{g} = (\gamma, \gamma-\mu, \Theta)$, $\gamma, \mu \in \mathbb{R}$, $\Theta = (-(k+1), 0]$, $k \in \mathbb{N} \cup \{\infty\}$, be the set of all operator functions of the form

$$\begin{aligned} a(y,\eta) := {} & \sigma r^{-\mu}\{\omega_\eta \mathrm{op}^{\gamma-n/2}_M(h)(y,\eta)\omega'_\eta + (1-\omega_\eta)\mathrm{Op}_r(p)(y,\eta)(1-\omega''_\eta)\}\sigma' \\ & + \varphi \mathrm{Op}_r(p_{\mathrm{int}})(y,\eta)\varphi' + (m+g)(y,\eta) \end{aligned} \tag{23}$$

for cut-off functions $\omega'' \prec \omega \prec \omega'$ and σ, σ' on the r half axis, $\varphi, \varphi' \in C^\infty_0(\mathbb{R}_+)$, h, p as in Theorem 5, $p_{\mathrm{int}}(r,y,\rho,\eta) \in C^\infty(\mathbb{R}_+ \times \Omega, L^\mu(X, \mathbb{R}^{1+q}_{\rho,\eta}))$, and $m + g \in \mathcal{R}^\mu_{M+G}(\Omega \times \mathbb{R}^q, \mathrm{g})$.

Remark 3 From Definitions 2 and 5, it follows that an $a(y,\eta) \in \mathcal{R}^{\mu}(\Omega \times \mathbb{R}^q, \mathrm{g})$ is a function in $(y,\eta) \in \Omega \times \mathbb{R}^q$, taking values in $L^{\mu}(X^{\wedge}, \mathrm{g})$.

Theorem 6

We have

$$\mathcal{R}^{\mu}(\Omega \times \mathbb{R}^q, \mathrm{g}) \subset S^{\mu}(\Omega \times \mathbb{R}^q; \mathcal{K}^{s,\gamma}(X^{\wedge}), \mathcal{K}^{s-\mu,\gamma-\mu}(X^{\wedge}))$$

and

$$\mathcal{R}^{\mu}(\Omega \times \mathbb{R}^q, \mathrm{g}) \subset S^{\mu}(\Omega \times \mathbb{R}^q; \mathcal{K}^{s,\gamma}_P(X^{\wedge}), \mathcal{K}^{s-\mu,\gamma-\mu}_Q(X^{\wedge}))$$

for every $s \in \mathbb{R}$ and every asymptotic type P for some resulting Q.

A proof of this theorem can be found in [19], see also [9, Theorem 4.4.20]. An inspection of the details shows that it suffices to verify the claimed properties of (23) separately for the involved summands. While the term containing p_{int} is simple, cf. also Remark 4, and the assertion for the smoothing Mellin term nearly evident (for the Green term it is a consequence of the definition anyway), the main issue is

$$r^{-\mu}\omega_\eta \operatorname{op}_M^{\gamma-n/2}(h)(y,\eta)\omega'_\eta \tag{24}$$

and

$$\sigma r^{-\mu}(1-\omega_\eta)\mathrm{Op}_r(p)(y,\eta)(1-\omega''_\eta)\}\sigma'. \tag{25}$$

The original factors σ, σ' in (24) are omitted since we may choose the cut-off functions ω, ω' in such a way that $\sigma\omega_\eta = \omega_\eta$, $\omega'_\eta\sigma' = \omega'_\eta$ for all η.

3 Composition Results

3.1 Properties of Edge Symbols

Remark 4 An element $a(y,\eta) \in \mathcal{R}^{\mu}(\Omega \times \mathbb{R}^q, \mathrm{g})$, first of the form (23), can always be written as

$$a(y,\eta) := \sigma r^{-\mu}\{\omega_\eta \operatorname{op}_M^{\gamma-n/2}(h)(y,\eta)\omega'_\eta + (1-\omega_\eta)\mathrm{Op}_r(p)(y,\eta)(1-\omega''_\eta)\}\sigma' + (m+g)(y,\eta) \tag{26}$$

for other cut-off functions σ, σ' and another Green symbol $g(y,\eta)$.

Remark 5 We have

$$a(y,\eta) = \sigma r^{-\mu}\mathrm{Op}_r(p)(y,\eta)\sigma' \quad \mathrm{mod}\ C^{\infty}(\Omega, L^{-\infty}(X^{\wedge}; \mathbb{R}^q_\eta)) \tag{27}$$

and

$$a(y,\eta) = \sigma r^{-\mu} \operatorname{op}_M^{\gamma-n/2}(h)(y,\eta)\sigma' \quad \mathrm{mod}\ C^{\infty}(\Omega, L^{-\infty}(X^{\wedge}; \mathbb{R}^q_\eta))$$

for every $\gamma \in \mathbb{R}$.

From the first representation in the latter remark, we obtain a homogeneous principal symbol of order μ

$$\sigma_\psi(a)(r,x,y,\rho,\xi,\eta) = r^{-\mu}\tilde{p}_{(\mu)}(r,x,y,r\rho,\xi,r\eta), \tag{28}$$

referring to the local splitting of variables $(r,x,y) \in \mathbb{R}_+ \times U \times \Omega$ for an open $U \subseteq \mathbb{R}^n$ belonging to a chart on X, and covariables (ρ,ξ,η). Here, $\tilde{p}_{(\mu)}(r,x,y,\tilde{\rho},\xi,\tilde{\eta})$ is the homogeneous principal symbol of order μ of a corresponding local symbol $\tilde{p}(r,x,y,\tilde{\rho},\xi,\tilde{\eta}) \in S^\mu_{\mathrm{cl}}(\overline{\mathbb{R}}_+ \times U \times \Omega \times \mathbb{R}^{1+n+q}_{\tilde{\rho},\xi,\tilde{\eta}})$. Clearly, the cut-off functions are included as a factor $(\sigma\sigma')(r)$. Moreover, we set

$$\tilde{\sigma}_\psi(a)(r,x,y,\tilde{\rho},\xi,\tilde{\eta}) = \tilde{p}_{(\mu)}(r,x,y,\tilde{\rho},\xi,\tilde{\eta}). \tag{29}$$

The twisted homogeneous principal edge symbol of $a(y,\eta) \in \mathcal{R}^\mu(\Omega \times \mathbb{R}^q, \mathrm{g})$, written in the form (23) is defined as

$$\sigma_\wedge(a)(y,\eta) := r^{-\mu}\{\omega_{|\eta|}\,\mathrm{op}_M^{\gamma-n/2}(h_0)(y,\eta)\omega'_{|\eta|} + (1-\omega_{|\eta|})\mathrm{Op}_r(p_0)(y,\eta)(1-\omega''_{|\eta|})\}$$
$$+\,\sigma_\wedge(m+g)(y,\eta), \tag{30}$$

$\eta \neq 0$, cf. formula (13).

Lemma 1 *The homogeneous principal edge symbol (30) can be computed as a limit*

$$\sigma_\wedge(a)(y,\eta) = \lim_{\delta\to\infty}\delta^{-\mu}\kappa_\delta^{-1}a(y,\delta\eta)\kappa_\delta,$$

cf. also Remark 8 *below, although $a(y,\eta)$ as an operator-valued symbol in $S^\mu(\Omega \times \mathbb{R}^q; \mathcal{K}^{s,\gamma}(X^\wedge), \mathcal{K}^{s-\mu,\gamma-\mu}(X^\wedge))$ is not classical, except for the $\mathcal{R}_{M+G}$-part, cf. relation (14).*

Proof Because of relation (14) it suffices to consider $a(y,\eta) - (m+g)(y,\eta)$, for brevity in this proof again denoted by $a(y,\eta)$. Then

$$\begin{aligned}\delta^{-\mu}\kappa_\delta^{-1}a(y,\delta\eta)\kappa_\delta &= \delta^{-\mu}(\kappa_\delta^{-1}\sigma\kappa_\delta)(\kappa_\delta^{-1}r^{-\mu}\kappa_\delta)\\ &\{(\kappa_\delta^{-1}\omega_{\delta\eta}\kappa_\delta)(\kappa_\delta^{-1}\,\mathrm{op}_M^{\gamma-n/2}(h)(y,\delta\eta)\kappa_\delta)(\kappa_\delta^{-1}\omega'_{\delta\eta}\kappa_\delta)\\ &+(\kappa_\delta^{-1}(1-\omega_{\delta\eta})\kappa_\delta)(\kappa_\delta^{-1}\mathrm{Op}_r(p)(y,\delta\eta)\kappa_\delta)(\kappa_\delta^{-1}(1-\omega''_{\delta\eta})\kappa_\delta)\}(\kappa_\delta^{-1}\sigma'\kappa_\delta).\end{aligned}$$

It suffices now to observe that $\delta^{-\mu}(\kappa_\delta^{-1}r^{-\mu}\kappa_\delta) = r^{-\mu}$, and for $\delta \to \infty$

$$\kappa_\delta^{-1}\sigma\kappa_\delta \to 1,\ \kappa_\delta^{-1}\omega_{\delta\eta}\kappa_\delta \to \omega_{|\eta|},$$

etc., and

$$\kappa_\delta^{-1}\,\mathrm{op}_M^{\gamma-n/2}(h)(y,\delta\eta)\kappa_\delta \to \mathrm{op}_M^{\gamma-n/2}(h_0)(y,\eta),\ \kappa_\delta^{-1}\mathrm{Op}_r(p)(y,\delta\eta)\kappa_\delta \to \mathrm{Op}_r(p_0)(y,\eta).$$

□

Remark 6 The results of [7] show that for any h as in Definition 5 we have

$$a(y,\eta) := \sigma r^{-\mu}\,\mathrm{op}_M^{\gamma-n/2}(h)(y,\eta)\sigma' \in \mathcal{R}^\mu(\Omega \times \mathbb{R}^q, \mathrm{g})$$

for $\mathrm{g} = (\gamma, \gamma - \mu, (-\infty, 0]))$ and

$$\sigma_\wedge(a)(y,\eta) = r^{-\mu}\, \mathrm{op}_M^{\gamma-n/2}(h_0)(y,\eta).$$

Another consequence of [7] is that for suitable σ, σ' every element of $\mathcal{R}^\mu(\Omega\times\mathbb{R}^q, \mathrm{g})$ for $\mathrm{g} = (\gamma, \gamma-\mu, \Theta))$, $\Theta = (-(k+1), 0]$, $k \in \mathbb{N}\cup\{\infty\}$, has the form

$$\sigma r^{-\mu}\, \mathrm{op}_M^{\gamma-n/2}(h)(y,\eta)\sigma'$$

modulo $\mathcal{R}^\mu_{M+G}(\Omega\times\mathbb{R}^q, \mathrm{g})$.

For purposes below, we ignore for the moment the $\mathcal{R}^\mu_{M+G}$-summand, and assume $a(y,\eta) \in \mathcal{R}^\mu(\Omega\times\mathbb{R}^q, \mathrm{g})$ in the form

$$a(y,\eta) = \sigma r^{-\mu}\big\{\omega_\eta\, \mathrm{op}_M^{\gamma-n/2}(h)(y,\eta)\omega'_\eta + (1-\omega_\eta)\mathrm{Op}_r(p)(y,\eta)(1-\omega''_\eta)\big\}\sigma',$$

cf. Remark 4. Let us split up $a(y,\eta)$ as a sum

$$a(y,\eta) = a_{[0]}(y,\eta) + a_{[1]}(y,\eta) \tag{31}$$

for

$$a_{[i]}(y,\eta) := \sigma r^{-\mu}\{\omega_\eta\, \mathrm{op}_M^{\gamma-n/2}(h_{[i]})(y,\eta)\omega'_\eta + (1-\omega_\eta)\mathrm{Op}_r(p)(y,\eta)(1-\omega''_\eta)\}\sigma',$$

where

$$h_{[0]}(r,y,z,\eta) := \sum_{j=0}^{N} r^j \frac{1}{j!}\Big\{\frac{\partial^j}{\partial r^j}\tilde{h}(r,y,z,r\eta)\Big\}\Big|_{r=0}, \tag{32}$$

for any fixed $N\in\mathbb{N}$ and $h_{[1]}(r,y,z,\eta) := h(r,y,z,\eta) - h_{[0]}(r,y,z,\eta)$. Note that

$$h_{[0]}(r,y,z,\eta) = \sum_{j=0}^{N} r^j \sum_{|\alpha|\le j} h_{j,\alpha}(y,z)\eta^\alpha$$

for some $h_{j,\alpha}(y,z) \in C^\infty(\Omega, M_{\mathcal{O}}^{\mu-|\alpha|}(X))$.

Definition 6 Given $a(y,\eta) \in \mathcal{R}^\mu(\Omega\times\mathbb{R}^q, \mathrm{g})$ in the form (23), we set

$$\sigma_M^{\mu-j}(a)(y,z,\eta) := \sum_{|\alpha|\le j} h_{j,\alpha}(y,z)\eta^\alpha + \sigma_M^{\mu-j}(m+g)(y,z,\eta),\ \ j=0,\ldots,N, \tag{33}$$

called conormal symbols of a.

In the following theorem, we assume $\mathrm{g} = (\gamma, \gamma-\mu, (-(k+1), 0])$, $k\in\mathbb{N}\cup\{\infty\}$.

Theorem 7 *For any $a(y,\eta) \in \mathcal{R}^\mu(\Omega\times\mathbb{R}^q, \mathrm{g})$, regarded as a family of operators in $L^\mu(X^\wedge, \mathrm{g})$, the conormal symbols $\sigma_M^{\mu-j}(a)(y,z,\eta)$, $j = 0,\ldots,k$, are uniquely determined by $a(y,\eta)$.*

The recovering process for conormal symbols can be found in [24, Sect. 1.3.1, Theorem 4]; it applies in the cone algebra over $X^\wedge$, not only over $\mathbb{R}_+$.

Applying Taylor's formula at $r = 0$, we can write

$$\tilde{h}_{[1]}(r, y, z, \tilde{\eta}) = \tilde{h}^L_{[1]}(r, y, z, \tilde{\eta}) + \tilde{h}_{[1],(L+1)}(r, y, z, \tilde{\eta})$$

for

$$\tilde{h}^L_{[1]}(r, y, z, \tilde{\eta}) := \sum_{j=0}^{L} \frac{1}{j!} r^j \left(\frac{\partial^j}{\partial r^j} \tilde{h}_{[1]} \right) (0, y, z, \tilde{\eta})$$

and

$$\tilde{h}_{[1],(L+1)}(r, y, z, \tilde{\eta}) \in C^\infty(\overline{\mathbb{R}}_+ \times \Omega, M^\mu_{\mathcal{O}}(X; \mathbb{R}^q_{\tilde{\eta}}))$$

vanishing at $r = 0$ of order $L + 1$. In a similar manner, we decompose (17) as

$$\tilde{p}(r, y, \tilde{\rho}, \tilde{\eta}) = \tilde{p}^L(r, y, \tilde{\rho}, \tilde{\eta}) + \tilde{p}_{(L+1)}(r, y, \tilde{\rho}, \tilde{\eta})$$

for

$$\tilde{p}^L(r, y, \tilde{\rho}, \tilde{\eta}) := \sum_{j=0}^{L} \frac{1}{j!} r^j \left(\frac{\partial^j}{\partial r^j} \tilde{p} \right) (0, y, \tilde{\rho}, \tilde{\eta})$$

and

$$\tilde{p}_{(L+1)}(r, y, \tilde{\rho}, \tilde{\eta}) \in C^\infty\big(\overline{\mathbb{R}}_+ \times \Omega, L^\mu_{\mathrm{cl}}(X; \mathbb{R}^{1+q}_{\tilde{\rho},\tilde{\eta}})\big),$$

vanishing at $r = 0$ of order $L + 1$. Setting

$$h^L_{[1]}(r, y, z, \eta) = \tilde{h}^L_{[1]}(r, y, z, r\eta),\ h_{[1],(L+1)}(r, y, z, \eta) = \tilde{h}_{[1],(L+1)}(r, y, z, r\eta),$$
$$p^L(r, y, \rho, \eta) = \tilde{p}^L(r, y, r\rho, r\eta),\ p_{(L+1)}(r, y, \rho, \eta)\tilde{p}_{(L+1)}(r, y, r\rho, r\eta)$$

it follows that

$$a_{[1]}(y, \eta) = a^L_{[1]}(y, \eta) + a_{[1],(L+1)}(y, \eta) \tag{34}$$

for

$$a^L_{[1]}(y, \eta) := \sigma r^{-\mu}\big\{\omega_\eta\, \mathrm{op}_M^{\gamma-n/2}(h^L_{[1]})(y, \eta)\omega'_\eta + (1 - \omega_\eta)\mathrm{Op}_r(p^L)(y, \eta)(1 - \omega''_\eta)\big\}\sigma', \tag{35}$$

$$\begin{aligned} a_{[1],(L+1)}(y, \eta) := \sigma r^{-\mu}\{&\omega_\eta\, \mathrm{op}_M^{\gamma-n/2}(h_{[1],(L+1)})(y, \eta)\omega'_\eta \\ &+ (1 - \omega_\eta)\mathrm{Op}_r(p_{L+1})(y, \eta)(1 - \omega''_\eta)\}\sigma'. \end{aligned} \tag{36}$$

This yields a decomposition

$$a(y,\eta) = \mathbf{a}_{[0]}(y,\eta) + \mathbf{a}_{[1]}(y,\eta) \tag{37}$$

for

$$\mathbf{a}_{[0]}(y,\eta) := a_{[0]}(y,\eta) + a^{L}_{[1]}(y,\eta),\ \ \mathbf{a}_{[1]}(y,\eta) := a_{[1],(L+1)}(y,\eta) \tag{38}$$

for any $L \in \mathbb{N}$, where

$$\mathbf{a}_{[0]}(y,\eta) \in S^{\mu}(\Omega \times \mathbb{R}^q; \mathcal{K}^{s,\gamma}(X^\wedge), \mathcal{K}^{s-\mu,\gamma-\mu}(X^\wedge)), \tag{39}$$

and

$$\mathbf{a}_{[1]}(y,\eta) \in S^{\mu-(L+1)}(\Omega \times \mathbb{R}^q; \mathcal{K}^{s,\gamma}(X^\wedge), \mathcal{K}^{s-\mu,\gamma-\mu}(X^\wedge)). \tag{40}$$

3.2 Compositions

Theorem 8 *Let*

$$a(y,\eta) \in \mathcal{R}^{\mu}(\Omega \times \mathbb{R}^q, \mathrm{g}),\ b(y,\eta) \in \mathcal{R}^{\nu}(\Omega \times \mathbb{R}^q, \mathrm{h})$$

for $\mathrm{g} = (\gamma - \nu, \gamma - (\mu + \nu), \Theta),\ \mathrm{h} = (\gamma, \gamma - \nu, \Theta),\ \mu, \nu, \gamma \in \mathbb{R}$. *Then*

$$(ab)(y,\eta) \in \mathcal{R}^{\mu+\nu}(\Omega \times \mathbb{R}^q, \mathrm{g} \circ \mathrm{h}) \tag{41}$$

for $\mathrm{g} \circ \mathrm{h} = (\gamma, \gamma - (\mu + \nu), \Theta)$, *and we have*

$$\sigma_\psi(ab) = \sigma_\psi(a)\sigma_\psi(b),\ \tilde{\sigma}_\psi(ab) = \tilde{\sigma}_\psi(a)\tilde{\sigma}_\psi(b), \tag{42}$$

and

$$\sigma_\wedge(ab) = \sigma_\wedge(a)\sigma_\wedge(b). \tag{43}$$

Moreover, if a *or* b *belongs to* $\mathcal{R}_{M+G}$ *or* $\mathcal{R}_G$, *then the same is true of* ab.

Proof Without loss of generality we assume that a, b are as in Remark 4. In addition, we first consider the case without summands in $\mathcal{R}_{M+G}$, i.e.,

$$a(y,\eta) := \sigma r^{-\mu}\big\{\omega_\eta \operatorname{op}_M^{\gamma-n/2}(h)(y,\eta)\omega'_\eta + (1-\omega_\eta)\mathrm{Op}_r(p)(y,\eta)(1-\omega''_\eta)\big\}\sigma'$$

and

$$b(y,\eta) := \sigma r^{-\nu}\big\{\omega_\eta \operatorname{op}_M^{\gamma-n/2}(k)(y,\eta)\omega'_\eta + (1-\omega_\eta)\mathrm{Op}_r(b)(y,\eta)(1-\omega''_\eta)\big\}\sigma'.$$

In this case, for the weight interval we take $\Theta = (-(N+1, 0)]$ for every $N \in \mathbb{N}$. Because of Theorem 6, the composition ab can be carried out as a composition of operator-valued symbols, and we obtain

$$a(y,\eta)b(y,\eta) \in S^{\mu+\nu}(\Omega \times \mathbb{R}^q; \mathcal{K}^{s,\gamma}(X^\wedge), \mathcal{K}^{s-(\mu+\nu),\gamma-(\mu+\nu)}(X^\wedge)). \quad (44)$$

There are

$$p_{\mathrm{cp}}(r,y,\rho,\eta) := \tilde{p}_{\mathrm{cp}}(r,y,r\rho,r\eta), \quad (45)$$

$$\tilde{p}_{\mathrm{cp}}(r,y,\tilde{\rho},\tilde{\eta}) \in C^\infty\big(\overline{\mathbb{R}}_+ \times \Omega, L^{\mu+\nu}_{\mathrm{cl}}(X; \mathbb{R}^{1+q}_{\tilde{\rho},\tilde{\eta}})\big) \quad (46)$$

and

$$h_{\mathrm{cp}}(r,y,z,\eta) := \tilde{h}_{\mathrm{cp}}(r,y,z,r\eta), \quad (47)$$

$$\tilde{h}_{\mathrm{cp}}(r,y,z,\tilde{\eta}) \in C^\infty(\overline{\mathbb{R}}_+ \times \Omega, M^{\mu+\nu}_{\mathcal{O}}(X; \mathbb{R}^q_{\tilde{\eta}})), \quad (48)$$

where $p_{\mathrm{cp}}(r,y,\rho,\eta)$ is related to $h_{\mathrm{cp}}(r,y,z,\eta)$ as in Theorem 5, such that for

$$c_{\mathrm{cp}}(y,\eta) := \sigma r^{-(\mu+\nu)}\{\omega_\eta \operatorname{op}_M^{\gamma-n/2}(h_{\mathrm{cp}})(y,\eta)\omega'_\eta + (1-\omega_\eta)\mathrm{Op}_r(p_{\mathrm{cp}})(y,\eta)(1-\omega''_\eta)\}\sigma'$$

we have

$$a(y,\eta)b(y,\eta) - c_{\mathrm{cp}}(y,\eta) \in C^\infty(\Omega \times \mathbb{R}^q, L^{-\infty}(X^\wedge)). \quad (49)$$

In the composition $(ab)(y,\eta)$, we may integrate the cut-off factor $\sigma\sigma'$ in the middle in the dependence of $\tilde{k}(r,y,z,r\eta) = k(r,y,z,\eta)$ and $\tilde{b}(r,y,r\rho,r\eta) = b(r,y,\rho,\eta)$ on the first r-variable as smooth functions up to $r=0$. In other words, without loss of generality we omit $\sigma\sigma'$ in the middle. By virtue of Theorem 4, the operator family $(ab)(y,\eta)$ belongs (y,η)-wise to $L^{\mu+\nu}(X^\wedge, \mathrm{g} \circ \mathrm{h})$. From Theorem 3(ii), we see that (49) (y,η)-wise belongs to $L_{M+G}(X^\wedge, \mathrm{g} \circ \mathrm{h})$. Therefore, it remains to show that (49) belongs to $\mathcal{R}^{\mu+\nu}_{M+G}(\Omega \times \mathbb{R}^q, \mathrm{g} \circ \mathrm{h})$. As noted before, it suffices to consider the case $\Theta = (-(N+1), 0]$ for some $N \in \mathbb{N}$. By virtue of Theorem 7, we find the sequence of conormal symbols of (49) in a unique way, in this case smoothly depending on $y \in \Omega$, and with the typical polynomial behaviour in η. The latter property follows by rephrasing the operator families a, b, and c_{cp} as $a_{[0]}+a_{[1]}$, $b_{[0]}+b_{[1]}$, and $c_{\mathrm{cp},[0]}+c_{\mathrm{cp},[1]}$, respectively, according to (31). Then we easily see that only $a_{[0]}b_{[0]}-c_{\mathrm{cp},[0]}$ contribute to the conormal symbols, those are smoothing and holomorphic in this case, and, applying the composition rules in the cone calculus we obtain polynomials in η, as claimed. Thus, we can form an element $c_M(y,\eta) \in \mathcal{R}^{\mu+\nu}_{M+G}(\Omega \times \mathbb{R}^q, \mathrm{g} \circ \mathrm{h})$, such that

$$(ab)(y,\eta) - (c_{\mathrm{cp}} + c_M)(y,\eta) \in S^{\mu+\nu}(\Omega \times \mathbb{R}^q; \mathcal{K}^{s,\gamma}(X^\wedge), \mathcal{K}^{s-(\mu+\nu),\gamma-(\mu+\nu)}(X^\wedge))$$

and takes values in $L_G(X^\wedge, \mathrm{g} \circ \mathrm{h})$. It remains to show the relation

$$c_g(y,\eta) := (ab)(y,\eta) - (c_{\mathrm{cp}} + c_M)(y,\eta) \in \mathcal{R}^{\mu+\nu}_G(\Omega \times \mathbb{R}^q, \mathrm{g} \circ \mathrm{h}).$$

Here we use decompositions of a, b, c_{cp} like (37), namely,

$$a = \mathbf{a}_{[0]} + \mathbf{a}_{[1]},\ b = \mathbf{b}_{[0]} + \mathbf{b}_{[1]},\ c = \mathbf{c}_{\mathrm{cp},[0]} + \mathbf{c}_{\mathrm{cp},[1]} \tag{50}$$

for any sufficiently large L. This gives us

$$(ab) - (c_{\mathrm{cp}} + c_M) = (\mathbf{a}_{[0]}\mathbf{b}_{[0]}) - (\mathbf{c}_{\mathrm{cp},[0]} + c_M)) + R_{L+1}. \tag{51}$$

From relations of the kind (39) and (40), we see that

$$R_{L+1} \in S^{\mu+\nu-(L+1)}\big(\Omega \times \mathbb{R}^q; \mathcal{K}^{s,\gamma}(X^\wedge), \mathcal{K}^{s-(\mu+\nu),\gamma-(\mu+\nu)}(X^\wedge)\big)$$

for all s. The expression on the left-hand side of (51) takes values in $L_G(X^\wedge, \mathrm{g} \circ \mathrm{h})$. Also $(\mathbf{a}_{[0]}\mathbf{b}_{[0]}) - (\mathbf{c}_{\mathrm{cp},[0]} + c_M))$ is $L_G(X^\wedge, \mathrm{g} \circ \mathrm{h})$-valued, since its conormal symbols vanish. In this construction, without loss of generality, we assume that the above-mentioned N in the weight interval Θ is equal to L. So, if we enlarge L, we also change the meaning of c_M. We do not explicitly indicate that in the notation of c_M. In any case, if we determine homogeneous components $(ab)_{(\mu+\nu-j)}(y,\eta)$, $j = 0, \ldots, L$, they will be independent of R_{L+1} and take values in $L_G(X^\wedge, \mathrm{g} \circ \mathrm{h})$. Here, we employ successively relations (73) of Remark 8, namely,

$$c_{g,(\mu+\nu)}(y,\eta) = \lim_{\delta\to\infty}\delta^{-\mu}\tilde{\kappa}_\delta^{-1}\{(\mathbf{a}_{[0]}\mathbf{b}_{[0]}) - (\mathbf{c}_{\mathrm{cp},[0]} + c_M)\}(y,\delta\eta)\kappa_\delta, \tag{52}$$

etc. Since L is arbitrary, we obtain the full sequence of homogeneous components $c_{g,(\mu+\nu-j)}(y,\eta)$, $j \in \mathbb{N}$. Then, for an excision function $\chi(\eta)$, we get a Green symbol $c_g(y,\eta) \sim \sum_{j=0}^{\infty} \chi(\eta) c_{g,(\mu+\nu-j)}(y,\eta)$. Using Remark 2 it follows that

$$(ab) - (c_{\mathrm{cp}} + c_M) = (\mathbf{a}_{[0]}\mathbf{b}_{[0]}) - (\mathbf{c}_{\mathrm{cp},[0]} + c_M + c_g)$$

is a Green symbol of order $-\infty$, also belonging to the admitted edge amplitude functions, and hence we proved relation (41) for the case that the smoothing Mellin plus Green symbols are omitted. However, the arguments for the desired composition behaviour including summands in $\mathcal{R}_{M+G}$ are similar to the previous ones. The composition rule (42) is a simple consequence of (27) and (41). Relation (43) can be deduced by applying Lemma 1 combined with

$$\begin{aligned}\sigma_\wedge(ab)(y,\eta) &= \lim_{\delta\to\infty}\delta^{-(\mu+\nu)}\kappa_\delta^{-1}(ab)(y,\delta\eta)\kappa_\delta \\ &= \lim_{\delta\to\infty}\delta^{-\mu}\kappa_\delta^{-1}(a)(y,\delta\eta)\kappa_\delta \lim_{\delta\to\infty}\delta^{-\nu}\kappa_\delta^{-1}(b)(y,\delta\eta)\kappa_\delta.\end{aligned}$$

The fact that the composition ab belongs to $\mathcal{R}_{M+G}$ or $\mathcal{R}_G$ as soon as a or b is of that kind can be proved in a similar manner as the result in general, since the characterisation of ab is a consequence of the corresponding (y,η)-wise behaviour in the cone algebra. □

3.3 Conclusions and Remarks

There is an extension of Definition 5 to the case of 2×2 block matrices with $a(y,\eta) \in \mathcal{R}^{\mu}(\Omega \times \mathbb{R}^q, \mathrm{g})$ in the upper-left corners, while the other entries contain trace and potential symbols as well as classical symbols in the lower-right corner. Similar to Theorem 8, there is then a composition result also including trace and potential entries. The picture is similar to Boutet de Monvel's calculus, where the extra trace and potential symbols are motivated by the task to formulate elliptic boundary problems and to express parametrices. In the edge calculus of [22], the situation is similar, see also the material in several monographs in this bibliography, e.g., [4, 8, 26], etc., or the article jointly with Coriasco [3]. Recall that Boutet de Monvel's calculus is entirely contained in the edge calculus of [22] as soon as orders and types are zero. In general, for Green and trace operators of positive type, there are symbols referring to the right of the respective weight lines. In the edge calculus version, this situation has been treated in a joint paper with Kapanadze and Seiler [10], and for the case of edge boundary value problems jointly with Wei in [29]. Also, here we have an algebra structure which is essential for the construction of parametrices. Note that it is important to find the inverses of corresponding operator-valued symbols within the respective algebras of symbols. For symbols taking values in Boutet de Monvel's calculus, this follows from [23]. A similar technique shows such a result for the invertibility of edge symbols and edge operators within the corresponding algebras.

Let us make a few remarks on the edge algebra itself. Similarly as the cone algebra consisting of operator spaces $L^{\mu}(X^{\wedge}, \mathrm{g})$, cf. Definition 2, or, more generally, on a manifold M with conical singularities, for a manifold B with smooth edge Y we have an edge algebra, consisting of spaces $L^{\mu}(B, \mathrm{g})$, of edge pseudo-differential operators of order μ. More generally, we have spaces of block matrix operators with trace and potential entries, but we do not elaborate the details here. A concise and update version of the edge calculus is given in a joint article with Flad and Harutyunyan [6]. Modulo smoothing operators in this calculus the operators $A \in L^{\mu}(B, \mathrm{g})$ are locally near Y in coordinates $y \in \Omega$, $\Omega \subseteq \mathbb{R}^q$ open, $q = \dim Y$, of the form $\mathrm{Op}_y(a)$ for an $a(y,\eta) \in \mathcal{R}^{\mu}(\Omega \times \mathbb{R}^q, \mathrm{g})$. For compact B, we have spaces $H^{s,\gamma}(B)$ of smoothness $s \in \mathbb{R}$ and weight $\gamma \in \mathbb{R}$, locally near Y modelled on spaces $\mathcal{W}^s(\mathbb{R}^q, \mathcal{K}^{s,\gamma}(X^{\wedge}))$, cf. Definitions 8, 9, and subspaces $H_P^{s,\gamma}(B)$ with asymptotics of type P, locally near Y modelled on spaces $\mathcal{W}^s(\mathbb{R}^q, \mathcal{K}_P^{s,\gamma}(X^{\wedge}))$, cf. formula (75). The continuity of operators $A \in L^{\mu}(B, \mathrm{g})$ for weight data $\mathrm{g} = (\gamma, \gamma - \mu, \Theta)$,

$$A : H^{s,\gamma}(B) \to H^{s-\mu,\gamma-\mu}(B), \quad H_P^{s,\gamma}(B) \to H_Q^{s-\mu,\gamma-\mu}(B)$$

can be proved in many ways, e.g., by means of arguments in [24, Sect. 3.2.1, Theorem 6], which employs a tensor product argument, or the ideas of Seiler's paper [31].

The discussion so far concerns the case of constant (in the edge variable y) discrete asymptotics. Note that from the very beginning in the cone and edge pseudo-differential calculus, since [17, 20, 22], there have been observed continuous and variable branching discrete asympototics, see also the corresponding material in [24,

Sects. 1.1.5, 3.3.6; 9, Sect. 2.3]. Concerning the subsequent development, see also the bibliography in [28]. Clearly, the structure of edge symbols, especially their composition behaviour, is in this context fundamental as well, but in the present overview, we do not go into further details.

4 Auxiliary Material

4.1 Kegel Spaces and Other Aspects of the Cone Calculus

Let us prepare some necessary tools. The Mellin transform is defined as

$$Mu(z) = \int_0^\infty r^{z-1}u(r)\mathrm{d}r, \tag{53}$$

first for $u \in C_0^\infty(\mathbb{R}_+)$ and later on extended to larger distribution spaces. Let

$$\Gamma_\beta := \{z \in \mathbb{C} : \mathrm{Re}\, z = \beta\} \tag{54}$$

for any real β. By $\mathcal{A}(U)$ for an open set $U \subset \mathbb{C}$, we denote the space of holomorphic functions in U. For $u \in C_0^\infty(\mathbb{R}_+)$, we have $Mu \in \mathcal{A}(\mathbb{C})$ and $Mu|_{\Gamma_\beta} \in \mathcal{S}(\Gamma_\beta)$, for every $\beta \in \mathbb{R}$. This allows us to form the weighted Mellin transform

$$M_\gamma u := Mu\big|_{\Gamma_{\frac{1}{2}-\gamma}}$$

for a weight $\gamma \in \mathbb{R}$. We employ the fact that M_γ extends to an isomorphism

$$M_\gamma : r^\gamma L^2(\mathbb{R}_+) \to L^2(\Gamma_{\frac{1}{2}-\gamma}),$$

and we have

$$M_\gamma^{-1}g(r) = \int_{\Gamma_{\frac{1}{2}-\gamma}} r^{-z}g(z)\mathrm{d}\!\!{}^{-}\!z.$$

For the weighted Mellin transform, we have an analogue of Schwartz spaces, namely,

$$\mathcal{T}^\gamma(\mathbb{R}_+) := M_\gamma^{-1}\mathcal{S}(\Gamma_{\frac{1}{2}-\gamma}).$$

From the above remarks, we have $C_0^\infty(\mathbb{R}_+) \subset \mathcal{T}^\gamma(\mathbb{R}_+)$ for every $\gamma \in \mathbb{R}$. There are also versions of Sobolev spaces based on the weighted Mellin transform. Let

$$\mathcal{H}^{s,\gamma}(\mathbb{R}_+) \quad \text{for} \quad s, \gamma \in \mathbb{R},$$

denote the completion of $\mathcal{T}^\gamma(\mathbb{R}_+)$ with respect to the norm

$$\|u\|_{\mathcal{H}^{s,\gamma}(\mathbb{R}_+)} := \left\{\int_{\Gamma_{\frac{1}{2}-\gamma}} \langle z\rangle^{2s}|Mu(z)|^2\mathrm{d}\!\!{}^{-}\!z\right\}^{1/2}.$$

The Mellin–Sobolev spaces will occur in combination with Sobolev spaces in $\mathbb{R}^n$. So, we define

$$\mathcal{H}^{s,\gamma}(\mathbb{R}_+ \times \mathbb{R}^n)$$

as the completion of $C_0^\infty(\mathbb{R}_+ \times \mathbb{R}^n)$ with respect to the norm

$$\|u\|_{\mathcal{H}^{s,\gamma}(\mathbb{R}_+\times\mathbb{R}^n)} := \left\{ \int_{\mathbb{R}^n} \int_{\Gamma_{\frac{n+1}{2}-\gamma}} \langle z, \xi \rangle^{2s} |(FMu)(z,\xi)|^2 đz đ\xi \right\}^{1/2}.$$

The choice $\Gamma_{\frac{n+1}{2}-\gamma}$ of the weight line in the latter definition is motivated by the shape of functions in $L^2(\mathbb{R}^{n+1})$ under substituting polar coordinates $\mathbb{R}^{n+1} \setminus \{0\} \ni \tilde{x} \to (r, x) \in \mathbb{R}_+ \times S^n$ with S^n being the unit sphere in $\mathbb{R}^{n+1}$. In fact, we then have

$$L^2(\mathbb{R}^{n+1}) = r^{-n/2} L^2(\mathbb{R}_+ \times S^n)$$

with respect to the measure $\mathrm{d}r\mathrm{d}x$ and $\mathrm{d}x$ defined by the restriction of the Riemannian metric of $\mathbb{R}^{n+1}$ to S^n.
Observe that

$$\mathcal{H}^{s,\gamma+\beta}(\mathbb{R}_+ \times \mathbb{R}^n) = r^\beta \mathcal{H}^{s,\gamma}(\mathbb{R}_+ \times \mathbb{R}^n)$$

for every $s, \gamma, \beta \in \mathbb{R}$.

Let X be a closed smooth manifold of dimension n. Choose an open covering $\{U_1, ..., U_N\}$ by coordinate neighborhoods and charts

$$\chi_j : U_j \to \mathbb{R}^n.$$

Moreover, let $\{\varphi_1, ..., \varphi_N\}$ be a subordinate partition of unity. Then $\mathcal{H}^{s,\gamma}(X^\wedge)$ for $X^\wedge = \mathbb{R}_+ \times X$ is defined as the completion of $C_0^\infty(\mathbb{R}_+ \times X)$ with respect to the norm

$$\|u\|_{\mathcal{H}^{s,\gamma}(X^\wedge)} = \left\{ \sum_{j=1}^{N} \|\varphi_j u \circ (\mathrm{id}_{\mathbb{R}_+} \times \chi_j)^{-1}\|^2_{\mathcal{H}^{s,\gamma}(\mathbb{R}_+\times\mathbb{R}^n)} \right\}^{1/2} \tag{55}$$

where $\mathrm{id}_{\mathbb{R}_+} \times \chi_j$ is defined by $(r, x) \mapsto (r, \chi_j(x))$. It can be easily verified that the choice of partition of unity only affects the norm (55) up to equivalence.
Recall that on X we fixed a Riemannian metric and the associated measure $\mathrm{d}x$. We then have an identification

$$\mathcal{H}^{0,0}(X^\wedge) = r^{-n/2} L^2(\mathbb{R}_+ \times X) \tag{56}$$

with $L^2(\mathbb{R}_+ \times X)$ referring to $\mathrm{d}r\mathrm{d}x$.
We also employ cylindrical Sobolev spaces $\mathcal{H}^s(\mathbb{R}^q \times X)$ defined as the completion of $C_0^\infty(\mathbb{R}^q \times X)$ with respect to the norm

$$\|u\|_{H^s(\mathbb{R}^q\times X)} = \left\{ \sum_{j=1}^{N} \|\varphi_j u \circ (\mathrm{id}_{\mathbb{R}^q} \times \chi_j)^{-1}\|^2_{H^s(\mathbb{R}^q\times\mathbb{R}^n)} \right\}^{1/2}$$

for $(\mathrm{id}_{\mathbb{R}^q} \times \chi_j)(y,x) := (y, \chi_j(x))$.

Lemma 2 *Let* $(S_{\gamma-n/2}u)(\mathbf{r},x) := e^{-(\frac{n+1}{2}-\gamma)\mathbf{r}}u(e^{-\mathbf{r}},x)$ *first for* $u \in C_0^\infty(\mathbb{R}_+ \times X)$,

$$S_{\gamma-n/2} : C_0^\infty(\mathbb{R}_+ \times X) \to C_0^\infty(\mathbb{R} \times X). \tag{57}$$

Then (57) extends to an isomorphism

$$S_{\gamma-n/2} : \mathcal{H}^{s,\gamma}(X^\wedge) \to H^s(\mathbb{R} \times X) \tag{58}$$

for every $s, \gamma \in \mathbb{R}$.

In the singular analysis, we interpret $X^\wedge$ as an open stretched cone with base X. It will be necessary to impose another control of distributions for $r \to \infty$ than that formulated by the spaces $\mathcal{H}^{s,\gamma}(X^\wedge)$. This will be done in terms of spaces which control the behaviour up to the conical exit $r \to \infty$.

Let $\chi : U \to V$ be a diffeomorphism from a coordinate neighborhood U on X to an open set $V \subset S^n$ with S^n being the unit sphere in $\mathbb{R}^{1+n}_{\tilde{x}}$, let

$$V^\wedge := \{\tilde{x} \in \mathbb{R}^{1+n} \setminus \{0\} : \tilde{x}/|\tilde{x}| \in V\},$$

and consider the diffeomorphism

$$\chi_{\mathrm{cone}} : \mathbb{R}_+ \times U \to V^\wedge, \quad (r,x) \to r\chi(x).$$

The following spaces will be often referred to as cone spaces of smoothness s.

Definition 7 We set

$$H^s_{\mathrm{cone}}(X^\wedge) := \big\{u \in H^s_{\mathrm{loc}}(\mathbb{R} \times X)\big|_{\mathbb{R}_+ \times X} : (1-\omega)\varphi u \circ \chi^{-1}_{\mathrm{cone}} \in H^s(\mathbb{R}^{1+n}) \tag{59}$$

$$\text{for every } \varphi \in C_0^\infty(U) \text{ and for any } \chi_{\mathrm{cone}} \text{ of the above-mentioned kind}\big\}, \tag{60}$$

where ω is some cut-off function on the r half axis.

Example 1 We have

$$H^s_{\mathrm{cone}}((S^n)^\wedge) = \big\{u \in H^s_{\mathrm{loc}}(\mathbb{R} \times S^n)\big|_{\mathbb{R}_+ \times S^n} : ((1-\omega)u) \circ p \in H^s(\mathbb{R}^{n+1})\big\},$$

for some cut-off function ω, where $p : \mathbb{R}^{n+1}_{\tilde{x}} \setminus \{0\} \to \mathbb{R}_+ \times S^n$, $p : \tilde{x} \mapsto (r,x)$, indicates polar coordinates in $\mathbb{R}^{n+1}_{\tilde{x}} \setminus \{0\}$.

Remark 7 There is the following equivalent variant of Definition 7. Without loss of generality, we consider coordinate neighbourhoods U on X which are diffeomorphic to a ball in $\mathbb{R}^n$. Let $B := \{\tilde{x} = (\tilde{x}_1, \dots, \tilde{x}_{n+1}) \in \mathbb{R}^{n+1} : \tilde{x}_1 = 1, |(\tilde{x}_2, \dots, \tilde{x}_{n+1})| < 1\}$, and let $\zeta : U \to B$ be a diffeomorphism. Moreover, set $B^\wedge := \{\lambda\tilde{x} : \tilde{x} \in B, \lambda \in \mathbb{R}_+\}$, and form the diffeomorphism

$$\zeta_{\mathrm{cone}} : \mathbb{R}_+ \times U \to B^\wedge, \quad (r,x) \to (r, r\zeta(x)).$$

Then we have

$$H^s_{\mathrm{cone}}(X^\wedge) := \big\{u \in H^s_{\mathrm{loc}}(\mathbb{R} \times X)\big|_{\mathbb{R}_+ \times X} : (1-\omega)\varphi u \circ \zeta^{-1}_{\mathrm{cone}} \in H^s(\mathbb{R}^{1+n})$$
$$\text{for every } \varphi \in C^\infty_0(U) \text{ and for any } \zeta_{\mathrm{cone}} \text{ of the above-mentioned kind}\big\}, \tag{61}$$

where ω is some cut-off function on the r half axis.

Next, we define Kegel spaces of smoothness $s \in \mathbb{R}$ and weight $\gamma \in \mathbb{R}$ for $r \to 0$ and $e \in \mathbb{R}$ for $r \to \infty$.

Definition 8 We set

$$\mathcal{K}^{s,\gamma}(X^\wedge) := \big\{\omega u + (1-\omega)v : u \in \mathcal{H}^{s,\gamma}(X^\wedge),\ v \in H^s_{\mathrm{cone}}(X^\wedge)\big\} \tag{62}$$

for any cut-off function ω on the r half axis. More generally, let

$$\mathcal{K}^{s,\gamma;e}(X^\wedge) := [r]^{-e}\mathcal{K}^{s,\gamma}(X^\wedge) \tag{63}$$

for any strictly positive smooth function $[r]$ on $\mathbb{R}_+$ such that $[r] = r$ for $r \geq c$ for some $c > 0$.

The space (62) is independent of the specific choice of ω.

4.2 *Edge Spaces and Symbols with Twisted Symbolic Estimates*

A Hilbert space H is said to be endowed with a group action $\kappa = \{\kappa_\delta\}_{\delta \in \mathbb{R}_+}$ if $\kappa_\delta : H \to H$ is an isomorphism and $\kappa_\delta\kappa_{\delta'} = \kappa_{\delta\delta'}$ for every $\delta, \delta' \in \mathbb{R}_+$ and if $\delta \to \kappa_\delta h$ defines an element in $C(\mathbb{R}_{+,\delta}, H)$ for every $h \in H$. Recall that then there are constants $c, M > 0$ such that

$$\|\kappa_\delta\|_{\mathcal{L}(H)} \leq c \max\{\delta, \delta^{-1}\}^M.$$

On $\mathcal{K}^{s,\gamma}(X^\wedge)$,we consider a family of isomorphisms

$$\kappa_\delta : \mathcal{K}^{s,\gamma}(X^\wedge) \to \mathcal{K}^{s,\gamma}(X^\wedge), \delta \in \mathbb{R}_+, \tag{64}$$

defined by

$$(\kappa_\delta u)(r,x) := \delta^{(n+1)/2}u(\delta r, x), \tag{65}$$

for $n := \dim X$.

Definition 9 [22, Sect. 2.2, Definition 1.1]. The abstract edge space

$$\mathcal{W}^s(\mathbb{R}^q, H), \quad s \in \mathbb{R}, \tag{66}$$

modelled on a Hilbert space H with group action $\kappa = \{\kappa_\delta\}_{\delta \in \mathbb{R}_+}$, is defined as the set of all $u \in \mathcal{S}'(\mathbb{R}^q, H)(= \mathcal{L}(\mathcal{S}(\mathbb{R}^q), H))$ such that $Fu(\eta)$ is locally integrable and

$$\|u\|_{\mathcal{W}^s(\mathbb{R}^q,H)} := \left\{ \int \langle\eta\rangle^{2s} \|\kappa^{-1}_{\langle\eta\rangle}(Fu)(\eta)\|^2_H d\!\!{}^-\!\eta \right\}^{1/2} \tag{67}$$

is finite.

Clearly, the space (66) depends on the choice of κ. Let H and $\tilde{H}$ be Hilbert spaces with group action κ and $\tilde{\kappa}$, respectively. By

$$S^{(\mu)}(\Omega \times (\mathbb{R}^q \setminus \{0\}); H, \tilde{H}) \tag{68}$$

for an open set $\Omega \subseteq \mathbb{R}^p$ and $\mu \in \mathbb{R}$, we denote the subspace of all

$$a_{(\mu)}(y, \eta) \in C^\infty(\Omega \times (\mathbb{R}^q \setminus \{0\}), \mathcal{L}(H, \tilde{H}))$$

satisfying the (twisted) homogeneity relation

$$a_{(\mu)}(y, \delta\eta) = \delta^\mu \tilde{\kappa}_\delta a_{(\mu)}(y, \eta)\kappa_\delta^{-1}$$

for all $\delta \in \mathbb{R}_+$. Note that for any excision function $\chi(\eta)$ and any $a_{(\mu)}(y, \eta) \in S^{(\mu)}(\Omega \times (\mathbb{R}^q \setminus \{0\}); H, \tilde{H})$, the function

$$a(y, \eta) := \chi(\eta) a_{(\mu)}(y, \eta)$$

satisfies the system of estimates

$$\|\tilde{\kappa}^{-1}_{\langle\eta\rangle}\{D^\alpha_y D^\beta_\eta a(y, \eta)\}\kappa_{\langle\eta\rangle}\|_{\mathcal{L}(H,\tilde{H})} \leq c\,\langle\eta\rangle^{\mu-|\beta|} \tag{69}$$

for all $(y, \eta) \in K \times \mathbb{R}^q$, $K \Subset \Omega, \alpha \in \mathbb{N}^p, \beta \in \mathbb{N}^q$, for constants $c = c(\alpha, \beta, K) > 0$.

Definition 10 [22, Sect. 2.3, Definition 1]. Let $\Omega \subseteq \mathbb{R}^p$ be open and $\mu \in \mathbb{R}$.

(i) The space

$$S^\mu(\Omega \times \mathbb{R}^q; H, \tilde{H}) \tag{70}$$

is defined as the set of all $a(y, \eta) \in C^\infty(\Omega \times \mathbb{R}^q, \mathcal{L}(H, \tilde{H}))$ satisfying the symbolic estimates (69). The elements are called operator-valued symbols of order μ (referring to the chosen Hilbert spaces with group action).

(ii) The space of classical symbols

$$S^\mu_{\mathrm{cl}}(\Omega \times \mathbb{R}^q; H, \tilde{H}) \tag{71}$$

is defined as the set of all $a(y, \eta)$ in (70) such that there is a sequence of homogeneous components $a_{(\mu-j)}(y, \eta) \in S^{(\mu-j)}(\Omega \times (\mathbb{R}^q \setminus \{0\}); H, \tilde{H})$, $j \in \mathbb{N}$, such that for any excision function χ

$$r_N(y, \eta) := a(y, \eta) - \sum_{j=0}^{N} \chi(\eta) a_{(\mu-j)}(y, \eta) \in S^{\mu-(N+1)}(\Omega \times \mathbb{R}^q; H, \tilde{H}) \tag{72}$$

for every $N \in \mathbb{N}$.

Remark 8 For a symbol $a(y,\eta) \in S^{\mu}_{\mathrm{cl}}(\Omega \times \mathbb{R}^q; H, \tilde{H})$, we can recover the sequence $a_{(\mu-j)}$, $j \in \mathbb{N}$, of homogeneous components. First, we form the limit

$$a_{(\mu)}(y,\eta) = \lim_{\delta\to\infty} \delta^{-\mu} \tilde{\kappa}_{\delta}^{-1} a(y, \delta\eta)\kappa_{\delta} \tag{73}$$

for $\eta \neq 0$. Then, for any excision function $\chi(\eta)$ in $\mathbb{R}^q$ it follows that

$$a_{\mu-1}(y,\eta) := a(y,\eta) - \chi(\eta) a_{(\mu)}(y,\eta) \in S^{\mu-1}_{(\mathrm{cl})}(\Omega \times \mathbb{R}^q; H, \tilde{H}).$$

Applying the procedure to $a_{\mu-1}(y,\eta)$ similarly as (73), we obtain $a_{(\mu-1)}(y,\eta)$, and then, successively, $a_{(\mu-j)}(y,\eta)$ for all j.

4.3 Asymptotics

A sequence $P := \{(p_j, m_j)\}_{j\in\mathbb{J}} \subset \mathbb{C} \times \mathbb{N}, \mathbb{C} = \mathbb{C}_z$, for some $\mathbb{J} \subseteq \mathbb{N}$ is called discrete asymptotic type if $\pi_{\mathbb{C}} P = \{p_j\}_{j\in\mathbb{J}}$ is contained in $\{\frac{n+1}{2} - \gamma + \vartheta < \mathrm{Re}\, z < \frac{n+1}{2} - \gamma\}$ for some $-\infty \leq \vartheta < 0$ where $\pi_{\mathbb{C}} P$ is finite for $\vartheta > -\infty$ and $\mathrm{Re}\, p_j \to -\infty$ as $j \to \infty$ when $\vartheta = -\infty$ and $\pi_{\mathbb{C}} P$ is infinite. If we have such a P, we say that P is associated with (γ, Θ) for $\Theta := (\vartheta, 0]$. An asymptotic type P is said to satisfy the shadow condition if $(p, m) \in P$ implies $(p - j, m) \in P$ for every $j \in \mathbb{N}$ such that $\mathrm{Re}\, p - j > \frac{n+1}{2} - \gamma + \vartheta$. For convenience, we always assume that the asymptotic types in consideration satisfy the shadow condition, although this is not really essential. Let

$$\mathcal{K}^{s,\gamma}_{\Theta}(X^{\wedge}) := \varprojlim_{l\in\mathbb{N}} \mathcal{K}^{s,\gamma-\vartheta-(l+1)^{-1}}(X^{\wedge})$$

endowed with the Fréchet topology of the projective limit. In the case of finite Θ and $\pi_{\mathbb{C}} P = \{p_0, \ldots, p_N\}$, we form

$$\mathcal{E}_P(X^{\wedge}) := \left\{ \omega(r) \sum_{j=0}^{N} \sum_{k=0}^{m_j} c_{jk}(x) r^{-p_j} \log^k r : c_{jk} \in C^{\infty}(X) \right\} \tag{74}$$

for some fixed cut-off function $\omega(r)$. The space (74) is Fréchet in a natural way. Then for $\pi_{\mathbb{C}} P \subset \{\mathrm{Re}\, z < \frac{n+1}{2} - \gamma\}$ we have

$$\mathcal{E}_P(X^{\wedge}) \subset \mathcal{K}^{\infty,\gamma}(X^{\wedge}).$$

This space is direct to $\mathcal{K}^{s,\gamma}_{\Theta}(X^{\wedge})$, and we set

$$\mathcal{K}^{s,\gamma}_{P}(X^{\wedge}) := \mathcal{K}^{s,\gamma}_{\Theta}(X^{\wedge}) + \mathcal{E}_P(X^{\wedge}) \tag{75}$$

in the Fréchet topology of the direct sum. We can also endow the space (75) with the group action (65), using the fact that $\mathcal{K}^{s,\gamma}_{P}(X^{\wedge})$ can be written as a projective limit of Hilbert spaces E^j, $j \in \mathbb{N}$, continuously embedded in $E^0 := \mathcal{K}^{s,\gamma}(X^{\wedge})$ (with

stronger topologies) where (65) also acts on E^j for every j. However, κ does not act on $\mathcal{E}_P(X^\wedge)$, since the involved cut-off function ω is changed.
By a discrete Mellin asymptotic type, we understand a sequence

$$R := \{(r_j, n_j)\}_{j\in\mathbb{Z}} \subset \mathbb{C} \times \mathbb{N}$$

such that $\pi_{\mathbb{C}} R := \{r_j\}_{j\in\mathbb{Z}}$ intersects every strip $\{c \le \operatorname{Re} z \le c'\}, c \le c'$, in a finite set. Then

$$M_R^{-\infty}(X)$$

is defined to be the set of all $f(z) \in \mathcal{A}(\mathbb{C} \setminus \pi_{\mathbb{C}} R, L^{-\infty}(X))$ which are meromorphic with poles at the points r_j of multiplicity $n_j + 1$ and with finite rank Laurent coefficients in $L^{-\infty}(X)$, for all j, and such that for any $\pi_{\mathbb{C}} R$-excision function $\chi(z)$ (i.e., $\chi \in C^\infty(\mathbb{C}), \chi(z) = 0$ for $\operatorname{dist}(z, \pi_{\mathbb{C}} R) < \varepsilon_0$, $\chi(z) = 1$ for $\operatorname{dist}(z, \pi_{\mathbb{C}} R) > \varepsilon_1$ for some $0 < \varepsilon_0 < \varepsilon_1$) we have $(\chi f)|_{\Gamma_\beta} \in \mathcal{S}(\Gamma_\beta, L^{-\infty}(X))$ for every $\beta \in \mathbb{R}$, uniformly in compact β-intervals.

References

1. L. Boutet de Monvel, Boundary problems for pseudo-differential operators, Acta. Math. **126**, 11–51 (1971)
2. H.O. Cordes, A global parametrix for pseudo-differential operators over $\mathbb{R}^n$, with applications. Reprint SFB 72, Universität Bonn (1976)
3. **{Cori2}** S. Coriasco, B.-W. Schulze, Edge problems on configurations with model cones of different dimensions. Osaka J. Math. **43**, 1–40 (2006)
4. Ju.V. Egorov, B.-W. Schulze, *Pseudo-Differential Operators, Singularities, Applications*, vol. 93. Oper. Theory: Adv. Appl. (Birkhäuser, Basel, 1997)
5. G.I. Eskin, Boundary value problems for elliptic pseudodifferential equations (Transl. of Nauka, Moskva, 1973). Math. Monographs, vol. 52 (American Mathematical Society, Providence, Rhode Island, 1980)
6. H.-J. Flad, G. Harutyunyan, B.-W. Schulze, Asymptotic parametrices of elliptic edge operators. arXiv:1010.1453 (2010)
7. J.B. Gil, B.-W. Schulze, J. Seiler, Cone pseudodifferential operators in the edge symbolic calculus. Osaka J. Math. **37**, 221–260 (2000)
8. G. Harutjunjan, B.-W. Schulze, *Elliptic Mixed, Transmission and Singular Crack Problems* (European Mathematical Society, Zürich, 2008)
9. D. Kapanadze, B.-W. Schulze, *Crack Theory and Edge Singularities* (Kluwer Academic Publishers, Dordrecht, 2003)
10. D. Kapanadze, B.-W. Schulze, J. Seiler, Operators with singular trace conditions on a manifold with edges. Integr. Equ. Oper. Theory. **61**, 241–279 (2008)
11. V.A. Kondratyev, Boundary value problems for elliptic equations in domains with conical points. Trudy Mosk. Mat. Obshch. **16**, 209–292 (1967)
12. C. Parenti, Operatori pseudo-differenziali in $\mathbb{R}^n$ e applicazioni. Annali Mat. Pura Appl. (4). **93**, 359–389 (1972)
13. S. Rempel, B.-W. Schulze, *Index Theory of Elliptic Boundary Problems* (Akademie-Verlag, Berlin, 1982)
14. S. Rempel, B.-W. Schulze, Parametrices and boundary symbolic calculus for elliptic boundary problems without transmission property. Math. Nachr. **105**, 45–149 (1982)

15. S. Rempel, B.-W. Schulze, *Pseudo-differential and Mellin Operators in Spaces with Conormal Singularity* (boundary symbols). Report R-Math-**01/84** (Karl-Weierstrass Institut, Berlin, 1984)
16. S. Rempel, B.-W. Schulze, Complete Mellin and Green symbolic calculus in spaces with conormal asymptotics. Ann. Glob. Anal. Geom. **4**(2), 137–224 (1986)
17. S. Rempel, B.-W. Schulze, *Asymptotics for elliptic mixed boundary problems* (pseudo-differential and Mellin operators in spaces with conormal singularity). Math. Res., vol. 50 (Akademie-Verlag, Berlin, 1989)
18. E. Schrohe, in *Spaces of weighted symbols and weighted Sobolev spaces on manifolds*. Springer Lecture Notes in Math. vol. 1256, "Pseudo-Differential Operators" (Springer, Berlin, Heidelberg, 1987), pp. 360–377
19. E. Schrohe, B.-W. Schulze, in *A Symbol Algebra for Pseudodifferential Boundary Value Problems on Manifolds with Edges*. Differential Equations, Asymptotic Analysis, and Mathematical Physics. Math. Research, vol. 100 (Akademie Verlag, Berlin, 1997), pp. 292–324
20. B.-W. Schulze, Ellipticity and continuous conormal asymptotics on manifolds with conical singularities, Math. Nachr. **136**, 7–57 (1988)
21. B.-W. Schulze, Corner Mellin operators and reduction of orders with parameters. Ann. Sc. Norm. Sup. Pisa Cl. Sci. **16**(1), 1–81 (1989)
22. B.-W. Schulze, in *Pseudo-differential operators on manifolds with edges*, Teubner-Texte zur Mathematik, vol. 112, Symp. "Partial Differential Equations, Holzhau 1988" (BSB Teubner, Leipzig, 1989), pp. 259–287.
23. B.-W. Schulze, in *Topologies and invertibility in operator spaces with symbolic structures*, Teubner-Texte zur Mathematik, vol. 111 "Problems and Methods in Mathematical Physics" (BSB Teubner, Leipzig, 1989), pp. 257–270
24. B.-W. Schulze, *Pseudo-differential Operators on Manifolds with Singularities* (North-Holland, Amsterdam, 1991)
25. B.-W. Schulze, *Pseudo-differential Boundary Value Problems, Conical Singularities, and Asymptotics* (Akademie Verlag, Berlin, 1994)
26. B.-W. Schulze, *Boundary Value Problems and Singular Pseudo-differential Operators* (Wiley, Chichester, 1998)
27. B.-W. Schulze, J. Seiler, The edge algebra structure of boundary value problems. Ann. Glob. Anal. Geom. **22**, 197–265 (2002)
28. B.-W. Schulze, L. Tepoyan, *The Singular Functions of Branching Edge Asymptotics*, arXiv:1202.0387v2 [math.AP]; ed. by S. Molahajloo, S. Pilipovič,, J. Toft, M.W. Wong. Pseudo-Differential Operators, Generalized Functions and Asymptotics, Oper. Theory: Adv. Appl., vol. 231 (Birkhäuser, Basel, 2013), pp. 27–53.
29. B.-W. Schulze, Y. Wei, Edge-boundary problems with singular trace conditions, Ann. Glob. Anal. Geom. **35**, 413–429 (2009)
30. M.A. Shubin, Pseudodifferential operators in $\mathbb{R}^n$. Dokl. Akad. Nauk SSSR. **196**, 316–319 (1971)
31. J. Seiler, Continuity of edge and corner pseudo-differential operators. Math. Nachr. **205**, 163–182 (1999)
32. M.A. Shubin, *Pseudodifferential Operators and Spectral Theory* (Springer, Berlin, 1987)

On Bifurcation for Semilinear Elliptic Dirichlet Problems on Shrinking Domains

Nils Waterstraat

Abstract We consider the Dirichlet problem for semilinear elliptic equations on a bounded domain which is diffeomorphic to a ball and investigate bifurcation from a given (trivial) branch of solutions, where the radius of the ball serves as bifurcation parameter. Our methods are based on well-known results from variational bifurcation theory, which we outline in a separate section for the readers' convenience.

1 Introduction

Let $\Omega \subset \mathbb{R}^n$ be a bounded domain and

$$\Phi : \bar{\Omega} \to B[0,1]$$

a diffeomorphism, where $B[0,1]$ denotes the closed unit ball around $0 \in \mathbb{R}^n$. We consider the Dirichlet boundary value problem:

$$\begin{cases} Lu(x) + g(x, u(x)) = 0, & x \in \Omega \\ u(x) = 0, & x \in \partial\Omega, \end{cases} \tag{1}$$

where

$$Lu(x) = -\sum_{i,j=1}^{n} \frac{\partial}{\partial x_j}\left(a_{ij}(x)\frac{\partial u}{\partial x_i}(x)\right), \quad x \in \Omega,$$

for some smooth functions $a_{ij} : \bar{\Omega} \to \mathbb{R}$, $a_{ij} = a_{ji}$, $1 \leq i, j \leq n$, which satisfy the ellipticity condition

$$\sum_{i,j=1}^{n} a_{ij}(x)\xi_i\xi_j > 0, \quad x \in \Omega, \ (\xi_1, \dots, \xi_n) \in \mathbb{R}^n \setminus \{0\}. \tag{2}$$

N. Waterstraat (✉)
Institut für Mathematik, Humboldt-Universität zu Berlin, 10099 Berlin, Germany
e-mail: waterstn@math.hu-berlin.de

J. Escher et al., (eds.), *Elliptic and Parabolic Equations,* Springer Proceedings in Mathematics & Statistics 119, DOI 10.1007/978-3-319-12547-3_12

Moreover, we assume that $g : \bar{\Omega}\times\mathbb{R} \to \mathbb{R}$ is continuously differentiable, $g(x, 0) = 0$, $x \in \Omega$, and we suppose that there are constants α, C such that

$$\left|\frac{\partial g}{\partial \xi}(x,\xi)\right| \leq C(1 + |\xi|^{\alpha-1}), \quad (x,\xi) \in \Omega \times \mathbb{R}, \tag{3}$$

where $1 \leq \alpha \leq \frac{n+2}{n-2}$ if $n \geq 3$ and $1 \leq \alpha < \infty$ if $n = 2$. Finally, in the case $n = 1$, that is, (1) is an ordinary differential equation, we do not impose a growth condition on the nonlinearity g.

Let us denote for $0 < r \leq 1$ by $B(0,r)$ the open ball of radius r around 0 in $\mathbb{R}^n$ and by

$$\Omega_r := \Phi^{-1}(B(0,r)) \tag{4}$$

the domain Ω shrinked by the factor r. We consider the corresponding boundary value problems

$$\begin{cases} Lu(x) + g(x, u(x)) = 0, & x \in \Omega_r \\ \qquad\qquad\qquad u(x) = 0, & x \in \partial\Omega_r. \end{cases} \tag{5}$$

Note that $u \equiv 0$ is a solution of (5) for all instants r. We call $r_0 \in (0, 1]$ a *bifurcation instant* for the boundary value problems (5), if there exists a sequence of radii $r_n \to r_0$ and functions $u_n \in H^1_0(\Omega_{r_n})$ such that u_n is a non-trivial weak solution of (5) on Ω_{r_n} and $\|u_n\|_{H^1_0(\Omega_{r_n})} \to 0$.

Let us point out that we exclude from the definition the limiting case $r_0 = 0$ in which the domain Ω_r degenerates to a point. The reason is that if $r_n \to 0$, then $\|u_n\|_{H^1_0(\Omega_{r_n})} \to 0$, for example, for any sequence of functions $\{u_n\}_{n\in\mathbb{N}} \subset H^1_0(\Omega_{r_n})$ such that $\|\frac{\partial u_n}{\partial x_i}\|_{L^\infty(\Omega_r)} < \infty$, $n \in \mathbb{N}$, $i = 1, \ldots, n$. Consequently, a bifurcation instant $r_0 = 0$ would not imply the existence of non-trivial solutions of (5) for small $r > 0$ which are arbitrarily close to the trivial solution $u \equiv 0$ in a suitable sense.

In what follows, we suppose that the function $f : \Omega \to \mathbb{R}$ defined by $f(x) = \frac{\partial g}{\partial \xi}(x, 0)$ is smooth, and we consider the linearised boundary value problems:

$$\begin{cases} Lu(x) + f(x)u(x) = 0, & x \in \Omega_r \\ \qquad\qquad\qquad u(x) = 0, & x \in \partial\Omega_r. \end{cases} \tag{6}$$

We call $r_0 \in (0, 1]$ a *conjugate instant* for (6) if the dimension of the space of classical solutions:

$$m(r_0) := \dim\{u \in C^2(\Omega_r) \cap C(\bar{\Omega}_r) : u \text{ solves } (6)\} \tag{7}$$

is non-zero, and from now on we assume that $m(1) = 0$. Our main result reads as follows:

Theorem 1 *The bifurcation instants of (5) are precisely the conjugate instants of (6).*

Our proof of Theorem 1 uses crossing forms from variational bifurcation theory as follows: Let H be a real separable Hilbert space and $\psi : I \times H \to \mathbb{R}$ a C^2-function, where $I = [0, 1]$ denotes the unit interval. We assume that $0 \in H$ is a critical point of all functionals $\psi_\lambda := \psi(\lambda, \cdot)$, $\lambda \in I$, and that the Riesz representations L_λ of the second derivatives $D_0^2\psi_\lambda : H \times H \to \mathbb{R}$ at this critical point are Fredholm operators and have finite Morse indices $\mu_-(L_\lambda)$. It is well known that, if L_a and L_b are invertible for some $0 \leq a < b \leq 1$ and $\mu_-(L_a) \neq \mu_-(L_b)$, then there exists a bifurcation instant $\lambda_0 \in (a, b)$ for critical points of ψ, that is, every neighbourhood of $(\lambda_0, 0)$ in $I \times H$ contains elements (λ, u), where $u \neq 0$ is a critical point of ψ_λ. In other words, a jump in the Morse index along the path $L = \{L_\lambda\}$ entails bifurcation of critical points of ψ. Besides this result, we recall below in Sect. 2 that, if L is continuously differentiable with respect to the parameter $\lambda \in I$, then jumps in the Morse index can be computed as follows: assume that L_{λ_0} has a non-trivial kernel for some $\lambda_0 \in (0, 1)$. The derivative $\dot{L}_{\lambda_0}$ of L at λ_0 plugged into the scalar product on H defines a quadratic form on H and λ_0 is called regular if its restriction to $\ker L_{\lambda_0}$ is nondegenerate. We recall from [7] that if λ_0 is a regular crossing of L, then there exists $\varepsilon > 0$ such that L_λ is invertible for all $0 < |\lambda - \lambda_0| < \varepsilon$, and the jump in the Morse index when λ passes λ_0 is given by the signature of the quadratic form $\langle \dot{L}_{\lambda_0}\cdot, \cdot\rangle$ on $\ker L_{\lambda_0}$. This result traces back to the seminal paper [19] of Robbin and Salamon.

The third section is devoted to the proof of Theorem 1. We introduce a family of functionals $\psi : I \times H_0^1(B(0, 1)) \to \mathbb{R}$ parametrised by the radius r of the ball in (4), such that critical points of $\psi_r := \psi(r, \cdot)$ correspond to weak solutions of (5) under an obvious rescaling mapping from $H_0^1(B(0, 1))$ to $H_0^1(\Omega_r)$. Since $u \equiv 0$ is a solution of all equations (5), $0 \in H_0^1(B(0, 1))$ is a critical point of all functionals ψ_r, $r \in I$. Moreover, $r_0 \in (0, 1]$ is a bifurcation instant for ψ if and only if it is a bifurcation instant for the equations (5). We explain below that the kernels of the associated operators L_r induced by the second derivative of ψ_r at the critical point 0 correspond to solutions of the linearised boundary value problems (6). The main part of our proof shows that each crossing $r_0 \in (0, 1)$ of this path $L = \{L_r\}$ is regular and that the signature of the restriction of $\langle \dot{L}_{r_0}\cdot, \cdot\rangle$ to $\ker L_{r_0}$ is given by the dimension $m(r_0)$ introduced in (7). Consequently, if $m(r_0) \neq 0$, then r_0 is a bifurcation instant for the equations (5). Since every bifurcation instant is easily seen to be conjugate by the implicit function theorem, this will prove Theorem 1.

In the final Sect. 4, we show at first a corollary that we obtain from our proof of Theorem 1 as outlined in the previous paragraph. Since each crossing $r_0 \in (0, 1)$ of L is regular, we deduce that there exist only finitely many instants at which $\ker L_r \neq \{0\}$, and hence $m(r) = \dim\ker L_r = 0$ for all but finitely many $r \in (0, 1)$. Let μ_- denote the Morse index of the linearised equation (6) on $\Omega = \Omega_1$, i.e. the number of negative eigenvalues

$$\begin{cases} Lu(x) + f(x)u(x) = \lambda\, u(x), & x \in \Omega \\ u(x) = 0, & x \in \partial\Omega, \end{cases} \tag{8}$$

counted according to their multiplicities. We derive from our proof of Theorem 1 that

$$\mu_- = \sum_{0<r<1} m(r). \tag{9}$$

Let us point out that this result was already obtained by Smale in [20] (cf. also [21]) for general strongly elliptic differential operators on vector bundles over compact manifolds with boundary, which is a generalisation of the Morse index theorem for geodesics in Riemannian manifolds to partial differential equations and motivates the definition of conjugate instants for (6). However, Smale's argument is based on the domain monotonicity of eigenvalues and consequently it is rather different from our way to equality (9).

We derive from (9) the following corollary of Theorem 1.

Corollary 1 *If $\mu_- \neq 0$, then there exist at least*

$$\left\lfloor \frac{\mu_-}{\max_{0<r<1} m(r)} \right\rfloor$$

distinct bifurcation instants in $(0, 1)$*, where* $\lfloor \cdot \rfloor$ *denotes the integral part of a real number.*

We conclude the fourth section by some examples of our theory including ordinary differential equations and a well-known equation from geometric analysis.
Finally, let us point out that Theorem 1 and Corollary 1 have appeared in various generality in our joint articles [15] and [16] with A. Portaluri from Università degli studi di Torino. In [15], we assume that L is the Laplacian on a star-shaped domain in $\mathbb{R}^n$, whereas in [16], we consider the Laplace–Beltrami operator on a geodesic ball of a Riemannian manifold. Of course, after introducing coordinates in the latter case, both results are special cases of the slightly more general setting that we consider here. A further aim of this chapter is to give a thorough exposition on crossing forms in bifurcation theory as outlined above, which should make the presentation self-contained in contrast to our previous works [15] and [16], which strongly rely on the article [7] of Fitzpatrick et al.

2 Bifurcation of Critical Points for Essentially Positive Functionals

The aim of this section is to explain the necessary basics from variational bifurcation theory that we need in order to prove Theorem 1 in the following section. Before we come to bifurcation theory, we recap in a first subsection some facts from spectral theory for self-adjoint Fredholm operators on real Hilbert spaces. Subsequently, we recall the well-known principle from nonlinear functional analysis that jumps in the Morse index of the Hessians entail bifurcation of critical points of families of functionals (cf. Theorem 2). Since this result can be found in common textbooks, we do not prove it here, but just quote some references. The final part of this section is concerned with crossing forms and their relation to jumps of the Morse index along paths of self-adjoint Fredholm operators. The main result is stated in Proposition 1, which finally leads to the bifurcation theorem 3 on which the proof of our main Theorem 1 relies. Since the only reference for Proposition 1 that we are aware of is the more general Theorem 4.1 in [7], we provide full details in this section and in particular include a proof of Lemma 3, which was left to the reader in [7].

2.1 Self-Adjoint Fredholm Operators and Their Spectra

Let H be a real Hilbert space. We denote by $\mathcal{L}(H)$ the space of bounded linear operators, by $GL(H) \subset \mathcal{L}(H)$ the open subset of invertible operators, and by $\mathcal{S}(H) \subset \mathcal{L}(H)$ the closed subset of self-adjoint operators. For $T \in \mathcal{S}(H)$, the spectrum

$$\sigma(T) = \{\lambda \in \mathbb{R} : \lambda - T \notin GL(H)\}$$

is non-empty, and our first aim is to introduce spectral projections for T. Before we do this, we briefly discuss complexifications of real Hilbert spaces and their operators.

The complexification $H^{\mathbb{C}}$ of H is the linear space consisting of all formal elements $u + \mathrm{i}v$, where $u, v \in H$. $H^{\mathbb{C}}$ is a Hilbert space with respect to the scalar product

$$\langle u_1 + \mathrm{i}v_1, u_2 + \mathrm{i}v_2\rangle_{H^{\mathbb{C}}} := \langle u_1, u_2\rangle_H + \langle v_1, v_2\rangle_H + \mathrm{i}\langle v_1, u_2\rangle_H - \mathrm{i}\langle u_1, v_2\rangle_H.$$

Each operator $T \in \mathcal{L}(H)$ induces canonically a bounded linear operator on $H^{\mathbb{C}}$ by $T^{\mathbb{C}}(u + iv) = Tu + \mathrm{i}Tv$, and conversely, if A is a bounded linear operator on $H^{\mathbb{C}}$, it is readily seen that there are unique operators T, S in $\mathcal{L}(H)$ such that $A = T^{\mathbb{C}} + iS^{\mathbb{C}}$. As a consequence, there exists a conjugation $\bar{\cdot}$ on the Banach space $\mathcal{L}(H^{\mathbb{C}})$ of all bounded linear operators on $H^{\mathbb{C}}$, and $A = T^{\mathbb{C}}$ for some $T \in \mathcal{L}(H)$ if and only if $\overline{A} = A$. This particularly holds for the real part $\mathrm{Re}(A) := \frac{1}{2}(A + \overline{A})$ of an operator $A \in \mathcal{L}(H^{\mathbb{C}})$.

Let now $T \in \mathcal{S}(H)$, and $a, b \notin \sigma(T)$ such that $[a, b] \cap \sigma(T)$ is a non-empty finite set consisting solely of eigenvalues of finite type; i.e., $0 < \dim\ker(\lambda - T) < \infty$ for all $\lambda \in [a, b] \cap \sigma(T)$. Let $D \subset \mathbb{C}$ be a disc such that $\partial D \cap \mathbb{R} = \{a, b\}$. It is readily seen that $\sigma(T) = \sigma(T^{\mathbb{C}})$ and that $(\ker(\lambda - T))^{\mathbb{C}} = \ker(\lambda - T^{\mathbb{C}})$. In particular, $\sigma(T^{\mathbb{C}}) \cap D$ consists only of eigenvalues of finite type and $\partial D \cap \sigma(T^{\mathbb{C}}) = \emptyset$. Let us recall that the spectral projection of $T^{\mathbb{C}}$ with respect to $D \cap \sigma(T^{\mathbb{C}})$ is the orthogonal projection

$$P(T^{\mathbb{C}}) = \frac{1}{2\pi \mathrm{i}} \int_{\partial D} (\lambda - T^{\mathbb{C}})^{-1}\, \mathrm{d}\lambda, \tag{10}$$

which projects onto the direct sum of all eigenspaces $\ker(\lambda - T^{\mathbb{C}})$ of $T^{\mathbb{C}}$ for eigenvalues $\lambda \in D$. Using once again that $\sigma(T) = \sigma(T^{\mathbb{C}})$ and $\ker(\lambda - T^{\mathbb{C}}) = (\ker(\lambda - T))^{\mathbb{C}}$, it follows that $P_{[a,b]}(T) := \mathrm{Re}(P(T^{\mathbb{C}}))$ is the orthogonal projection in H onto the direct sum of the eigenspaces of T with respect to the eigenvalues in $[a, b]$. We deduce from the construction of $P_{[a,b]}(T)$ and the corresponding result for linear operators on complex Hilbert spaces (cf. e.g. [9], Theorem II.4.2), the following continuity property for isolated eigenvalues of finite multiplicity.

Lemma 1 *Let $T \in \mathcal{S}(H)$ and $a,b \in \mathbb{R}$ such that $a,b \notin \sigma(T)$. Assume that $(a,b)\cap\sigma(T)$ consists of isolated eigenvalues of finite multiplicity. Then there exists $\varepsilon > 0$ such that $a,b \notin \sigma(S)$ and*

$$\dim\operatorname{im} P_{[a,b]}(S) = \dim\operatorname{im} P_{[a,b]}(T)$$

for all $S \in \mathcal{S}(H)$ such that $\|T - S\|_{\mathcal{L}(H)} < \varepsilon$.

In what follows, we denote by $\mathcal{FS}(H)$ the set of all elements in $\mathcal{S}(H)$ which are Fredholm, and we note that an operator in $\mathcal{S}(H)$ belongs to $\mathcal{FS}(H)$, if and only if its kernel is of finite dimension and its image is closed.

Lemma 2 *If $T \in \mathcal{FS}(H)$, then either 0 belongs to the resolvent set of T or it is an isolated eigenvalue of finite multiplicity.*

Proof Since the set of Fredholm operators is open in $\mathcal{L}(H)$, there exists $\varepsilon > 0$ such that $\lambda - T$ is Fredholm for all $|\lambda| < \varepsilon$. Hence, if $|\lambda| < \varepsilon$, then either $\lambda - T$ is invertible or it has a finite dimensional kernel. It remains to show that 0 is isolated in $\sigma(T)$. Since T is self-adjoint and an Fredholm operator, there is an orthogonal decomposition $H = \ker T \oplus \operatorname{im} T$, and the restriction T' of T to the closed subspace $(\ker T)^{\perp} = \operatorname{im} T$ is an isomorphism onto $\operatorname{im} T$. Since $GL(\operatorname{im} T) \subset \mathcal{L}(H)$ is open, there exists $\kappa > 0$ such that $\sigma(T')\cap(-\kappa,\kappa) = \emptyset$. Let now $0 < |\lambda| < \min\{\kappa,\varepsilon\}$ and let $u = u_1 + u_2 \in \ker T \oplus \operatorname{im} T$ be an element of $\ker(\lambda - T)$. Then $0 = \lambda u - Tu = \lambda u_1 + \lambda u_2 - Tu_2$ and so

$$\lambda u_1 = (T - \lambda)u_2 = (T' - \lambda)u_2.$$

Since the left-hand side of this equality is in $\ker T$ and the right-hand side is in $\operatorname{im} T = (\ker T)^{\perp}$, we conclude that $\lambda u_1 = (T' - \lambda)u_2 = 0$. From $\lambda \neq 0$ and the invertibility of $T' - \lambda$, it follows that $u = u_1 + u_2 = 0$. □

As a final piece of notation, we let $\mathcal{FS}_+(H)$ be the set of all elements $T \in \mathcal{FS}(H)$ such that $\sigma(T)\cap(-\infty,0)$ consists of a finite number of eigenvalues of finite type. The Morse index

$$\mu_-(T) = \dim\left(\bigoplus_{\lambda<0}\ker(\lambda - T)\right)$$

is finite for all $T \in \mathcal{FS}_+(H)$ and we note for later reference the following elementary properties:

(i) If $U : H \to H$ is an orthogonal operator, then $UTU^{-1} \in \mathcal{FS}_+(H)$ and

$$\mu_-(UTU^{-1}) = \mu_-(T).$$

(ii) If T is reduced by a splitting $H = H_0 \oplus H_1$, that is, $T(H_i) \subset H_i$, $i = 0, 1$, then

$$\mu_-(T) = \mu_-(T\,|_{H_0}) + \mu_-(T\,|_{H_1}).$$

(iii) If T_0, T_1 belong to the same component of $\mathcal{FS}_+(H) \cap GL(H)$, then

$$\mu_-(T_1) = \mu_-(T_0).$$

Note that the first two properties follow immediately from the definition, whereas the last one is a consequence of Lemma 1.

Finally, let us mention for the sake of completeness that Atiyah and Singer proved in [1] that $\mathcal{FS}(H)$ consists of three components

$$\mathcal{FS}(H) = \mathcal{FS}_+(H) \cup \mathcal{FS}_*(H) \cup \mathcal{FS}_-(H),$$

where an operator $L \in \mathcal{FS}(H)$ belongs to $\mathcal{FS}_\pm(H)$ if and only if $\mu_-(\pm L) < \infty$. Moreover, the spaces $\mathcal{FS}_\pm(H)$ are contractible, whereas $\mathcal{FS}_*(H)$ is a classifying space for the KO-theory functor KO^{-7}.

2.2 *The Bifurcation Theorem*

A Classical Bifurcation Theorem

As in the previous section, let H be a real Hilbert space and recall from the introduction that we denote by I the unit interval $[0, 1]$. Let $\psi : I \times H \to \mathbb{R}$ be a continuous function such that the derivatives $D_u\psi_\lambda$ and $D_u^2\psi_\lambda$ of $\psi_\lambda := \psi(\lambda, \cdot) : H \to \mathbb{R}$ exist and depend continuously on $(\lambda, u) \in I \times H$. In what follows, we assume that $0 \in H$ is a critical point of all functionals ψ_λ, $\lambda \in I$.

Definition 1 An instant $\lambda_0 \in I$ is called a bifurcation point of critical points of ψ if any neighbourhood of $(\lambda_0, 0) \in I \times H$ contains elements (λ, u) such that $u \neq 0$ is a critical point of ψ_λ.

The bilinear forms $D_0^2\psi_\lambda : H \times H \to \mathbb{R}$ given by the second derivative at the critical point $0 \in H$ defined by the Riesz representation theorem a unique path $L : I \to \mathcal{S}(H)$ such that

$$D_0^2\psi_\lambda(u, v) = \langle L_\lambda u, v \rangle_H, \quad u, v \in H.$$

The following theorem is a standard result in bifurcation theory.

Theorem 2 *Assume that $L_\lambda \in \mathcal{FS}_+(H)$ for all $\lambda \in I$, so that in particular the Morse index $\mu_-(L_\lambda)$ is finite for all $\lambda \in I$. Let $a, b \in I$, $a < b$, such that L_a, L_b are invertible. If $\mu_-(L_a) \neq \mu_-(L_b)$, then there exists a bifurcation point of critical points in (a, b).*

Different proofs of this theorem can be found for example in [13], Sect. 8.9, by using the continuity of the critical groups in Morse theory, and in [12], Sect. II.7.1, where the argument is based on Conley index theory for flows of vector fields.

Note that a corresponding assertion holds for families of functionals ψ such that $L_\lambda \in \mathcal{FS}_-(H)$ and $\mu_-(-L_a) \neq \mu_-(-L_b)$, which clearly follows by applying

Theorem 2 to $-\psi_\lambda$, $\lambda \in I$. However, the situation turns out to be much more involved if $L_\lambda \in \mathcal{FS}_*(H)$, $\lambda \in I$, since in this case neither $\mu_-(L_\lambda)$ nor $\mu_-(-L_\lambda)$ is finite. Atiyah et al. constructed in [2] in connection with spectral asymmetry and the η-invariant an integer-valued homotopy invariant for paths in any component of $\mathcal{FS}(H)$ which is called *spectral flow*. Roughly speaking, the spectral flow $\mathrm{sf}(L,[a,b])$ of a path $L:[a,b]\to\mathcal{FS}(H)$ is the number of negative eigenvalues of L_a that become positive as the parameter λ travels from a to b minus the number of positive eigenvalues of L_a that become negative. If $L_\lambda \in \mathcal{FS}_+(H)$, $\lambda \in [a,b]$, then $\mathrm{sf}(L,[a,b]) = \mu_-(L_a) - \mu_-(L_b)$ and so the non-vanishing of the spectral flow entails bifurcation in Theorem 2 if L is induced by the second derivative of a family of functionals ψ as above. Accordingly, one may ask if the assertion of Theorem 2 remains to be true if $L_\lambda \in \mathcal{FS}_*(H)$, $\lambda \in I$ and $\mathrm{sf}(L,[a,b]) \neq 0$. This question was answered in the affirmative by Fitzpatrick et al. in [7] under the additional assumption that the entire map $\psi: I \times H \to \mathbb{R}$ is C^2, and it was improved recently in the joint work [14] of Pejsachowicz and the author to continuous maps $\psi: I\times H\to\mathbb{R}$ that satisfy the same differentiability assumptions as in Theorem 2, that is, each ψ_λ is C^2 and its derivatives depend continuously on the parameter $\lambda \in I$.

Morse Index and Crossing Forms

Let us now consider a continuously differentiable path $L: I \to \mathcal{FS}_+(H)$ and let us denote by $\dot{L}_{\lambda_0}$ the derivative of L with respect to λ at λ_0. We follow the notation of Robbin and Salamon in [19] and denote by $\Gamma(L,\lambda_0)$ the quadratic form

$$\Gamma(L,\lambda_0)[u] = \langle \dot{L}_{\lambda_0}u, u\rangle_H, \quad u \in \ker L_{\lambda_0},$$

on the finite dimensional space $\ker L_{\lambda_0}$, which may be zero dimensional.

Definition 2 An instant $\lambda_0 \in I$ is called a crossing of the path L if $\ker L_{\lambda_0} \neq 0$, and a crossing λ_0 is called regular if $\Gamma(L,\lambda_0)$ is nondegenerate.

We begin by showing that regular crossings are isolated, which is an immediate consequence of the following lemma.

Lemma 3 *If λ_0 is a regular crossing of L, then there exist $\varepsilon, C > 0$ such that for $|\lambda-\lambda_0|<\varepsilon$*

$$\|L_\lambda u\| \geq C\,|\lambda-\lambda_0|\,\|u\|, \quad u \in H.$$

Proof Since L_{λ_0} is a self-adjoint Fredholm operator, there is an orthogonal decomposition $H = \ker L_{\lambda_0} \oplus \mathrm{im} L_{\lambda_0} =: H_0 \oplus H_1$. Let us assume at first that $u \in H_1$ and let $P^\perp_{\lambda_0}$ denote the orthogonal projection onto $H_1 = H_0^\perp$. Since $P^\perp_{\lambda_0} L_{\lambda_0} P^\perp_{\lambda_0}$ is invertible and $GL(H_1) \subset \mathcal{L}(H_1)$ is open, there exists $\varepsilon' > 0$ such that $P^\perp_{\lambda_0} L_\lambda P^\perp_{\lambda_0} \in \mathcal{L}(H_1)$ is invertible for all $|\lambda-\lambda_0| < \varepsilon'$. From $\|P^\perp_{\lambda_0}\| = 1$ and $P^\perp_{\lambda_0}u = u$ for all $u \in H_1$, we conclude that there is $C > 0$ such that

$$\|L_\lambda u\| \geq \|P^\perp_{\lambda_0} L_\lambda P^\perp_{\lambda_0} u\| \geq C\,\|u\| \geq C\,|\lambda-\lambda_0|\|u\|, \quad u \in H_1,$$

for all $|\lambda - \lambda_0| < \varepsilon$, where $0 < \varepsilon < \min\{1, \varepsilon'\}$.

Let us now assume that $u \in H_0$ and let us denote by P_{λ_0} the orthogonal projection onto H_0. From

$$\Gamma(L, \lambda_0) = \langle P_{\lambda_0} \dot{L}_{\lambda_0} P_{\lambda_0} u, u \rangle, \quad u \in H_0,$$

and the assumption that λ_0 is regular, we conclude that $P_{\lambda_0} \dot{L}_{\lambda_0} P_{\lambda_0}$ is invertible on the finite dimensional space H_0. Consequently, there exists $\varepsilon > 0$ such that

$$\frac{P_{\lambda_0} L_\lambda P_{\lambda_0}}{\lambda - \lambda_0} = \frac{P_{\lambda_0} L_\lambda P_{\lambda_0} - P_{\lambda_0} L_{\lambda_0} P_{\lambda_0}}{\lambda - \lambda_0} : H_0 \to H_0$$

is invertible for all $|\lambda - \lambda_0| < \varepsilon$. Since $\|P_{\lambda_0}\| = 1$ and $P_{\lambda_0} u = u$, $u \in H_0$, there exists $C > 0$ such that

$$\|L_\lambda u\| \geq \|P_{\lambda_0} L_\lambda P_{\lambda_0} u\| \geq C\,|\lambda - \lambda_0|\,\|u\|, \quad u \in H_0.$$

This finishes the proof. □

In what follows, we denote by $\operatorname{sgn}\Gamma(L, \lambda_0)$ the signature of the quadratic form $\Gamma(L, \lambda_0)$. Let us recall that $\operatorname{sgn}\Gamma(L, \lambda_0)$ is the difference of the number of positive and negative eigenvalues of the symmetric operator $P_{\lambda_0} \dot{L}_{\lambda_0} P_{\lambda_0}$ on the finite dimensional space $\ker L_{\lambda_0}$, where P_{λ_0} denotes the orthogonal projection onto $\ker L_{\lambda_0}$ as in the previous proof. In other words,

$$\operatorname{sgn}\Gamma(L, \lambda_0) = \mu_-(-P_{\lambda_0} \dot{L}_{\lambda_0} P_{\lambda_0}) - \mu_-(P_{\lambda_0} \dot{L}_{\lambda_0} P_{\lambda_0}).$$

Let us point out that the following proposition holds for any path L, irrespective if it is induced by the Hessians of a family of functionals as in the previous section.

Proposition 1 *If L_a, L_b are invertible for some $a, b \in I$, $a < b$, and L has only regular crossings in (a, b), then*

$$\mu_-(L_a) - \mu_-(L_b) = \sum_{\lambda \in (a,b)} \operatorname{sgn}\Gamma(L, \lambda).$$

Proof By Lemma 3 and property (iii) of the Morse index, we can clearly assume that there is only a single regular crossing λ_0 of L in (a, b). Then $\ker L_{\lambda_0} \neq \{0\}$, and we infer from Lemma 2 that there exists $\varepsilon > 0$ such that 0 is the only eigenvalue of L_{λ_0} in the interval $[-\varepsilon, \varepsilon]$. Moreover, by Lemma 1, there exists $\rho > 0$ such that $\pm\varepsilon$ is not in the spectrum of L_λ for all $\lambda \in [\lambda_0 - \rho, \lambda_0 + \rho] \subset [a, b]$. For simplicity of notation, we write $P_\lambda := P_{[-\varepsilon,\varepsilon]}(L_\lambda)$ for the orthogonal projection onto the sum of the eigenspaces of L_λ with respect to eigenvalues in $[-\varepsilon, \varepsilon]$ as introduced in Sect. 2.1. By (10) it is easily seen that that path $P = \{P_\lambda\} : [\lambda_0 - \rho, \lambda_0 + \rho] \to \mathcal{L}(H)$ of bounded projections is continuously differentiable, and according to [11], Sect. VI.2, there is an interval $[c, d] \subset [\lambda_0 - \rho, \lambda_0 + \rho]$ around λ_0 and a continuously differentiable path U_λ of orthogonal operators on H such that

$$U_{\lambda_0} = I_H, \quad U_\lambda P_\lambda U_\lambda^{-1} = P_{\lambda_0}, \quad \lambda \in [c, d]. \tag{11}$$

Since L_λ commutes with P_λ, we deduce from (11) that $U_\lambda L_\lambda U_\lambda^{-1}$ is reduced by the decomposition $H = H_0 \oplus H_0^\perp$, where $H_0 = \operatorname{im} P_{\lambda_0} = \ker L_{\lambda_0}$, and moreover, $U_\lambda L_\lambda U_\lambda^{-1}\mid_{H_0^\perp}: H_0^\perp \to H_0^\perp$ is an isomorphism, $\lambda \in [c,d]$. Consequently, we obtain from the properties (i)–(iii) of the Morse index

$$\begin{aligned}\mu_-(L_a) - \mu_-(L_b) &= \mu_-(L_c) - \mu_-(L_d) = \mu_-(U_c L_c U_c^{-1}) - \mu_-(U_d L_d U_d^{-1})\\ &= \mu_-(U_c L_c U_c^{-1}\mid_{H_0}) + \mu_-(U_c L_c U_c^{-1}\mid_{H_0^\perp})\\ &- (\mu_-(U_d L_d U_d^{-1}\mid_{H_0}) + \mu_-(U_d L_d U_d^{-1}\mid_{H_0^\perp}))\\ &= \mu_-(U_c L_c U_c^{-1}\mid_{H_0}) - \mu_-(U_d L_d U_d^{-1}\mid_{H_0}).\end{aligned} \tag{12}$$

Let us now consider $\ell_\lambda := U_\lambda L_\lambda U_\lambda^{-1}\mid_{H_0}$, $\lambda \in [c,d]$, which is a continuously differentiable path of symmetric operators on the finite dimensional space H_0. Clearly,

$$\begin{aligned}\dot{\ell}_\lambda &= \frac{\mathrm{d}}{\mathrm{d}\lambda}(U_\lambda)\circ L_\lambda \circ U_\lambda^{-1}\mid_{H_0} + U_\lambda \circ \frac{\mathrm{d}}{\mathrm{d}\lambda}(L_\lambda)\circ U_\lambda^{-1}\mid_{H_0}\\ &+ U_\lambda \circ L_\lambda \circ \frac{\mathrm{d}}{\mathrm{d}\lambda}(U_\lambda^{-1})\mid_{H_0}, \quad \lambda \in [c,d],\end{aligned}$$

and from $U_{\lambda_0} = I_H$, we see that

$$\begin{aligned}\langle \dot{\ell}_{\lambda_0} u, u\rangle &= \langle \dot{L}_{\lambda_0} u, u\rangle + \langle L_{\lambda_0} \dot{U}_{\lambda_0}^{-1} u, u\rangle = \langle \dot{L}_{\lambda_0} u, u\rangle + \langle \dot{U}_{\lambda_0}^{-1} u, L_{\lambda_0} u\rangle\\ &= \langle \dot{L}_{\lambda_0} u, u\rangle = \Gamma(L,\lambda_0)[u], \quad u \in H_0 = \ker L_{\lambda_0}.\end{aligned} \tag{13}$$

Consequently, by (12) the proposition is shown once we have proven that

$$\mu_-(\ell_c) - \mu_-(\ell_d) = \operatorname{sgn}\langle \dot{\ell}_{\lambda_0}\cdot,\cdot\rangle.$$

From (13) and the assumption that λ_0 is regular, we conclude that the operator $\dot{\ell}_{\lambda_0}$ is invertible on the finite dimensional space H_0. Thus, there is a constant $\alpha > 0$ such that $\dot{\ell}_{\lambda_0} + B$ is invertible for all linear operators $B : H_0 \to H_0$ of norm less than α. We define for $\lambda = c, d$ two straight paths by

$$T_\lambda : [0,1] \to \mathcal{L}(H_0), \quad T_\lambda(t) = t\ell_\lambda + (1-t)(\lambda - \lambda_0)\,\dot{\ell}_{\lambda_0}.$$

From $\ell_{\lambda_0} = L_{\lambda_0}\mid_{H_0} = 0$, we see that

$$T_\lambda(t) = (\lambda - \lambda_0)\left(t\left(\frac{\ell_\lambda - \ell_{\lambda_0}}{\lambda - \lambda_0} - \dot{\ell}_{\lambda_0}\right) + \dot{\ell}_{\lambda_0}\right),\ t \in [0,1].$$

By choosing $c < \lambda_0$ and $d > \lambda_0$ in (11) closer to λ_0, we can assume that the norm of $\frac{\ell_\lambda - \ell_{\lambda_0}}{\lambda - \lambda_0} - \dot{\ell}_{\lambda_0}$ is less than the constant α for $\lambda = c, d$, and consequently, $T_\lambda(t)$ is invertible for $t \in [0,1]$ and $\lambda = c, d$. We obtain

$$\mu_-(\ell_c) = \mu_-((c-\lambda_0)\,\dot{\ell}_{\lambda_0}) = \mu_-(-\,\dot{\ell}_{\lambda_0}) \text{ and } \mu_-(\ell_d) = \mu_-((d-\lambda_0)\,\dot{\ell}_{\lambda_0}) = \mu_-(\dot{\ell}_{\lambda_0}),$$

and finally

$$\operatorname{sgn}\langle \dot{\ell}_{\lambda_0}\cdot,\cdot\rangle = \mu_-(-\dot{\ell}_{\lambda_0}) - \mu_-(\dot{\ell}_{\lambda_0}) = \mu_-(\ell_c) - \mu_-(\ell_d).$$

This finishes the proof. □

Finally, we come back to bifurcation theory and consider a family of functionals $\psi : I \times H \to \mathbb{R}$ such that $0 \in H$ is a critical point of all ψ_λ, $\lambda \in I$, as in Sect. 2.2. Moreover, we denote as before by L_λ the Riesz representation of $D_0^2\psi_\lambda$, and we assume that $L_\lambda \in \mathcal{FS}_+(H)$, $\lambda \in I$. We now obtain from Theorem 2 and Proposition 1 the following bifurcation result, on which our proof of Theorem 1 is based.

Theorem 3 *If* $\lambda_0 \in (0,1)$ *is a regular crossing of* L *and* $\operatorname{sgn}\Gamma(L,\lambda_0) \neq 0$, *then* λ_0 *is a bifurcation point of critical points for* ψ.

3 Elliptic Dirichlet Problems on Shrinking Domains

The aim of this section is to prove Theorem 1. Let us denote by $\nu = (\nu_1,\ldots,\nu_n)$ the outer normal unit vector field on $\partial\Omega$ and recall that

$$\int_\Omega \frac{\partial u}{\partial x_i}(x)\,v(x)\mathrm{d}x = -\int_\Omega u(x)\frac{\partial v}{\partial x_i}(x)\mathrm{d}x + \int_{\partial\Omega} u(x)\,v(x)\nu_i(x)\,\mathrm{d}S \tag{14}$$

for all $u, v \in C^1(\Omega)\cap C(\bar{\Omega})$. From this integration by parts formula, we clearly see that a weak solution of the boundary value problem (5) is a function $\hat{u} \in H_0^1(\Omega_r)$ such that

$$\int_{\Omega_r}\sum_{i,j=1}^n a_{ij}(x)\frac{\partial \hat{u}}{\partial x_i}(x)\frac{\partial \hat{v}}{\partial x_j}(x)\mathrm{d}x + \int_{\Omega_r} g(x,\hat{u}(x))\,\hat{v}(x)\,\mathrm{d}x = 0, \quad \hat{v} \in H_0^1(\Omega_r). \tag{15}$$

In what follows, we denote for the sake of simplicity by $B := B(0,1)$ the unit ball around 0 in $\mathbb{R}^n$. It is readily seen that, after a change of coordinates, (15) is equivalent to

$$\int_B\sum_{k,l=1}^n \tilde{a}_{kl}(r\cdot x)\frac{\partial u}{\partial x_k}(x)\frac{\partial v}{\partial x_l}(x)\mathrm{d}x + r^2\int_B \tilde{g}(r\cdot x, u(x))\,v(x)\,\mathrm{d}x = 0, \;\; v \in H_0^1(B), \tag{16}$$

where

$$u(x) = \hat{u}(\Phi^{-1}(r\cdot x)), \quad x \in B, \tag{17}$$

$\tilde{g}$ is continuously differentiable on $\overline{B}\times\mathbb{R}$, and $\tilde{a}_{kl} \in C^\infty(\overline{B})$ are such that $\tilde{a}_{kl} = \tilde{a}_{lk}$, $1 \le k,l \le n$, and

$$\sum_{k,l=1}^n \tilde{a}_{kl}(x)\xi_k\xi_l > 0, \quad x \in B,\ (\xi_1,\ldots,\xi_n) \in \mathbb{R}^n\setminus\{0\}. \tag{18}$$

Let us consider the family of functionals $\psi : I \times H_0^1(B) \to \mathbb{R}$ defined by

$$\psi(r,u) = \frac{1}{2}\int_B \sum_{k,l=1}^{n} \tilde{a}_{kl}(r\cdot x)\,\frac{\partial u}{\partial x_k}(x)\frac{\partial u}{\partial x_l}(x)\,\mathrm{d}x + r^2\int_B G(r\cdot x, u(x))\,\mathrm{d}x,$$

where

$$G(x,t) = \int_0^t \tilde{g}(x,\xi)\,\mathrm{d}\xi, \quad (x,t)\in B\times\mathbb{R},$$

is a primitive of g. According to [18], Proposition B.10, each ψ_r is C^2 and the derivative of ψ_r at $u \in H_0^1(B)$ is given by (16), i.e.,

$$\begin{aligned}(D_u\psi_r)v = &\int_B \sum_{k,l=1}^{n} \tilde{a}_{kl}(r\cdot x)\frac{\partial u}{\partial x_k}(x)\frac{\partial v}{\partial x_l}(x)\mathrm{d}x\\ &+ r^2\int_B \tilde{g}(r\cdot x, u(x))\,v(x)\,\mathrm{d}x, \ \ v\in H_0^1(B).\end{aligned}$$

Hence, $D_u\psi_r = 0 \in (H_0^1(B))^*$ if and only if the function $\hat{u}$ defined by (17) is a weak solution of the boundary value problem (5) on Ω_r. In particular, $0 \in H_0^1(B)$ is a critical point of all ψ_r, and $r^* \in (0,1]$ is a bifurcation instant for ψ if and only if it is a bifurcation instant for the boundary value problems (5). Consequently, the study of bifurcation for (5), as defined in the introduction, boils down to an investigation of bifurcation for ψ in the sense of Definition 1.
By [18], Proposition B.34, the second derivative of ψ_r at $0 \in H_0^1(B)$ is given by

$$\begin{aligned}D_0^2\psi_r(u,v) = &\int_B \sum_{k,l=1}^{n} \tilde{a}_{kl}(r\cdot x)\frac{\partial u}{\partial x_k}(x)\frac{\partial v}{\partial x_l}(x)\,\mathrm{d}x\\ &+ r^2\int_B \tilde{f}(r\cdot x)\,u(x)\,v(x)\,\mathrm{d}x, \quad u,v\in H_0^1(B),\end{aligned} \tag{19}$$

where $\tilde{f}(x) = \frac{\partial\tilde{g}}{\partial\xi}(x,0)$, $x \in B$. We now define in accordance with the notation in the previous section a bounded symmetric operator L_r by

$$\langle L_r u, v\rangle_{H_0^1(B)} = D_0^2\psi_r(u,v), \quad u,v\in H_0^1(B),$$

which depends continuously differentiable on the parameter r. Clearly, $u \in \ker L_r$ if and only if the rescaled function $\hat{u}$ defined by (17) satisfies

$$\int_{\Omega_r} \sum_{i,j=1}^{n} a_{ij}(x)\frac{\partial\hat{u}}{\partial x_i}(x)\frac{\partial\hat{v}}{\partial x_j}(x) + \int_{\Omega_r} f(x)\,\hat{u}(x)\,\hat{v}(x)\,\mathrm{d}x = 0, \quad \hat{v}\in H_0^1(\Omega_r),$$

which means that $\hat{u}$ is a weak solution of the linearised boundary value problem (6). Since weak solutions of (6) are smooth by classical regularity theory (cf. [8], Corrollary 8.11), we conclude in particular that

$$\dim\ker L_r = m(r), \quad r\in(0,1], \tag{20}$$

where $m(r)$ is the number introduced in (7).

In summary, Theorem 1 is proven once we have shown that the bifurcation instants r of ψ in $(0, 1]$ are precisely the instants for which $\ker L_r \neq 0$.

Lemma 4 *The operators L_r, $r \in [0, 1]$, are Fredholm.*

Proof We set $L_r = L_{r,1} + L_{r,2} \in \mathcal{L}(H_0^1(B))$, where

$$\langle L_{r,1}u, v\rangle_{H_0^1(B)} = \int_B \sum_{k,l=1}^{n} \tilde{a}_{kl}(r \cdot x)\frac{\partial u}{\partial x_k}(x)\frac{\partial v}{\partial x_l}(x)\,\mathrm{d}x, \quad u, v \in H_0^1(B),$$

$$\langle L_{r,2}u, v\rangle_{H_0^1(B)} = r^2 \int_B \tilde{f}(r \cdot x)\, u(x)\, v(x)\,\mathrm{d}x, \quad u, v \in H_0^1(B).$$

From the ellipticity condition (18) and the boundedness of the functions a_{kl} on B, it clearly follows that $L_{r,1}$ is invertible. $L_{r,2}$ extends to a bounded operator on $L^2(B)$ and hence its restriction to $H_0^1(B)$ is compact, because of the compactness of the embedding $H_0^1(B) \hookrightarrow L^2(B)$. Consequently, L_r is a compact perturbation of an invertible operator and hence Fredholm. □

We now split the proof of Theorem 1 into two parts.

Every Bifurcation Instant r_0 of Critical Points for ψ is a Crossing of L

The assertion follows easily from the implicit function theorem in Banach spaces (cf. [3], Sect. 3). Here we only use a special case, which reads as follows.

Theorem 4 *Let X, Y be Banach spaces and $F : I \times X \to Y$ continuous. Assume that the equation*

$$F(\lambda, x) = 0 \tag{21}$$

has a solution $(\lambda_0, x_0) \in (0, 1) \times X$, and that the derivative $D_x F_\lambda$ of $F_\lambda := F(\lambda, \cdot)$ with respect to $x \in X$ exists and depends continuously on $(\lambda, x) \in I \times X$. If $D_{x_0} F_{\lambda_0} \in GL(X, Y)$, then there exists a neighbourhood $U \times V \subset I \times X$ of (λ_0, x_0) and a continuous map $f : U \to V$ such that $f(\lambda_0) = x_0$, $F(\lambda, f(\lambda)) = 0$ for all $\lambda \in U$ and every solution of (21) in $U \times V$ is of the form $(\lambda, f(\lambda))$.

Indeed, let us assume on the contrary that $\ker L_{r_0} = \{0\}$. We define a map

$$F : I \times H_0^1(B) \to (H_0^1(B))^*, \quad F(r, u) = D_u \psi_r$$

and note that by assumption $F(r, 0) = 0$ for all $r \in I$. Since $(D_0 F_{r_0} u)v = \langle L_{r_0} u, v\rangle$, $u, v \in H_0^1(B)$, and $\ker L_{r_0} = 0$ by assumption, we obtain from Lemma 4 that $D_0 F_{r_0} \in GL(H_0^1(B), (H_0^1(B))^*)$. Consequently, all solutions of the equation $F(r, u) = 0$ in a neighbourhood of $(r_0, 0) \in I \times H_0^1(B)$ are of the form $(r, 0)$.

Finally, let us mention that, since $m(1) = 0$ by assumption, bifurcation can only occur in the interior of I.

Every Crossing r_0 of L is a Bifurcation Instant of Critical Points of ψ

Let r_0 be a crossing of L and note that $r_0 \in (0,1)$ by (20) and the assumption that $m(1) = 0$. Our aim is to show that r_0 is regular and that the signature of the corresponding crossing form does not vanish.

Let $u \in \ker L_{r_0}$ and let us note for later reference that

$$\begin{aligned}\Gamma(L,r_0)[u] = &\int_B \sum_{k,l=1}^{n} \langle \nabla \tilde{a}_{kl}(r_0 \cdot x), x\rangle \frac{\partial u}{\partial x_k}\frac{\partial u}{\partial x_l}\,\mathrm{d}x\\ &+\int_B \frac{\mathrm{d}}{\mathrm{d}r}\mid_{r=r_0}(r^2 \tilde{f}(r\cdot x))u(x)^2\,\mathrm{d}x.\end{aligned} \tag{22}$$

As already observed above, from $u \in \ker L_{r_0}$, we obtain that u is smooth and so

$$\begin{aligned}0 &= -\sum_{k,l=1}^{n} \frac{\partial}{\partial x_k}\left(\tilde{a}_{kl}(r_0\cdot x)\frac{\partial u}{\partial x_l}(x)\right) + r_0^2 \tilde{f}(r_0\cdot x)u(x)\\ &= -\sum_{k,l=1}^{n} r_0 \frac{\partial \tilde{a}_{kl}}{\partial x_k}(r_0\cdot x)\frac{\partial u}{\partial x_l}(x) + \tilde{a}_{kl}(r_0\cdot x)\frac{\partial^2 u}{\partial x_k \partial x_l}(x) + r_0^2 f(r_0\cdot x)u(x)\end{aligned} \tag{23}$$

for all $x \in B$. We now introduce a new function by $v_r(x) := u(\frac{r}{r_0}\cdot x)$, $r \in (0, r_0]$, $x \in B$, and write

$$\dot{u}(x) := \frac{\mathrm{d}}{\mathrm{d}r}\mid_{r=r_0} v_r(x) = \frac{1}{r_0}\langle \nabla u(x), x\rangle, \quad x \in B. \tag{24}$$

From (23), we infer that

$$\begin{aligned}&-\sum_{k,l=1}^{n} \frac{\partial}{\partial x_k}\left(\tilde{a}_{kl}(r\cdot x)\frac{\partial v_r}{\partial x_l}(x)\right) + r^2 f(r\cdot x)v_r(x)\\ &= \frac{r^2}{r_0^2}\left(-\sum_{k,l=1}^{n}\left(r_0\frac{\partial \tilde{a}_{kl}}{\partial x_k}\left(r_0\frac{r}{r_0}x\right)\frac{\partial u}{\partial x_l}\left(\frac{r}{r_0}x\right) + \tilde{a}_{kl}\left(r_0\frac{r}{r_0}x\right)\frac{\partial^2 u}{\partial x_k\partial x_l}\left(\frac{r}{r_0}x\right)\right)\right.\\ &\left.+ r_0^2 \tilde{f}\left(r_0\frac{r}{r_0}x\right)u\left(\frac{r}{r_0}x\right)\right) = 0,\end{aligned}$$

and by differentiating with respect to r at $r = r_0$ we get

$$\begin{aligned}0 = &-\sum_{k,l=1}^{n}\frac{\partial}{\partial x_k}\left(\langle \nabla\tilde{a}_{kl}(r_0\cdot x), x\rangle \frac{\partial u}{\partial x_l}\right) - \sum_{k,l=1}^{n}\frac{\partial}{\partial x_k}\left(\tilde{a}_{kl}(r_0\cdot x)\frac{\partial \dot{u}}{\partial x_l}\right)\\ &+\frac{\mathrm{d}}{\mathrm{d}r}\mid_{r=r_0}(r^2\tilde{f}(r\cdot x))u(x) + r_0^2\tilde{f}(r_0\cdot x)\dot{u}(x), \quad x \in B.\end{aligned} \tag{25}$$

Now, we multiply (25) by u and integrate over B:

$$0 = -\int_B \sum_{k,l=1}^n \frac{\partial}{\partial x_k}\left(\langle \nabla \tilde{a}_{kl}(r_0 \cdot x), x\rangle \frac{\partial u}{\partial x_l}\right) u(x)\,\mathrm{d}x$$
$$-\int_B \sum_{k,l=1}^n \frac{\partial}{\partial x_k}\left(\tilde{a}_{kl}(r_0 \cdot x)\frac{\partial \dot{u}}{\partial x_l}\right) u(x)\,\mathrm{d}x + \int_B \frac{\mathrm{d}}{\mathrm{d}r}\,|_{r=r_0}\,(r^2 \tilde{f}(r \cdot x))u(x)^2\,\mathrm{d}x$$
$$+\int_B r_0^2 \tilde{f}(r_0 \cdot x)\dot{u}(x)u(x)\,\mathrm{d}x.$$

Since u vanishes on ∂B, we obtain from (14)

$$0 = \int_B \sum_{k,l=1}^n \langle \nabla \tilde{a}_{kl}(r_0 \cdot x), x\rangle \frac{\partial u}{\partial x_l}\frac{\partial u}{\partial x_k}\,\mathrm{d}x$$
$$-\int_{\partial B}\left(\sum_{k,l=1}^n \langle \nabla \tilde{a}_{kl}(r_0 \cdot x), x\rangle \nu_k(x)\frac{\partial u}{\partial x_l}\right) u(x)\,\mathrm{d}S$$
$$+\int_B \sum_{k,l=1}^n \tilde{a}_{kl}(r_0 \cdot x)\frac{\partial \dot{u}}{\partial x_l}\frac{\partial u}{\partial x_k}\,\mathrm{d}x - \int_{\partial B}\left(\sum_{k,l=1}^n \tilde{a}_{kl}(r_0 \cdot x)\nu_k(x)\frac{\partial \dot{u}}{\partial x_l}\right) u(x)\,\mathrm{d}S$$
$$+\int_B \frac{\mathrm{d}}{\mathrm{d}r}\,|_{r=r_0}\,(r^2 \tilde{f}(r \cdot x))u(x)^2\,\mathrm{d}x + \int_B r_0^2 \tilde{f}(r_0 \cdot x)\dot{u}(x)u(x)\,\mathrm{d}x$$
$$= \int_B \sum_{k,l=1}^n \langle \nabla \tilde{a}_{kl}(r_0 \cdot x), x\rangle \frac{\partial u}{\partial x_l}\frac{\partial u}{\partial x_k}\,\mathrm{d}x - \int_B \sum_{k,l=1}^n \frac{\partial}{\partial x_l}\left(\tilde{a}_{kl}(r_0 \cdot x)\frac{\partial u}{\partial x_k}\right)\dot{u}(x)\,\mathrm{d}x$$
$$+\int_{\partial B}\left(\sum_{k,l=1}^n \tilde{a}_{kl}(r_0 \cdot x)\nu_l(x)\frac{\partial u}{\partial x_k}\right)\dot{u}(x)\,\mathrm{d}S$$
$$+\int_B \frac{\mathrm{d}}{\mathrm{d}r}\,|_{r=r_0}\,(r^2 \tilde{f}(r \cdot x))u(x)^2\,\mathrm{d}x + \int_B r_0^2 \tilde{f}(r_0 \cdot x)\dot{u}(x)u(x)\,\mathrm{d}x.$$

Now, we use the first equality in (23), as well as (22) and (24) to conclude that

$$\Gamma(L, r_0)[u] = -\frac{1}{r_0}\int_{\partial B}\left(\sum_{k,l=1}^n \tilde{a}_{kl}(r_0 \cdot x)\nu_l(x)\frac{\partial u}{\partial x_k}\right)\langle \nabla u(x), x\rangle\,\mathrm{d}S,$$

which can be written as

$$\Gamma(L, r_0)[u] = -\frac{1}{r_0}\int_{\partial B}\langle A(r_0 \cdot x)x, \nabla u(x)\rangle\,\langle \nabla u(x), x\rangle\,\mathrm{d}S,$$

where $A(x) := \{\tilde{a}_{kl}(x)\}$, $x \in B$, and we use that $\nu(x) = x$ for all $x \in \partial B$. Denoting by $(A(r_0 \cdot x)x)^T$, $x \in \partial B$, the component of the vector $A(r_0 \cdot x)x$ tangential to ∂B, we have

$$\langle A(r_0 \cdot x)x, \nabla u(x)\rangle = \langle \nabla u(x), x\rangle\,\langle A(r_0 \cdot x)x, x\rangle + \langle \nabla u(x), (A(r_0 \cdot x)x)^T\rangle$$

and it follows that

$$\begin{aligned}\Gamma(L,r_0)[u] &= -\frac{1}{r_0}\int_{\partial B}\langle\nabla u(x),x\rangle^2\,\langle A(r_0\cdot x)x,x\rangle\,\mathrm{d}S\\ &\quad-\frac{1}{r_0}\int_{\partial B}\langle\nabla u(x),x\rangle\,\langle\nabla u(x),(A(r_0\cdot x)x)^T\rangle\,\mathrm{d}S.\end{aligned}$$

From

$$\begin{aligned}&\operatorname{div}(u(x)\langle x,\nabla u(x)\rangle(A(r_0\,x)x)^T) = \langle x,\nabla u(x)\rangle\langle\nabla u(x),(A(r_0\,x)x)^T\rangle\\ &+u(x)\langle\nabla\langle x,\nabla u(x)\rangle,(A(r_0\,x)x)^T\rangle + u(x)\langle x,\nabla u(x)\rangle\operatorname{div}(A(\lambda_0 x)x)^T,\end{aligned}$$

and $u\mid_{\partial\Omega}=0$, we see that

$$\langle\nabla u(x),x\rangle\,\langle\nabla u(x),(A(r_0\cdot x)x)^T\rangle = \operatorname{div}(u(x)\langle x,\nabla u(x)\rangle(A(r_0\cdot x)x)^T),\quad x\in\partial B,$$

and now Stokes' theorem gives

$$\Gamma(L,r_0)[u] = -\frac{1}{r_0}\int_{\partial B}\langle\nabla u(x),x\rangle^2\,\langle A(r_0\cdot x)x,x\rangle\,\mathrm{d}S\leq 0. \tag{26}$$

Here we use that $A(x)$ is positive definite for all $x\in B$, which follows from the ellipticity condition (18). Finally, even the strict inequality holds in (26) if $u\neq 0$. For otherwise,

$$\langle\nabla u(x),x\rangle = \langle\nabla u(x),\nu(x)\rangle = \frac{\partial u}{\partial\nu}(x) = 0,\quad x\in\partial B,$$

and since $u\mid_{\partial B}=0$, this implies $u\equiv 0$ by the uniqueness of the Cauchy problem for elliptic second-order operators (cf. [6, 10]).

In summary, we have shown that $\Gamma(L,r_0)$ is negative definite, and so in particular nondegenerate with the nonvanishing signature:

$$\operatorname{sgn}\Gamma(L,r_0) = -\dim\ker L_{r_0} = -m(r_0), \tag{27}$$

where the second equality was already shown in (20). By Theorem 3, r_0 is a bifurcation instant, and so Theorem 1 is proven.

4 Corollaries and Examples

The first aim of this final section is to prove Corollary 1, which is a rather immediate consequence of the equality (9). In order to derive (9) from the proof of Theorem 1, we note at first that the Morse index μ_- of (8) is given by the Morse index $\mu_-(L_1)$ of the operator $L_1\in\mathcal{FS}_+(H^1_0(B))$. Moreover, since L_0 is positive by (18), we see that $\mu_-(L_0)=0$. We have obtained in (27) that all crossings r_0 of L are regular and $\operatorname{sgn}\Gamma(L,r_0) = -m(r_0)$. Consequently, Lemma 3 and Proposition 1 show that $m(r) = \dim\ker L_r = 0$ for all but finitely many $r \in (0,1]$ and

$$\mu_- = \mu_-(L_1) - \mu_-(L_0) = -\sum_{0<r<1} \operatorname{sgn}\Gamma(L,r) = \sum_{0<r<1} m(r)$$

which is (9).

Let us point out that (9) was proven by Smale in [20] by studying the monotonicity of eigenvalues under shrinking of domains. Hence, we have obtained a new proof of Smale's theorem for the boundary value problems (6), and moreover, Corollary 1 is now an immediate consequence of (9) and Theorem 1.

Let us now conclude this final section by some examples and remarks. For applying Corollary 1, it is necessary to have an upper bound on the kernel dimensions $m(r), r \in (0,1]$, which may be difficult to determine in general. A particular example is given by semilinear ordinary differential equations, where (5) and (6) reduce to

$$\begin{cases} -(a(x)u'(x))' + g(x,u(x)) = 0, & x \in [0,r] \\ u(0) = u(r) = 0, \end{cases} \tag{28}$$

and

$$\begin{cases} -(a(x)u'(x))' + f(x)u(x) = 0, & x \in [0,r] \\ u(0) = u(r) = 0. \end{cases} \tag{29}$$

Here $a : [0,1] \to \mathbb{R}$ is positive and smooth, $g : [0,1] \times \mathbb{R} \to \mathbb{R}$ is any C^1 function such that $g(x,0) = 0$, $x \in [0,1]$, and $f(x) = \frac{\partial g}{\partial \xi}(x,0)$ is assumed to be smooth on $[0,1]$. Since clearly $0 \leq m(r) \leq 1$, we not only conclude that every conjugate instant of (29) is a bifurcation instant of (28) but also obtain from Corollary 1 that there are exactly μ_- distinct bifurcation instants, where μ_- is the number of negative eigenvalues of (29) for $r = 1$, i.e. the number of $\lambda < 0$ such that

$$\begin{cases} -(a(x)u'(x))' + f(x)u(x) = \lambda u(x), & x \in [0,1] \\ u(0) = u(1) = 0 \end{cases}$$

has a non-trivial solution.

Let us mention in passing that the computation of the crossing forms of L in the proof of Theorem 1 not only simplifies in the case of ordinary differential equations but also yields a new proof of the Morse index theorem for geodesics in Riemannian manifolds, which can be found in the recent joint work [17] of the author with A. Portaluri.

In higher dimensions, many semilinear equations come from problems in geometric analysis. Let us refer to [4, 5] and just mention as an example on compact Riemannian manifolds (M,g) of dimension $n \geq 3$, the equation:

$$4\,\frac{n-1}{n-2}\,\Delta u(p) + s(p)u(p) = \mu\,|u(p)|^{\frac{n+2}{n-2}}, \quad p \in M, \tag{30}$$

where Δ denotes the Laplace–Beltrami operator, $s : M \to \mathbb{R}$ the scalar curvature function and μ the Yamabe invariant of the metric g on M. The significance of (30) is that if $u \in C^\infty(M)$ is a positive solution, then $\tilde{g} = u^{\frac{4}{n-2}} g$ is a metric of constant scalar curvature on M.

Let us now assume that $\Phi : \mathcal{U} \to \mathbb{R}^n$ is a chart of M and consider $\Omega := \Phi^{-1}(B(0,1))$. In this case, the boundary value problems (5) are

$$\begin{cases} 4\dfrac{n-1}{n-2}\,\Delta u(p) + s(p)u(p) = \mu\,|u(p)|^{\frac{n+2}{n-2}}, & p \in \Omega_r \\ u(p) = 0, & p \in \partial\Omega_r \end{cases} \tag{31}$$

and we obtain as corresponding linear equations (6):

$$\begin{cases} 4\dfrac{n-1}{n-2}\,\Delta u(p) + s(p)u(p) = 0, & p \in \Omega_r \\ u(p) = 0, & p \in \partial\Omega_r. \end{cases} \tag{32}$$

We see that the bifurcation instants do not depend on the Yamabe invariant μ. Moreover, if g is already of constant scalar curvature, then the bifurcation instants are entirely determined by the spectrum of the Laplace–Beltrami operator. In particular, there is no bifurcation for (31) on manifolds of constant negative scalar curvature.

References

1. M.F., Atiyah, I.M. Singer, Index theory for skew-adjoint Fredholm operators. Inst. Hautes Etudes Sci. Publ. Math. **37**, 5–26 (1969)
2. M.F. Atiyah, V.K. Patodi, I.M. Singer, Spectral asymmetry and Riemannian geometry III. Proc. Cambridge Philos. Soc. **79**, 71–99 (1976)
3. A. Ambrosetti, G. Prodi, *A Primer of Nonlinear Analysis* (Cambridge University Press, Cambridge, 1993)
4. T. Aubin, *Nonlinear Analysis on Manifolds. Monge-Ampère Equations* (Springer, New York, 1988)
5. Besse, A.L., *Einstein Manifolds* (Springer, Berlin Heidelberg, 1987)
6. A.P. Calderon, Uniqueness in the Cauchy problem for partial differential equations. Amer. J. Math. **80**, 16–36 (1958)
7. P.M. Fitzpatrick, J. Pejsachowicz, L. Recht, Spectral flow and bifurcation of critical points of strongly-indefinite functionals Part I: General theory. J. Funct. Anal. **162**, 52–95 (1999)
8. D. Gilbarg, N.S. Trudinger, *Elliptic Partial Differential Equations of Second Order* (Springer, Berlin, 2001)
9. I. Gohberg, S. Goldberg, M.A. Kaashoek, *Classes of Linear Operators*, vol. I (Birkhäuser, Basel, 1990)
10. L. Hörmander, *Linear Partial Differential Operators* (Springer, New York, 1969)
11. T. Kato, *Perturbation Theory for Linear Operators* (Springer, Berlin-New York, 1976)
12. H. Kielhöfer, *Bifurcation Theory. An Introduction with Applications to Partial Differential Equations* (Springer, New York, 2012)
13. J. Mawhin, M. Willem, *Critical Point Theory and Hamiltonian Systems* (Springer, New York, 1989)

14. J. Pejsachowicz, N. Waterstraat, Bifurcation of critical points for continuous families of C^2 functionals of Fredholm type. J. Fixed Point Theory Appl. **13**, 537–560 (2013)
15. A. Portaluri, N. Waterstraat, On bifurcation for semilinear elliptic Dirichlet problems and the Morse-Smale index theorem. J. Math. Anal. Appl. **408**, 572–575 (2013)
16. A. Portaluri, N. Waterstraat, On bifurcation for semilinear elliptic Dirichlet problems on geodesic balls. J. Math. Anal. Appl. **415**, 240–246 (2014)
17. A. Portaluri, N. Waterstraat, Yet another proof of the Morse index theorem. Expo. Math. (to appear)
18. P.H. Rabinowitz, *Minimax Methods in Critical Point Theory with Applications to Differential Equations*. (American Mathematical Society, Providence, RI 1986)
19. J. Robbin, D. Salamon, The spectral flow and the Maslov index. Bull. London Math. Soc. **27**, 1–33 (1995)
20. S. Smale, On the Morse index theorem. J. Math. Mech. **14**, 1049–1055 (1965)
21. S. Smale, Corrigendum: "On the Morse index theorem". J. Math. Mech. **16**, 1069–1070 (1967)

MIX
Papier aus verantwortungsvollen Quellen
Paper from responsible sources
FSC® C105338
FSC
www.fsc.org

If you have any concerns about our products,
you can contact us on
ProductSafety@springernature.com

In case Publisher is established outside the EU,
the EU authorized representative is:
Springer Nature Customer Service Center GmbH
Europaplatz 3, 69115 Heidelberg, Germany

Printed by Libri Plureos GmbH
in Hamburg, Germany